# Communication and Tourism

## Reflecting on the Construction of the Tourist Image of Greece

# Communication and Tourism

## Reflecting on the Construction of the Tourist Image of Greece

**Michael Tsangaris**
*University of Piraeus, Greece*

CABI

**CABI is a trading name of CAB International**

CABI
Nosworthy Way
Wallingford
Oxfordshire OX10 8DE
UK

Tel: +44 (0)1491 832111
E-mail: info@cabi.org
Website: www.cabi.org

CABI
200 Portland Street
Boston
MA 02114
USA

T: +1 (617)682-9015
E-mail: cabi-nao@cabi.org



A catalogue record for this book is available from the British Library, London, UK.

ISBN-13: 9781800626010 (hardback)
9781800626027 (ePDF)
9781800626034 (ePub)

DOI: 10.1079/9781800626034.0000

Commissioning Editor: Claire Parfitt
Editorial Assistant: Helen Elliott
Production Editor: James Bishop

Typeset by Exeter Premedia Services Pvt Ltd, Chennai, India

# Contents

# About the Author

Michael Tsangaris (PhD) is a senior teaching fellow at the School of Economics, Business and International Studies at the University of Piraeus in Greece. His recent works include the books *Mobility, Tourism, and Society* (Kallipos, 2022) and *Radical Communications: Rebellious Expressions on Urban Walls* (Lexington Books, 2022). His research interests focus on visual representations, tourism, festivals, gender, alternative media and more generally on issues related to mass media, culture and society. For more information see: https://miketsag.wixsite.com/website.
E-mail: mtsang@unipi.gr

# Figures

# Introduction

## Communication: an Important Driving Force in Tourism

According to the French philosopher Henri Bergson (1859–1941), animal life, in contrast to plant life, is characterized by mobility in space, and there is an obvious relationship between mobility and consciousness. In the context of this relationship, he raises the question of whether consciousness is the cause or the causality of movement. At a certain point, he even argues that in one sense we can assume that consciousness is the cause of movement because its role is to direct it, but in another sense it is the result because consciousness is sustained by mobility, and once all movement ceases, consciousness atrophies or fossilizes (Bergson, 2014).

Consciousness, mobility and tourism seem to be naturally connected because travelling is a way of changing and expanding one's consciousness. Be that as it may, the most important factor that expands human consciousness is communication.

When we talk about communication, we generally refer to the exchange of meaning through the encoding and decoding of messages, but of course we can also refer to mobility (i.e. the flow of information, objects or people between different places). Communication can take place at different levels, between individuals, between social groups or between individuals and social groups, within a society or between societies, etc., and it can be supported by technological tools or simply take place without them, (i.e. through the media or through direct face-to-face contact between people) (Hill *et al.*, 2000).

Furthermore, we can say that a fundamental element of communication is our body, because it is the mirror of our existence and conveys messages about our particular characteristics and peculiarities to the outside world, to others and to society. The way we use our body, with movements, gestures, facial expressions, the way we dress, how we comb our hair, which tattoos we choose, how we position ourselves in space, etc., conveys certain social messages (Alexias, 2011).

Sometimes we can even speak of communication in terms of our personal perception of the outside world (Rogers, 1995).

Since tourism involves personal experiences with the outside world, contact, interaction and the exchange of messages between tourists, the inhabitants of the places visited and the actors of the tourism industry, as well as the transport of people, objects and information, it is obvious that it is closely linked to the field of communication.

From a phenomenological point of view, we can see that each actor in tourism uses their body in different ways to express themselves socially. The tourist, for example, has his or her own characteristic way of dressing, speaking or moving in space, expressions that together convey the social message that he or she is a tourist. Not to mention that tourism also includes the dissemination of tourists' narratives about their travel experiences through social media or other forms of communication, as well as the advertising practices of the tourism industry.

In any case, it can be said that communication as an activity in itself is one of the most important driving forces of tourism. It would not be an exaggeration to say that this book, as a form of communication that reaches a wide audience, is also capable of indirectly (i.e. without the interference of the tourism industry and without our intention) shaping tourist images in relation to the places mentioned.

Thus, the narrative that reproduces Homer's account of the *Odyssey* and Ithaca in Chapter 8, has the remarkable ability to fire


DOI: 10.1079/9781800626034.INTRO

the reader's imagination in relation to this island. The information woven into the narrative has the power to paint a vivid picture of Ithaca as a captivating Mediterranean Greek island with mythological and historical charm. Consequently, it has the potential to pique the interest of tourists. However, it is important to point out that the aim of this book is not to actively promote a holiday to Ithaca or Greece.

In one way or another, communication awakens or mobilizes human movements through the construction of images of places.

### A country's tourist image

The construction, reconstruction and dissemination of a country's tourist image is a long but dynamic process that is influenced by many factors, such as social and historical memory, education, literature, cinema and art in general, diplomacy, tourists' narratives about their travel experiences on social media and, of course, the advertising practices of the tourism industry.

If we want to simplify schematically the gradual realization of the idea of a country's tourist image, we can start from two main points: (i) the historical and natural geographic narratives on the one hand; and (ii) the narratives related to contemporary constructions that attract tourists on the other.

The narratives about the historical and natural geographical features emphasize the ancient civilizations, the historical achievements of the inhabitants of these regions and their culture, as well as the natural environments, geographical location, ecosystems, landscapes, climate and geology of these places. The narratives about contemporary constructions that attract tourists, on the other hand, reflect the postmodern character of these places, their tourist infrastructure and the realization that all tourist sites are essentially social constructs.

The tourist image of a country can therefore be based on its ancient origins, its history or the fascinating landscapes within its borders, or at least be an extension of these elements. But it can also be seen as a multi-layered neo-eclectic construct that encompasses various elements of contemporary culture, including postmodern architecture, fashion, entertainment, subcultures and their art, technological advances, ecological sensitivity, tourist facilities, theme parks, the tourist education level of the inhabitants and a distinct tourist culture.

However, we should emphasize that this assumption is only a simplification of reality, as there are many other elements that need to be taken into account when shaping a country's tourism image. And of course, we should always bear in mind that there is a dichotomy between 'appearance' and 'being' (i.e. between 'image' and 'reality', if we can of course accept that there is a stable objective reality).

## Human Need to Travel and Explore

Another driving force that has led to the development of tourism is the innate human need to move through geographical space. It is not difficult to assume that a person who wants to visit a particular place or travel in general already has some mediated ideas about that destination, which may be the motivation for the attraction. But it may also be that he or she is simply seized by a 'pure urge to move', an unmediated inner need to change surroundings, to wander, to explore and to expand consciousness.

Throughout history, humans have always had a fundamental urge to expand their knowledge, create a comprehensive understanding of the world and classify various aspects of life, including themselves. This innate desire for exploration and adaptability could stem from an inner compulsion to develop consciousness or be an evolutionary adaptation resulting from the extensive migrations of early nomadic humans.

The vital characteristics of nomadic life, namely mobility, freedom, unpredictability, self-sufficiency and diversity, were essential for the survival of primitive groups. Remarkably, it is precisely these characteristics that tourism is embracing today. In the postmodern age, tourism has undergone an evolution that emphasizes these facets and reflects the changing preferences and lifestyles of modern travellers. It could be argued that from this perspective, tourism represents the resurgence of a fundamental trait of our species, perhaps even a deeply ingrained life process that was once crucial to survival.

Tourism could indeed be what Jung (2014) calls a 'primitive tribal lore' linking contemporary societies to prehistoric societies. And if we want to be a little more poetic, mobility in the form of travelling could be, as Campbell (1988) very lyrically puts it, the eternal repetition of Adam and Eve's original expulsion from paradise and is part of the cycle of primal experience with 'birth, reunion and death'.

## Aim of the Book

Regarding the theme of this book, it should first be noted that it relates to communication and its impact on the so-called construction of social thought and, in particular, ideas about tourism and images of places as they have developed in the context of Western culture. One of the main objectives is to rethink and 'deconstruct' the concept of tourism as it is understood today in order to identify possible changes in its diachronic understanding and the role that the media has played in this process.

In other words, the aim is to highlight the various tensions and contradictions between the elements of the concept of tourism in order to emphasize the heterogeneity of this phenomenon. At the same time, importance is given to the functionalities and dysfunctionalities of tourism as they have been perceived or understood over time through experience and theoretical thinking.

Theories of communication and tourism are used to show that everything is rhizomatically connected. With the aim of looking at and understanding tourism diachronically, the contemporary commercial dimension of this phenomenon is critically analysed. At the same time, an attempt is made to interpret the word 'tourism', both in its colloquial use and when we encounter it in academic texts.

The term tourism in its current common understanding is the result of a social construction invented by some creative entrepreneurs of the 19th century, while it continues to evolve in this direction both in the business world and in most academic tourism studies programmes.[1]

We find it very unfortunate that terms such as travelling, hospitality or tourism are constantly restricted to meanings with an exclusively economic content, while terms such as marketing and branding also encompass politics and all areas of human life. Individual and collective identity, history, personal and social representations and, to a large extent, ideology and political positions coexist, interact and produce or reconstruct mnemonic products and concepts such as tourism, for the understanding of which we must turn to the study of the process, the interactions and, if possible, the elements that originally created them.

The British sociologist John Urry (1946–2016) traces the emergence of the so-called 'tourist gaze' (which is in some ways related to the commercial perspective of tourism) back to the 1840s, when the innate desire to travel coincided with the expansion of public transport, the establishment of the first travel agencies and the advance of photography, which depicted the landscape in a particular way (Urry and Larsen, 2011).[2]

When these surprising events coincided in time and space, they merged seamlessly. This merging led to a codification that would replace the meanings of 'leisure travel' and 'broadening horizons', which had previously been subsumed under the term 'tourism'. The new codification played a crucial role in creating the cornerstone on which the transformation of tourists into customer-consumers took place in a commercial give and take, ultimately fuelling the emergence of a 'tourism industry' in the wider context of 'mass society'.

International organizations and the tourism business have supported the expansion of this phenomenon into an 'industry', adding additional elements to give it a measurable economic dimension.

In addition to all these aspects, however, the will of the individual to travel, discover new places and pass on the so-called 'image of a place' to others also plays an important role in the emergence of today's concept of tourism.

In other words, this book explores the relationship between communication and tourism. In particular, it shows how the idea of tourism has been portrayed by the media at different historical moments and how it is portrayed today. It can be argued that the 'media reality' is not just a simple reflection of society, but also a 'constructed reality' created by various agencies, and furthermore an 'imaginary reality' fed by the fears, expectations and dreams of the public, which we can argue that are influenced one way or another by media producers.

In this context, the central issue explored in this book revolves around the question if the concept of tourism is presented exclusively as a commodified facet of human mobility, essentially reduced to the consumption of services; is it a matter of mere economic transactions, especially in the context of tourism's balance of payments.

We therefore wonder whether tourism emerged as an independent concept when various activities that have an inherent value were commercialized in the realm of the market, or whether this term can still describe something that has always been inherent to human nature and is a universal feature of human existence.

We assume that the essence of tourism has existed in rudimentary forms since ancient times and stems from an innate human urge that is closely linked to the desire to enjoy life to the full. Consequently, tourism can be seen as an outgrowth of humanity's constant endeavour to expand its understanding of the world, a constant quest to expand consciousness.

As Varvaresos (2017) characteristically states, travelling makes people young by transforming them into children, helping them to experience countless discoveries and revelations and driving them to seek the true and overcome the false and superficial. Mobility through space opens people's eyes, moulds and refines their souls, cultivates the art of remembering and gives them the affirmation of travelling to places and times of serenity, inner fullness and peace.

Words only mean what they mean because we agree that they do. As a linguist would put it, the relationship between a particular signifier and a signified is conventional. The main message of the book, then, is that the commercial dimension of tourism in our time and the extensive use of the media by the tourism industry have limited its meaning. The meaning of the term as an element of consumer society has marginalized its value as a vital human function. As with many other definitions of social phenomena, what we see today as the definition of the phenomenon of 'tourism' (especially the way tourism is defined by the United Nations World Tourism Organization (UNWTO)) reflects the prevailing ideologies. Therefore, tourism as an industry is today presented as a neutral phenomenon, although it is an imperative of the dominant ideology.

In fact, diachronically speaking, the tourist is not a stable figure; the tourist is not a manageable subject of 'typification' that could easily be defined uniformly over time, but a dynamic character that takes on its form and dimension each time depending on the values, opinions and general conditions of society. The 'typification' and study of tourists began to gain importance in the field of tourism and sociology in the mid-20th century. This coincided with the growth of the 'tourism industry' and the increasing interest in understanding the behaviour and motivations of modern tourists. The fact is, however, that the word 'tourist' has a different meaning in a late 18th-century text, in a 19th-century newspaper, in 20th-century academic discussions, or in UNWTO declarations.

Thus, in the various definitions of tourism, different characteristics and attributes are added over time as criteria, such as cultural education, recreation, distance, territoriality, temporality, working conditions, labour laws and regulations, and since they reflect the prevailing ideology, they influence our thinking and lead us to accept the current conditions as natural.

Tourism is indeed a blurred concept, the boundaries between the universe of touristic and non-touristic roles are fluid and there are many intermediate categories (Cohen, 1974).

The opinions that refer to the great difference between tourists, travellers, visitors, drifters, explorers, backpackers, new age travellers, the hobos, wanderers, vagabonds or nomads are also very vacillating and questionable in our logic.[3]

Although many researchers have tried to do so, we do not want to give tourism a specific historicity, so that it is the product of certain socio-economic conditions of a certain time. To make a long story short: we do not believe that there is a specific historical moment after which tourism began to exist, whereas it did not exist before.

Undoubtedly, there were certain historical conditions that we consider necessary for the emergence of the 'tourism industry'. However, the focus of our analysis is not the tourism industry that is identified with the word tourism today, but the voluntary mobility of people for the purpose of pleasure and the development of consciousness. This can be expressed more or less in many words, be it the word 'tourism', which dates back to the time of the British Enlightenment, or the ancient Greek word *periigisi*, or other terms such as 'travelling', 'wandering', 'nomadism', and so on. In reality, the word we are going to use is not that important, but we believe that the word 'tourism' is one of the most appropriate.

So, when we use the word 'tourism' in this book, we refer to 'travelling for pleasure' and 'travelling as an inner need to get to know the world', while when we talk about tourism as a commodity or as an industry, we use terms such as 'tourism product' or 'tourism industry'.

Be that as it may, to summarize, in this book we examine tourism not just as a business-oriented phenomenon or industry, but as an inherent activity that has been commodified, or more precisely, as an aspect of the general 'objectification' of human behaviour.

Indeed, as early as the 1970s, Dean MacCannell (1976) stated in the introduction to his book *The Tourist: a New Theory of the Leisure Class* that he wondered whether he was inadvertently portraying an ancient phenomenon as if it were a new invention.

In their philosophical explorations, some scholars likened 'travelling' to a contagion that, once it has broken out, infects the desire to explore and discover new destinations. Others compared 'travelling' to a cure that can heal the lifelessness of a stagnant existence. Still others saw tourism as a therapeutic tool for the alienated bourgeoisie to rediscover forgotten places and reawaken a sense of wholeness that was often missing in the routine of modern city life (Rimbaud, 1873; MacCannell, 1973; Baudelaire, 2012; Varvaresos, 2017).

From our perspective, the excessive commercial form that tourism has taken today bypasses and undermines the 'creative will' of life and the 'sense of freedom', those inner states associated with travel that bring completion and self-realization, and replaces them with an emptiness that springs from the purely speculative logic that expresses the interests of the tourism industry.

Indeed, one can convincingly argue that this represents a classic scenario in which the intangible inner experiences of the individual are transformed into tangible consumer goods. In today's world, tourists not only buy 'pleasure' but also 'creativity', 'safe adventure' and a sense of 'freedom'. This is achieved by: (i) tailoring alternative but standardized tourist experiences; (ii) exploring different but identical and fragmented urban environments; (iii) creating their own personalized tourist adventures from a selection of options offered by tourism platforms; (iv) exercising extensive control over their holiday planning via online agencies; and (v) anticipating the escapism of their trip with remarkable predictability.

It can therefore be said that the tourism industries of each country, through their homogenization, have transformed all destinations worldwide into familiar and comfortable places with identical tourist infrastructure. Although destinations, tourist choices and the variability of tourism products have increased, people enjoy standardized forms of tourism consumption in various combinations.

In this context, we will look at the phenomenon of tourism in the field of communication, consciously attempting to transcend the boundaries of a purely economic perspective. In addition, we will sometimes critically question procedures that commercialize the general phenomenon of tourism, especially in the context of Greece.

In order to streamline our approach, we often use a linear chronological sequence when exploring our topic. However, the frequent references to earlier and later points make it clear that tourism functions like a dynamic, interconnected network. It constantly builds a bridge between the past and contemporary perspectives and takes a central role in discussions that require continuous reflection and provide room for different interpretations and unforeseen developments.

In this context, it encourages us to reflect on the peculiarities that characterize the modern Greek state in the areas of tourism, communication and culture, which makes the study of Greek tourism extremely fascinating. Essentially, this fascination can be traced back to the compelling interplay between the nascent Greek state and the pre-existing, positive perception of ancient Greece by Europeans and the global community during its formative years.

Ancient Greek civilization, which largely developed in the region where the modern Greek state is located today, is regarded by many scholars as one of the foundations of Western thought. On closer inspection, Western thought, the foundation of today's 'tourism industry', stems from European civilization, which is rooted in Greco-Roman heritage, Christianity and the influence of the Germanic warriors who invaded the Roman Empire (Hirst, 2012).

Nevertheless, the eternal question that preoccupies us remains and challenges us to solve the riddle of the magnetic attraction that Greece exerts on the world, like a siren song that goes beyond the simple charm of the country's natural beauty and the tenderness of its mild climate.

Be that as it may, it is fair to say that Greece's irresistible appeal is based on the rich blend of historical, philosophical, political, social, artistic and architectural elements that it brings together.

Here, in this ancient cradle of civilization, human thought gave birth to a multitude of diverse ideas, through which reflections on the origins of humanity gave rise to myths and set in motion the eloquent dance of philosophy. The remarkable flowering of Greek philosophy was a great harmony that reverberated everywhere: in social and political discourse, in literature, in theatre, in science, but also in the famous Athenian democracy and in architectural wonders and art. Indeed, it was the works of the people who lived in these places that helped to ignite the fire of human enlightenment and show the Western world the way to a bliss that has never been equalled. Furthermore, Greece is characterized by a rich diversity of landscapes: countless islands, golden beaches, majestic mountains and timeless villages, all surrounded by a Mediterranean climate. Summers offer sun-soaked warmth, while mild winters provide lush greenery. This diversity makes Greece a captivating year-round destination for nature and culture lovers.

Ancient Greek civilization enjoyed the admiration of various cultures. In the period after the Middle Ages, Europeans studied ancient Greek philosophy extensively and made it their own, being profoundly influenced by its art. The European understanding of the ancient Greek cosmos predates the founding of the modern Greek state and shows a long-standing resonance that precedes today's diplomatic and cultural promotion efforts.

Without going into the period of the Roman Empire and its particular relationship with ancient Greek culture, by the time of the Renaissance, a 'core' of images had already emerged that formed positive social representations of the Greek world, which were disseminated more or less throughout the Western world through painting, sculpture, architecture, literature, poetry, etc.

If we look at this public attention that ancient Greece offered to the new Greek state in terms of tourism from an economic point of view, and especially from a marketing point of view, we would speak of a 'comparative advantage'. However, it is very unclear when the new Greek state realized this advantageous position and how effectively it was used for the promotion of Greek tourism; this is another important issue that we will not address here.

Therefore, in the study of tourism in terms of communication and culture, we will rely on examples relating mainly to the geographical area of Greece, whose history, culture and particular natural geography have attracted people since classical Athens, the Roman Empire, the medieval pilgrimages, the 'Grand Tour' and later the growing modern tourism in the 19th and 20th centuries. However, as this is necessary for a broader understanding of the subject, many examples from other parts of the world will also be presented.

As we will see in this book, modern Greece has based its tourism promotion and image building on the tradition of historical and natural geographical features. However, the tourist image of a country cannot be frozen in time like a photograph with an immortal existence, but is the result of various associations, interactions, borrowings and additions that occur from antiquity to the present. As tourist demand becomes more complicated and international competition in tourism increases at the beginning of the new millennium, numerous attempts are being made to reconstruct the image of Greece or to add to its original image.

Despite the increasing mobility of people and the great exchange of travel-related information worldwide, driven by various factors in both the West and the East, this book is primarily concerned with the field of tourism communication, especially with regard to people travelling to Greece to spend their leisure time, to be fascinated by the culture and to explore the coast. These travellers are fascinated by Greece's ancient heritage, picturesque landscapes, modern culture and the lure of a sun-drenched holiday by the sea.

Contemporary tourists, immigrants and refugees are different groups of people with different intentions and circumstances that lead them to move to a foreign country, but they can all contribute to the cultural exchange in the host country and have different impacts on the local community. However, the trend is that promoting integration and understanding between tourists, immigrants, refugees and local society can create a more inclusive and harmonious environment that benefits everyone.[4]

Undoubtedly, this study has a Eurocentric perspective, not by chance, but because the author's cultural background entails this view, and we cannot ignore the fact that the modern tourism industry emerged in the capitalist world. Comparable situations in non-Western societies and other cultures remain relatively unexplored. Our understanding of social groups that are often overshadowed by history and historians, such as the lower classes, socio-economically marginalized people, women, ethnic minorities and more, is also unfortunately limited.

Nevertheless, we approach our subject with a certain relativism, always recognizing that each culture has its own truths. There is no all-encompassing, universal philosophy, history, sociology or economics, let alone a definitive 'science of tourism'. There is no guarantee that all principles can transcend the boundaries of a particular culture. Their validity is inherently limited by factors such as time, place, political ideology, gender or other dimensions. Each society and subculture has its own structured knowledge, which includes its own philosophy, history, psychology and sociology, and thus its own truth, which is determined by its internal logic.

This relativistic perspective applies to the ancient Athenians, the modern Westerners, the Soviets, the Indians, the Bedouins, the Hobos, and countless others. If we adopt this perspective, we must recognize that our ideas make no claim to universal objectivity. Their validity depends on the perspective of the observer and the context in which they are analysed.

## Structure of the Book

Part One of the book provides a rich theoretical background to the exploration of the interface between media and tourism. This background consists of historical perspectives and foundational theories that are essential to understanding the symbiotic relationship between media and the world of travel.

Chapter 1 deals with the complex role of communication in tourism. It is a tapestry woven from threads of interpersonal communication, the importance of the human body in conveying emotions and the suggestive link between art and tourism communication. Advertising takes its place in this narrative, as does the landscape of mass media and the evolving transformative influence of new digital technologies. The emergence of 'digital tourism' and 'smart tourism' invites us to reflect on their potential and offers a glimpse of a future in which virtual reality will be an important factor in tourism marketing.

In Chapter 2, a historical odyssey is undertaken to trace the development of human mobility and its complex interaction with communication, travel and tourism. The story unfolds from antiquity to the present day and emphasizes the multidimensionality of the tourism discourse.

Chapter 3 examines the historical development of social and cultural dynamics intertwined with the concept of travelling. Social and cultural changes have decisively shaped the terminology of the traveller and the tourist. Definitions of tourism emerge as a result of historical changes, scientific perspectives and the pervasive influence of mass media.

Chapter 4 unfolds a complex relationship between culture and tourism that challenges us to abandon hierarchical judgements and embrace the kaleidoscope of cultural diversity. The intricate interplay of global forces, encompassing issues such as 'cultural preservation', 'cultural exchange' and 'cultural heritage' is analysed in the context of tourism.

Chapter 5, introduces the basic elements of mass media and tourism theories. It examines the processes of communication, the dissemination of information and the profound influence it exerts on public perception, as well as the way modern tourism has been understood and typified by scholars. The main theories running through both areas are uncovered and possible links and parallel developments between them are highlighted.

Chapter 6 explores the complex area of the construction of tourism images. Media, art, cultural diplomacy, marketing and the ubiquitous influence of social media shape perceptions of destinations and form narratives that influence public understanding and travel decisions.

Part Two of the book goes deeper, focusing on the role of static and dynamic media messages in shaping and communicating tourism experiences. From the tactile embrace of printed publications to the ever-evolving landscape of new media technologies, this section unfolds a variety of media.

Chapter 7 takes us back to a time long before the digital age and explores the historical influence of painting on travel motifs. From ancient cave paintings to contemporary digital art, this chapter paints a vivid picture of the role of visual art in shaping images of place and concepts of travel.

Chapter 8 explores the influential role of the written word and literature. It highlights the historical contributions of writers who have left an indelible mark on the perception of destinations and inspired countless journeys.

Chapter 9 looks at the intertwined history of tourism and photography. Its profound influence on how we perceive, experience and remember travel becomes clear. The chapter also examines the role of the press, tourism journalism, posters, postcards, brochures, guidebooks and maps in the emergence of the tourism industry.

Chapter 10 focuses on media with dynamic messages, with cinema, radio and television being the main players. These modern media captivate and inspire travellers through compelling narratives and audio-visual content.

Chapter 11 is dedicated to the digital age, where wireless technologies and new media applications have transformed the tourism landscape. The transition from the analogue to the digital world is explored, highlighting the 'democratizing' influence of the Internet on destination design and the growing influence of user-generated content. The focus is on the importance of communication tools and wireless networks in enhancing the travel experience.

Part Three of the book shifts the focus from the general to the particular in the relationship between communication and travelling. Taking Greece as an example, the historical development of the country's image and its profound effects on tourism are presented. It examines the important role that various actors, institutions and media have played in shaping the image of the region at the southern end of the Balkan Peninsula in different historical periods, from antiquity to the modern digital age. This section also examines the evolution of the concept of tourism in Greece and its complicated relationship with capitalism and the reification of human life.

Chapter 12 takes a journey through the visual arts of ancient Greece, oral traditions, epic poets, representational art and myths. Their influence on travel leaves a lasting legacy, furthering the notion that travel expands human consciousness and shapes the experience of the intrepid traveller.

Chapter 13 takes us back to the time of the ancient Romans and their civilization, which spread across Europe, Asia and Africa. Greco-Roman civilization, the Christian religion and Byzantine culture, conveyed through the media of the time, influenced the way people perceived travel and helped shape their ideas of distant lands and specific destinations.

Chapter 14 explores the European fascination with ancient Greek civilization during the Renaissance and Enlightenment through literature, art, philosophy and cultural movements such as humanism and classicism. The focus is on how this fascination shaped perceptions of the southern Balkans, even during the rule of the Ottoman Empire, and the culture of travel. The narrative involves important European travellers and Greek personalities of the time who play a crucial role in portraying the land where Greek civilization flourished.

Chapter 15 deals with the early 19th century, when the term 'tourism' first appeared in Western dictionaries. The shaping of the image of Greece during this period is examined through the lens of scholars, travel literature, paintings, newspapers, photographs, advertising and international events.

Chapter 16 draws attention to the first half of the 20th century and illustrates the emergence of the tourist image of Greece. Collective memory, interwoven with the media and literature, shapes the perception of modern Greece, which, apart from some efforts by the state tourism services, does not yet have an adequately organized system to support the Greek tourism industry.

Chapter 17 takes us to the period after the middle of the 20th century, when Greece's interest in developing tourism as an economic sector increased significantly. It was during this period that the traditional tourist image of 'sun, sea and sand' emerged, which today is a basic element of the tourist myth of Greece.

The final chapter in Part Three, Chapter 18, analyses the impact of various factors and events such as the unstoppable progress of communication technology, the 2004 Summer Olympics, the economic crisis, and the global coronavirus disease 2019 (COVID-19) pandemic. It is acknowledged that these and other factors have had a significant impact on Greece's ever-changing tourism image in recent times.

After a comprehensive exploration of theories, cultural and technological developments, historical events related to media, travel and tourism, and vivid examples from the region where Greek culture flourished, readers are invited to use their imagination to consider and construct counterfactual scenarios. They are encouraged to consider what might have happened in an alternative reality in which human society would have developed differently and tourism would have taken on different values and meanings. The book assumes that even in such a hypothetical universe, the media would play a central role in shaping the history of tourism and guiding its development. This assertion is based on the realization that communication is an essential aspect of the phenomenon of tourism and that tourism itself is a form of communication.

## Notes

[1] Conventionally, the origins of modern tourism are often linked to specific events, with authors giving different interpretations depending on the aspects of the phenomenon they wish to emphasize. Some link it to the Grand Tour, others refer to the first organized trip by Thomas Cook in 1841, while a broader view associates it with the Industrial Revolution and the transformative impact on labour practices and society (Inglis, 2000). In this book, however, we challenge these views by looking at tourism not as a by-product of cultural appropriation or as an industry emerging from social systems, but as an inherent need or characteristic of our species.

[2] Most dictionaries define the verb 'gaze' as to look fixedly, attentively or deliberately at something, but also as to look blankly or curiously at something. The concept of the 'gaze' has occupied philosophical, sociological and psychoanalytical theories about the relationship between the 'way of seeing' and systems of power, gender, colonization, imperialism or post-colonization, etc. In relation to tourism, John Urry has developed the well-known theory of the 'tourist gaze', based on the dominance of the visual dimension among the other senses, suggesting that the tourist experience involves a particular way of seeing (Urry, 1992, 2008). However, Urry argues that there are many different tourist gazes, such as: (i) the 'romantic tourist gaze', which emphasizes solitude, privacy and a personal, intellectual relationship between the individual and the object of their observation; (ii) the 'collective tourist gaze', which results from the satisfaction of visiting a famous destination or participating in a tourist event; (iii) the 'spectatorial tourist gaze', which concerns the momentary (collective) view of a framed tourist landscape, for example from the window of a bus; (iv) the 'reverential tourist gaze', which concerns the consumption of images in sacred places; (v) the 'anthropological tourist gaze', which examines space on the basis of meanings and symbols; (vi) the 'environmental tourist gaze', which is linked to tourist respect for the environment; and (vii) the 'mediated tourist gaze', which is based on tourist representations essentially created by the media. At this point, it can be argued that some of these types of 'tourist gaze' are more associated with commerce, while others point to a non-commercial view of the concept of tourism (i.e. as an inherent inner human need) (Urry and Larsen, 2011).

[3] Many researchers in this field regard the tourist as a temporary traveller who has a fixed abode during his/her journey, which is also his/her permanent place of residence, and this distinguishes him/her from the nomad, wanderer, vagrant, hobo and other types of permanent traveller. They also see the tourist experience as an exceptional and often unusual condition, a role that someone lives only occasionally or even rarely, whereas for the traveller-passenger, whether nomad, wanderer or vagrant, it is usually a complete way of life. In this sense, they see the tourist as a traveller on a tour, a round trip, so that his/her starting point is also his/her final destination. However, we believe that all these forms of mobility are much closer to tourism as we understand it in this book than the categories that today's tourism industry has established, such as business tourism, shopping tourism and medical tourism. This exclusion is probably close to the view that the poorest minority groups have always been the 'people without history' because they do not contribute much to the statistical categorizations that are made for economic reasons to count 'economic development'.

[4] There is no doubt that the current geopolitical situation in the Middle East, Asia and Africa in general is leading to a great mobility of people to Greece. However, we must emphasize at this point that the regulators of the international economic system today are, on the one hand, trying to remove all barriers that hinder the free movement of capital, commercial enterprises, products and skilled labour, and strongly encourage the free access of tourists to any destination, while at the same time erecting walls and reinforced borders against immigration from poor countries and persecuted refugees. We hope that this book can also inspire us to rethink or change the way we perceive and experience human mobility in general. In a way, it should remind us of the archaeology of certain human values and their true meaning, such as hospitality, altruism, liberty, equality and fraternity for all, which have been forgotten or transformed into opportunistic commercial actions with the rise of capitalism and mass society.

## References


Alexias, G. (2011) *Sociology of the Body*. Pedio Publications, Athens. (in Greek)

Baudelaire, C. (2012) *Journaux Intimes*. Deluxes, Paris. (in French)

Bergson, H. (2014) *Creative Evolution*. Dover Publications Inc., Mineola, New York.

Campbell, M.B. (1988) *The Witness and the Other World: Exotic European Travel Writing, 400–1600*. Cornell University Press, Ithaca, New York.

Cohen, E. (1974) Who is a tourist? A conceptual clarification. *The Sociological Review* 22(4), 527–555.

Hill, S., Abercrombie, N. and Turner, B. (2000) *The Penguin Dictionary of Sociology*. Penguin, London.

Hirst, J. (2012) *The Shortest History of Europe*. Black Inc., Collingwood, Victoria, Australia.

Inglis, F. (2000) *The Delicious History of the Holiday*. Routledge, London.

Jung, C.G. (2014) *The Archetypes and the Collective Unconscious*. Routledge, London.

MacCannell, D. (1973) Staged authenticity: arrangements of social space in tourist settings. *American Journal of Sociology* 79(3), 589–603.

MacCannell, D. (1976) *The Tourist: a New Theory of the Leisure Class*. University of California Press, Berkeley, California.

Rimbaud, A. (1873) *Une Saison en Enfer.* Pierre Mardaga, Bruxelles. (in French)

Rogers, C.R. (1995) *A Way of Being*. Houghton Mifflin Harcourt, Boston, Massachusetts.

Urry, J. (1992) The tourist gaze 'revisited'. *American Behavioral Scientist* 36(2), 172–186.

Urry, J. (2008) Globalising the tourist gaze. In: Babu, S.S., Mishra, S. and Parida, B.B. (eds) *Tourism Development Revisited: Concepts, Issues and Paradigms*. SAGE Publications India, New Delhi, pp. 150–160.

Urry, J. and Larsen, J. (2011) *The Tourist Gaze 3.0*. Sage, London.

Varvaresos, S. (2017) *The Lost Art of Travel*. Papazisis, Athens. (in Greek)

# Part One – Conceptual Framework

The complicated and symbiotic relationship between media and tourism is a captivating topic for both media and tourism studies. Within this dynamic interplay, an area emerges where these two disciplines intersect and illuminate the profound influence each has on the other. Exploring these intricacies provides a deep understanding of how these two fields influence each other and offers valuable insights into their shared role in shaping today's tourism industry.

As diverse and interdisciplinary fields, media studies and tourism studies weave together a rich tapestry of theories, methodologies and perspectives drawn from different academic traditions.

Much like the ever-evolving sciences that adapt to changing times, the paradigms that underpin these fields can change in response to societal shifts, as mass media and tourism carry significant economic weight and growth in today's consumer society. This is leading the academic community to accept these areas as important fields of study in their own right and to examine the complex relationships between them.

The constant evolution of digital technologies brings with it an ever-changing dimension of tourism communication that drives academic research on the interplay between these two fields.

Part One of this book focuses on establishing a conceptual framework for the topic, including basic definitions and parameters for the relationship between communication and tourism. This framework serves as a guide to explore how communication influences and is influenced by different aspects of tourism.

The aim is to promote a common understanding and systematic approach to complex topics and to help the reader navigate seamlessly through the text, organizing their thoughts, identifying variables, providing clarity and recognizing links between the different parts of the book.

Historical reviews, questions raised, proposed definitions and identification of relevant theories form the basis for developing new perspectives on the phenomenon of tourism. The chapters in Part One also serve to identify and highlight issues such as environmental concerns and gender equality that have received more attention in recent decades, which will have a positive impact on future research.

In order to establish a solid conceptual basis for studying the interaction between communication and tourism, important tasks need to be completed. These include: (i) identifying core components; (ii) describing terms, concepts and expressions in detail; and (iii) conducting retrospective analyses to gain valuable insights.

In Chapter 1, the key concepts of communication and tourism are identified and described step by step, tracing the evolution of their relationship over time. The role of communication in sharing travel experiences, promoting country images, creating travel opportunities and influencing perceptions is explored from ancient times to the digital revolution.

The study of the relationship between communication and travel emphasizes the central role it has played in shaping people's travel experiences throughout history is the focus of the discussion in Chapter 2. The complexity of this relationship plays a key role in the development and promotion of today's tourism industry.

Communication serves as a bridge that connects tourists with destinations, service providers and residents, enabling new cultural experiences and enriching the overall tourism experience.


DOI: 10.1079/9781800626034.PART1

Before discussing the relationship between media and tourism, it is important to define communication and tourism. Different aspects of communication in relation to tourism are presented in Chapter 3, as well as different opinions on tourism in relation to communication, taking into account the temporal dimension; the concept of culture will also be briefly discussed in Chapter 4.

Basic theories of media studies and tourism are discussed and form the basis of the study described in Chapter 5. Communication theories include topics such as the strong influence of the media, the two-phase model of media flow, cultivation theory, critical media theory, framing analysis, the model of 'uses and gratifications' and constructivism. In tourism research, sociological and anthropological theories are explored by scholars such as Daniel Boorstin, Dean MacCannell, Erik Cohen, Stanley Plog and John Urry (Boorstin, 1964; Cohen, 1972; Plog, 1974; MacCannell, 1976; Urry, 1992; Kaitatzi-Whitlock, 2012).

The concept of 'destination image' and its formation are explored in Chapter 6 and relate to how a place is perceived by potential tourists. Factors influencing this perception include collective memory, education, marketing, media portrayal, word of mouth and personal experience. Simplified models analyse the key elements, relationships and factors that contribute to tourists' perceptions and their impact on a country's image. The tourism industry's efforts to shape and improve this image are discussed, offering insights into how a country's image can be actively managed and improved.

## References


Boorstin, D. (1964) *The Image: a Guide to Pseudo-events in America*. Harper and Row, New York.

Cohen, E. (1972) Toward a sociology of international tourism. *Social Research* 39(1), 164–182.

Kaitatzi-Whitlock, S. (2012) *Communication: Theory in Practice*. A.A. Livani, Athens. (in Greek)

MacCannell, D. (1976) *The Tourist: a New Theory of the Leisure Class*. University of California Press, Berkeley, California.

Plog, S.C. (1974) Why destination areas rise and fall in popularity. *Cornell Hotel and Restaurant Administration Quarterly* 14(4), 55–58.

Urry, J. (1992) The tourist gaze 'revisited'. *American Behavioral Scientist* 36(2), 172–186.

# 1 Tourism Communication

**Abstract**
This chapter deals with the various facets of communication in tourism, which includes both the transmission of messages and the interactive processes. The initial focus is on the central role of interpersonal communication in maintaining relationships and exchanging information in tourism. The dynamic role of the human body in conveying emotions and shaping tourism experiences is emphasized. The chapter further explores the complex relationship between art and tourism communication, and examines the historical and contemporary importance of advertising in attracting tourists, as well as the evolving role of mass media in the construction of tourist images, and the ability of new media to facilitate transactions and information exchange. The narrative traces the journey of communication from primitive methods to the digital age, emphasizing the transformative impact of digital technologies on two-way interaction and information exchange.

## 1.1 What Is Communication?

The pioneering American sociologist Cooley (1909) argued that communication encompasses all symbols of the mind as well as the means of their transmission in space and time. This includes body movements and postures, facial expressions, gestures, tone of voice, words, writing, typography, railways, telegraph, telephone and all other achievements in the process of conquering space and time.

In fact, there are many different understandings of what communication means. They range from the simple assumption that it is the intentional transmission of a message to a recipient, as in writing a letter or watching television, to the view that it is a kind of interaction and synchronicity in which participants cooperate and coordinate to create shared meanings and reach an understanding.

Focusing more on the relational aspects of the process, we can say that communication is the mutual contact between individuals or groups or, more precisely, the transmission and exchange of messages and information from someone understood as a sender to someone understood as a receiver and vice versa, their interaction through a common system of signals, symbols or behaviours in order to create shared meanings and understand each other (Babiniotis, 1998, p. 655).

It is clear, then, that communication can be regarded as a fundamental part of life. Based on this position, it has been argued that in social life 'one cannot not communicate' and that all human behaviour is a form of communication (Watzlawick *et al.*, 2011).

That is, just as there can be no society without the formation of groups or without human gatherings, analogously there can be no social life without communication. Even a conscious lack of communication manifests itself as such and thus communicates something (Krippendorff, 1994).

Scholars and experts have classified communication into different categories based on various criteria leading to types, styles, etc., such as interpersonal communication, intrapersonal communication, face-to-face communication, mediated communication, verbal communication, non-verbal communication, written communication, visual communication, personal communication, mass communication, passive communication, aggressive communication, business communication, political communication, artistic communication, and so on.

To illustrate some important connections between communication and tourism, certain types of communication and topics are briefly discussed below, such as: (i) face-to-face interpersonal communication; (ii) the role of the human body; (iii) art as a form of communication; and


DOI: 10.1079/9781800626034.0001

(iv) communication and modern mass media, tourism marketing and communication in the new digital age.

## 1.2 Face-to-Face Communication

It makes no sense to speak of personal or social relationships without interpersonal communication, because face-to-face conversations, phone calls, face-to-face video chats, e-mail and letter exchanges, etc. are some of the practical human activities that connect people (Kirchmajer and Patterson, 2004).

Face-to-face interpersonal communication can be seen as the human act of conveying a message to others and ensuring that they understand it. It involves the exchange of information face to face, in the form of voice, facial expressions, body language and gestures, etc., or the absence of all these elements. It takes place when two or more people are in the same environment and can see each other during communication. In online meetings, face-to-face communication can take place even if the people are in a different location, but this type of communication can be described as mediated by technology.

Face-to-face communication promotes the engagement and participation of all parties in the dialogue, reduces misunderstandings, is quick, direct and more efficient, and usually creates a climate of trust. This type of communication is often more effective than text-only or audio-only conversations because when we are in the same natural environment and within reach, we can more easily give verbal and body language cues.

Face-to-face communication and language are interdependent components of effective human interaction. While language is the primary means of transmitting information and ideas, face-to-face communication enriches the communication process by adding non-verbal cues, emotional expressions, immediate feedback and cultural context. The way people use language influences their understanding of the world, their interactions with others and the way they organize and define their society. Language is the most common means of constructing and developing social reality. It not only reflects existing social structures and norms, but also actively shapes and influences them. In other words, language serves to construct and maintain the social world (Assmann, 2011).

In this respect, the discourse on interpersonal communication, especially face-to-face interpersonal communication, is fundamental to any study dealing with social and personal relationships, and in our field, tourism, it certainly occupies a special place. Effective interpersonal communication is very important for building and maintaining relationships. It enables people to resolve conflicts, exchange information, express feelings and understand each other. Interactions and communication skills in the professional environment of the tourism industry are issues that have preoccupied tourism studies from the beginning, both practically and theoretically, and continue to do so today (Burton and Burton, 1994).

## 1.3 The Human Body in Communication

However, as we have already noted, interpersonal communication also includes non-verbal elements (i.e. tone of voice, facial expressions, body language, gestures, etc.). The combination of body movement and dialogue in the form of language helps individuals to express themselves fully and to interpret the intentions of others. When body movement and language are in harmony, communication becomes more effective and promotes better understanding and cooperation between people. In short, the body is also central means to face-to-face interpersonal communication and social interaction in general, and as Goffman (1963) argued, it is an active element and not just a passive biological existence.

Since the body always plays an important role in initiating and establishing social relations, it is thus an essential element in the development of communication in tourism. In this sense, the body of the traveller or any other person participating in the tourist phenomenon, through a series of factors such as appearance, physical condition, experiences, emotions, passions, expressions and movements, forms its own dynamic communication system that receives, decodes, encodes and transmits messages and also reflects the attitude of society as a whole towards this phenomenon.

We can argue that the body of the traveller, or of any other person participating in the phenomenon of modern tourism, is repeatedly transformed to adapt to different forms of consumption or, more generally, to the transactions carried out in the developing consumer society, each time following different forms of habitus and assigning the desired social status to its owner (Featherstone, 1982; Bourdieu, 2018).

It must be emphasized, then, that in tourism, as in all areas of human life, the body can serve as a signifier in the semiotic sense. Thus, a body dressed in a certain casual manner, tanned by the sun, constantly holding a camera (or nowadays rather a mobile phone) and photographing tourist sights (i.e. adopting a certain posture and following certain typical movements and behaviours) corresponds to the signifier 'tourist'.

The bodily practices associated with tourism, such as the body tanned in a certain way from bathing in the sea or from the sun of the snow-covered mountain, can be explained as an attempt to show a social differentiation or even a norm within the mechanisms of social disciplining vis-à-vis the consumer culture of the bourgeoisie; in any case, many other interpretations can be put forward, depending on the intention of the researcher.

But apart from that, we must always keep in mind that the body is also a biological entity that has biological needs, feelings, passions, diseases, etc. that can lead to irrational behaviour. Moreover, the body is also active. It can shape social relations in tourism and contribute to their development, but also to their reconstruction and renewal. Finally, the body can be represented in the media and lead to social constructions and myths.

An important parameter of social analysis in relation to communication and tourism can thus be to examine how the tourist's body has been constructed by the tourism industry and how it has functioned over time, how it has been portrayed by the media and how the body itself has influenced tourism.

The tourist's body has been socially constructed as a means of freedom but also of control in relation to human activity, and the media has played an important role in this construction. In this sense, the tourist's body is often portrayed in the media as free from the conventions of everyday life, but it is obvious that it is constrained and controlled by the protocols of the tourism industry and the restrictions imposed by tourism products.

According to this logic, visitors to Disneyland, for example, can give free rein to their imagination, wear typical costumes of Disney characters bought in souvenir shops, paint their faces, express themselves with cartoon-like behaviour, etc., but within the limits imposed by the theme park in question.

The tourist's body as portrayed in the mass media also reflects the values, beliefs and stereotypes of society. In this context, the tourist's body has been the subject of many debates among social scientists on hedonism and sensuality, the objectification of women, self-discipline and self-management in the tourism industry, etc. (Pritchard and Morgan, 2011; Pemayun and Brata, 2018).

It is true that tourism is a universal phenomenon nowadays, and when we think about it, we can always ask ourselves if there is a similarity in the way tourists or hotel workers, etc., use their bodies in different countries or different eras (i.e. if there are certain diachronic techniques of the body that characterize the tourist or the hotel worker globally and are thus common to all people when they practice tourism). In other words, when we travel for pleasure, for example, are there certain connotations of our bodies that are common to all of us, whether through our desires as tourists or through the structures of tourism production, its regulation, etc., or is there at least a common image in the way tourism is portrayed internationally in the mass media?

These are all complex questions that are very difficult to answer unilaterally, as they are characterized by many dimensions and many logical answers can be given.

Be that as it may, let us return to our discussion of communication and its development. As early humans developed cognitive abilities and more complex kinship systems, the storage of thoughts and information became a crucial function for survival, social cohesion and cultural development. In other words, humans sought communication spaces outside of face-to-face interpersonal communication; spaces where they could store their messages and communicate across time, or space, to preserve knowledge, ensure survival, maintain cultural identity, facilitate communication and advance in various fields. This required the invention of methods to encode, store, transmit and retrieve messages. So after direct interpersonal communication came mediated communication.

## 1.4 Art as a Form of Communication

Since prehistoric times, people have used numerous means of communication, through various media such as artistic representations and symbols on cave walls, stone tablets, papyri and tarpaulins, or with the help of fire smoke, which was used to transmit messages over great distances from mountain to mountain. Later, the written word on paper, the press, the telegraph, the voice through the telephone, the modern means of mass communication, digital media, etc. were added.

Mediated communication fulfilled essential human needs, including bridging distances, acquiring information, efficiency, education, expression, social contact and access to resources. But mediated communication also contributed to the development of artistic expression through the creative use of various communication technologies. And one cannot deny that artistic expression has the power to stir emotions, challenge perspectives, stimulate critical thinking and inspire action.

It is therefore not difficult to assume that works of art such as painted ceramics, statues, mosaics, novels, paintings, art photographs, cinema films or graffiti, as well as various other media products that presuppose a certain artistic sensibility (e.g. television series, video productions), can easily convey a strong image of a place, and have the potential to attract people, and thus form strong tourist representations. In this sense, one could argue that artistic communication is closely linked to tourism.

However, in order to understand this link between artistic communication and tourism, we need to briefly look at some concepts of art.

From ancient times to the present, there have been different understandings of the meaning of art. In particular, art has often been seen as an imitation of nature and life, a revelation of a deeper reality, an expression of the creator's feelings, or a free play of the imagination. It is also often argued that art can aim at the self-realization of the creator, but also at economic, cognitive-pedagogical, moral and sociopolitical purposes. On the other hand, it can be argued that art is an autonomous activity, a 'purposeless expediency', as the 18th-century German philosopher Immanuel Kant (1724–1804) put it.

Arthur Schopenhauer (1788–1860), another German philosopher, held that art is an escape from the world of daily life, which is essentially violent and frustrating. He believed that when we look at works of art (i.e. paintings, sculptures and other forms of art) but especially when we listen to music, we feel that the weight of our personal existence and the pressure of our will that weighs us down in our lives is lifted, so that we forget our worries and unite with the eternal ideas that art expresses (Pellegrinis, 1998).

The question, then, is what defines something as an art product (i.e. what makes it be considered beautiful by the world) and consequently, considered in the context of this book, what makes it possible to involuntarily and easily form mental images about a place that create strong tourist representations.

If we think about it, the appreciation and attractiveness of a work of art is a question directly related to the definition of art, and as we have already mentioned, there are many variants of interpretation.

Another German philosopher, Friedrich Schiller (1759–1805), who may have been influenced by Kant's ideas about art, considered the aesthetic experience to be entirely selfless. That is, a work of art has value for itself and not as a means to an end to achieve some other end. This contrast is reflected in the difference between work and play. Work is seen as a means to an end (e.g. to save money) but play is in itself a pleasure and is the model for various activities such as sport, walking, travel, art and so on (Scruton, 1994).

At this point we can recall the basic idea that we often elaborate in this study, namely tourism as an inner need. As we have already mentioned, both art and travel can be seen, in Schiller's sense, as derivatives of what we call 'play' in the most general pedagogical sense.

Play is an inner need, an activity that is not tied to a material interest. Children do not play to achieve a material gain. Of course, there are goals in play, but these goals are seen as part of the play itself; children just want to play and perhaps win their game, not to profit from it. Play promotes a sense of autonomy, develops creative thinking and is seen as the root of all human expression (Huizinga, 1955).

In other words, we can say that tourism, like art, belongs to the realm of play associated with free expression and the development of individual consciousness, as opposed to work, which always has a rational purpose.

Although travel as an art has been discussed by scholars for many centuries, these discussions have never been part of an organized and systematic body of knowledge (Adler, 1989).

It can be said that tourism was mostly tied to the economic perspective and treated by the tourism industry as a means to make profit, which was only one dimension of tourism. However, as Adler (1989) states, tourism can also be practised as art. It is a means of expression, and it can also be said that when we travel we use certain classical aesthetic means, framing and isolating our impressions and emphasizing some particular feelings at the expense of others.

To continue with the descriptions of art: from a materialist point of view, according to Karl Marx (1818–1883), the aesthetic impulse is a historical product, the result of successive changes in the specific relations of material production, rather than a special sense or gift that can be attributed to a teleological, supernatural force or at least a biological endowment of the artist (Lifshitz, 2018).

That is, according to this view, art ultimately has a dialectical-historical dimension as part of the superstructure and, in particular, the institutions, principles, perceptions and ideas of a society based on its particular economic structure.

It should also be noted that several recent studies from the field of social sciences consider art as a form of social activity that, due to its 'nature', is not characterized by inherent, intrinsic values. Instead, they examine the production of art products as a social construction that develops within mapped historical and social contexts. In other words, this view holds that no artwork has a natural artistic value in itself.

According to Becker (1982), artistic value is not an intrinsic and inalienable substance of an object, but a 'label' attached to it by members of a social group that belongs to the 'art worlds' and has an interest in considering this object as 'art' (Tsangaris, 2016).

More or less in the same train of thought, MacCannell (1973) stated that the artwork becomes 'authentic' with its reproduction, in reversal of the view of Benjamin (1969), who believed that the authenticity of an artwork derives from its aura, which is associated with a kind of original ritual.

MacCannell therefore held that it is the reproductions of a work of art that ultimately constitute its aura and its socially constructed value. And given this position, he argued that tourist attractions in modern society are constructed in this way, including artistic attractions, and that the reason for the Grand Canyon's tourist 'aura' is due to this process of reproduction and construction (MacCannell, 1973).

Be that as it may, whether art is seen as a divine favour, a social construct or something else, experience with the relationship between art or aesthetic representation in general and tourism shows that it can make a decisive contribution to shaping tourism representations. Moreover, it is generally recognized that, in addition to the visual arts, all cultural derivatives and applied arts of a particular place, such as traditional artefacts, fashion, cuisine, folk tales and myths, can also contribute to this process.

At this point it would not be superfluous to mention that all these artistic expressions are usually not the work of voluntary efforts by the tourism industry to promote places. In other words, the attractiveness of art, together with all other external factors and internal needs of people, has played a very important role since ancient times in motivating people to visit faraway places without being supported by an existing tourism industry.

Nevertheless, the rationalization of the social relations of tourism after the Industrial Revolution in modern Western society was deemed necessary to prevent this act from breaking out of the social norms and rules of capitalism.

In order to become a commodity, the act of 'travelling for pleasure' had to be rationalized (i.e. specific procedures and protocols had to be put in place that would define with certainty the characteristics and limits of the tourist and all others involved in the process). And in this sense, the disciplining of the body that participates in the tourism process by adopting appropriate cultural behaviours and expressions as tourist acts could also be seen as part of the necessary process for the production and reproduction of capitalist society.

Thus, with the advent of industrial society and capitalism, tourism communication essentially referred only to the ways in which information and messages were transmitted to promote and facilitate tourist activities within an economic system characterized by private property and profit, while other organic forms of communication about travel for pleasure were generally ignored.

And it is worth mentioning at this point, in relation to the theme of this book, that tourism communication can be allegorically linked to the ancient Greek myth of the sirens, who were seductive creatures with the body of a bird and the head of a woman. These creatures sang so beautifully that no one could resist approaching them. They lived on rocky islands and whenever they saw passing sailors, they sang loudly and seduced them, causing their ships to crash against the rocks.

In a way, then, the sirens can be metaphorically linked to the idea of persuasive communication and advertising in tourism. Tourism promotion often aims to create persuasive and irresistible messages or images that entice consumers to visit a destination or buy a tourism product or service. The sirens' song, which is impossible to resist, can therefore be well compared to the persuasive techniques used in advertising to attract consumers' attention and interest.

It is important to note, however, that this connection is metaphorical and in any case not a complete correspondence, as it can only be made in retrospect, since the sirens were never involved in modern tourism, commerce or advertising. The myth apparently served as a cautionary tale in ancient Greece about the power of persuasion and the possible consequences of seduction by enticing messages or offers intended to lure the public.

Be that as it may, the ethical implications of tourism advertising today are of course not as dangerous as the sirens' song for sailors, but they are a worrying, complex and constantly evolving issue in terms of consumer deception, the transparency of tourism offers and, more generally, the impact of tourism advertising on society.

## 1.5 Communication and Mass Media

Returning to our exploration of the intersection between mass media and tourism, it is instructive to recognize that the rise of modern mass media – consisting of affordable newspapers, cinema, radio, television and the like – in the 20th century was a major factor in shaping the prevailing understanding of the concept of mass communication. In particular, it directed popular perception towards the one-way model of mass communication. This model, characterized by the transmission of messages from a single sender to numerous receivers, gained prominence and overshadowed the broader, more nuanced meaning of communication.

The technical structure and organizational framework of these mass communication channels inherently limited them to enabling only a vertical and unidirectional flow of information. Consequently, they prevented the possibility of a dialectical form of mass communication in which a more interactive and two-way exchange of information could take place.

As a result, many understood communication to be merely the transmission of information by a group of a few people to larger populations. This emphasis on the concept of communication as the simple transmission of a message in one direction to a large audience became the standard of mass communication that has prevailed in modern times, supported by theoretical considerations such as the Shannon-Weaver model of communication (Shannon and Weaver, 1949).

At that time, people informed themselves about tourist attractions through what magazines and newspapers wanted to present to them, they went to the cinema and dreamed about the spectacular locations a director had chosen for the film, or admired in front of the television the enchanting places shown in documentaries.

As the communication model went in one direction, from one sender to many receivers, information was completely controlled by the media industry, although, as already mentioned, the most important elements of comprehensive communication are exchange and interaction. On the other hand, one could say that some kind of real communication about tourism also took place when people exchanged information about places they had visited in their immediate environment such as family, friends, the workplace, or at least on an interpersonal level in the tourist office.

But the recent transformation of the mass communication landscape through the development of digital network systems has changed everything.

Undoubtedly, new relationships between people and media have emerged today, bringing back to the forefront the basic and primary meaning of the word communication as a two-way process where more people can interact equally.

Nowadays, tourist destinations are no longer served up before the viewer's eyes in a limited way, but anyone can search the Internet for audio-visual material and information about the places that interest them, post pictures or written texts about travel experiences, or share photos showing the sights of their country.

There is no doubt that the Internet offers a wealth of opinions and access to research and published tourism information. At this point, of course, it should be noted that the claim about the omnipotence of the individual in the communication universe of the network society has been questioned by several scholars due to certain factors.

Although there is a belief among scholars that mass communication via the Internet, which is currently two way, is equitable because the 'one sender for many receivers' media model has essentially transformed into a 'many for many communicators' model, some researchers believe that egalitarian communication has not really been successful. This is due to the architectural features of the Internet, which do not provide all users with the same opportunities to disseminate information widely. Moreover, social media applications can be controlled, censored, etc., so things have not changed as extremely as we would like compared to the previous situation (Lovink and Rasch, 2013).

Be that as it may, thanks to new technologies, the possibility of truly participative and two-way communication by technological means on a large scale and worldwide is closer to realization today than in any other period of history.

Although we have said that we will deal as little as possible with the economic side of tourism, at this point we would like to mention briefly, in a parenthesis, another variant of communication, namely 'tourism advertising', which is one of the most important activities for the good functioning of the tourism industry. In any case, tourism marketing has developed greatly in the age of new media, and therefore, in a book on the relationship between communication and tourism, it is in some way necessary to make some reflections on it.

'Tourism promotion' belongs exclusively to the field of business administration and is a form of communication whose main aim is to arouse the public's interest in certain tourist products or destinations. In business, the term promotion generally refers to the totality of activities aimed at making a product, brand or service known.

The organized establishment of tourism agencies to promote the tourism industry through mass media belongs to the field of tourism marketing. In this context, it should be noted that the tourism industry's promotional mechanism belongs to one of the variables of tourism marketing, also known as the 4Ps, a distinction derived from the initial letters of the words Product, Price, Place and Promotion. Promotion in turn includes advertising, sales techniques, sales promotion and public relations (Borden, 1964; Kotler, 1997).

To make things a little clearer, some of the important tools used by the tourism industry for promotion are: (i) travel posters; (ii) tourist maps; (iii) travel guides; (iv) travel agency brochures; (v) mail letters; (vi) telemarketing; (vii) tourism reports; (viii) newspaper and magazine advertisements; (ix) radio and television commercials; (x) sponsored radio and television programmes; (xi) travel journalism; (xii) sponsored cinema and television film productions; (xiii) travel websites; (xiv) e-mail offers; (xv) social media posts; (xvi) travel platforms; and (xvii) tourism exhibitions (Masouleh and Morteza, 2002).

It is clear that all these forms of tourism promotion are closely linked to the media and communications industry, and much of it is purely professional advertising.

Specifically, advertising, which is part of sales promotion, is a common form of informative, but above all manipulative, promotional communication that takes place in exchange for money. To make it even clearer, advertising is the unilateral communication of a company, organization or other entity with the public through promotional strategies to publicize and promote a product or service through the mass media at a certain financial cost. It is often said that advertising has a great influence on the preferences of the products offered and that it is one of the key factors of symbolic consumption.

It would be an omission not to emphasize that marketers very soon discovered that potential tourists based their decisions for tourist products on the representations offered by the media as well as on symbolic consumption.

Thus, advertising and publicity were already recognized as important promotional tools in the tourism industry in the early years of the development of modern mass tourism, and one of the most frequently used media was the poster.

As will be explained later, some communication products typically in the arts or leisure sector, such as films or television series, have been recognized and used in some countries for many years (albeit sporadically and slowly) as indirect promotional tools of organized tourism promotion. Their productions have sometimes been co-financed by destination management organizations, but also by tourism businesses.

Today, the tourism industry's advertising messages are mainly disseminated through the new digital media. However, billboards, newspapers, magazines, radio, cinema and television are still considered appropriate channels through which the tourism industry can produce attractive images to promote specific places, airlines, ferry companies, tourist boards, hotels, theme parks, etc.

Be that as it may, according to marketing research, each of these media has its own strengths and weaknesses when it comes to reproducing different types of advertisements and reaching different audiences (Assael, 1981; Batra *et al.*, 1995; Chaudhuri and Buck, 1995).

Therefore, it is easy to understand that in tourism advertising, choosing the appropriate communication channel is a key issue to reach the target audience and potential tourists, even in the digital age (Kim *et al.*, 2005).

Of course, there are differences in the process of constructing, transmitting, receiving and understanding a message in relation to the medium used. That is, a literary text, a newspaper, a song, a radio programme, a commercial, a TV series, a film or a multimedia presentation on a platform on the Internet are based on different production processes, while their messages are also transmitted based on different technologies and do not always appeal to the same senses. Therefore, each of the media used to present a topic shapes the impressions and perceptions of the recipients in different ways.

However, it is important to note that in most cases the information tourists receive today about a particular place comes from a combination of different media, not just one form.

Another aspect concerns the ethics of tourism advertising, which we have already discussed in the context of the representation of the tourist body in the media, has to do with the degradation of women through one-sided representations of the female body in tourism advertising. Thus, tourism advertising can be a mechanism that legitimizes and perpetuates hierarchies and inequalities between genders, but also differences between social classes, between nationalities, sexual orientations, etc. in the way they are portrayed in tourism activities.

In recent years, several studies have been carried out in the field of tourism on the representation of gender relations and the gender roles assigned to them in advertising. The results of most studies in the field of tourism, but also in other professional fields represented in advertising and the media, show that women are indeed portrayed as inferior to men and are always associated with outdated and stereotypical roles. It is worth noting that these representations sometimes portray gender relations through old and outdated notions and in this way not only contribute to supporting sexist stereotypes in tourism, but as untrue representations, distort reality and reinforce gender inequality (Sirakaya and Sonmez, 2000; Pritchard, 2001; Khan, 2011; Morgan and Pritchard, 2018).

As portrayed in tourism advertisements, women usually do not even get to speak, while men do all the talking and decide everything about the family, couple or group's destination. The basic image of the protagonist tourist, as often portrayed in Western media, corresponds to a certain type of person, namely the white, heterosexual male, who is seen as privileged in his decisions and actions over all other social groups that appear in the representations (Anderson, 1996; Duncan, 1996; Morgan and Pritchard, 1998; Khan, 2011).

On the other hand, women are usually used by the artistic directors of advertising to embellish the aesthetic aspect of a destination's image, or as indicators of courtesy and luxury in the services offered by tourism agencies such as airlines or hotel resorts (Kinnaird and Hall, 1994).

Other studies on tourism advertising have found that women are portrayed as objects waiting to be explored by the male viewer-tourist along with the destination (Bowen, 2002).

In this context, the question arises as to whether the image of tourism projected by the media portrays this growing phenomenon in a biased and outdated way that is incompatible with a progressive vision of society. It is also an important argument to reflect on whether tourism, as it is perceived in the world today, is absolutely characterized by the marketing methods of the tourism industry.

Finally, it is important to note at this point that the broader sector of mainstream media, which developed through entrepreneurship in a competitive market, has evolved in recent decades into the model of 'infotainment', which is essentially based on the combination of information and entertainment.

Although issues of quality in communication and regulation of mass media content have led to much debate and rivalry in mass media studies and discussions at the state level on issues such as language norms, censorship, fake news, violence and adult material, it could be argued that since the communication product is freely available, it must have a certain minimum level of respect for the public.

It can be argued that nowadays news programmes and light entertainment programmes as well as advertising use more and more the same narrative codes. Due to the increasing competition and commercialization that prevails in the world of communication, the main objective of most media is to attract the largest possible audience.

As a result, all programmes today master the same narrative elements of communication products, such as dramatization, personification, sound effects, music and fragmentation, in order to create suspense, pleasure and enjoyment and to hold the viewers' attention. For news programmes in particular, the increasing competition means that media producers prefer to create news in a way that appeals to viewers. This includes narrative structures such as those used in cinema or television films, often at the expense of objectivity or the seriousness of the subject matter covered.

Similarly, in entertainment programmes, apart from minor social issues and 'gossip', serious news of public interest is treated in a very light manner, which undermines its importance. In this way, all forms of programming are mixed, while elements of the aesthetic code of cinema, television series and advertising are applied to every kind of media text, forming a hybrid model of documentary and entertainment.

The expression of critical thinking in the media and journalism are no longer serious components of the programme, but aesthetically pleasing narratives whose main purpose is to entertain the audience rather than to seriously inform and impart knowledge (Pleios, 2011).

Given this, it is not surprising that the public is confused about the formats of programmes, as it has become difficult to distinguish pure news from light entertainment, where commercial announcements such as advertising, sponsorship, product placement and public relations (PR) announcements are allowed at any point in the communication process.

Apart from the changes in the format of mainstream media programmes, the entire field of communication has changed drastically in the last 20 years due to the dynamic development of new digital media technology and wireless networks.

The functions of the media have expanded from simply providing infotainment and advertising to directly engaging the public. The 'age of access' has now dawned, an age that is, as Rifkin (2000) puts it, radically different from the previous mass media age, as 'timely' access and participation in any kind of service is enforced and made the norm by vast commercial networks in cyberspace.

Today, people can plan their holidays at any time through easy-to-use tourism brokerage platforms or direct online transactions with tourism businesses of all kinds, but they need to keep abreast of what is on offer because everything is in flux on the Internet.

As a result of the changing communication landscape, the role of media has evolved from a simple infotainment provider to a crucial factor in the entire tourism process, including transactions that used to be carried out by other services (e.g. travel agencies, and car hire companies).

In today's hyperreal media world, it is difficult to separate seriousness from wit, the important from the tragic, the real from the virtual. Even catastrophic images presented in cinematic form on multimedia news platforms enhance the emotional charge of the confused audience and can be a means to arouse curiosity (i.e. tourist attractions). In the various forms of Internet communication (such as news and event websites, online forums and social media), posts on political issues, reports on violent crimes or celebrity weddings, advertisements for products and holiday resorts, and film trailers can be found side by side in the same visual space.

The 'user-generated content' of Web 2.0, combined with the 'infotainment model' and the commercial orientation of the Internet, leads to heterogeneous portfolios in which the real is presented with the fake, the serious with the silly, the moral with the corrupt, the general with the domestic, the political with the commercial, the private with the public, etc., all coming together in an 'eternal, illusory party' (Ludovico and Cirio, 2013).

In parallel, new definitions of tourism have been introduced, such as digital tourism, e-travel, e-tourism or smart tourism, most of which refer more or less to the introduction of the new information and communication technologies in tourism and the digital support of travellers before, during and after their trip.

In addition, 'virtual reality' technology is constantly evolving and its applications in the tourism sector have many very important implications. This technology allows access to a virtual world of attractions, destinations, museums, etc., where an actual visit is impossible in many cases, and yet is a dynamic tool for tourism marketing.[1]

Countless monuments and cultural assets around the world, already digitized as virtual three-dimensional (3D) models and accessible to the public via the Internet, have the potential to shape the tourist experience and the 'image of a destination' in a unique way before the tourist visits the site.

It could be argued that all the fanciful images projected by the mass media, the superficial and flashy stories, combined with the mainstream practices of tourism promotion taking place in the fragmented but symbiotic environment of the Internet, promote a strange postmodern state of tourism.

The infotainment model of the mainstream media and the content portals of the Internet, which combine information from various and diverse subject areas, such as news, gossip, advertising, reviews or stories, construct a perceptual pattern in the form of a mosaic that is able to create attractive images for an unusual range of new and old tourism products that are constantly reinventing themselves. Tourism representations shaped by this new model include hyperreal places promoted by the media, such as the mass-appeal destinations Europa Park, Galaxie Express Space Centre, Disneyland, Land of the Hobbit, Michael Jackson's Neverland Ranch and Hollywood Studio Tours. But also these tourism representations include enticing offers to visit geologically dangerous areas or other places affected by disasters (Mount St Helens, Chernobyl, Ground Zero, Aogashima, etc.) or even places where heinous acts have been committed (Mansonland, NYC Crime Tours, the Killing Fields in Cambodia, etc.).

In addition to the basic idea of tourism, new forms such as dark tourism, hyperreality tourism, adventure tourism, nostalgia tourism, smart tourism and robot tourism are being introduced or reintroduced, resulting in a mix of traditional forms of tourism and diverse, eclectic and personal experiences. And it must be stressed that in all these cases, in this new game of hyperreality, people have the opportunity to intervene as active players in the tourism phenomenon by presenting their own ideas and interacting in real time through social media.

## 1.6 Summary

To summarize, the tourist experience is profoundly shaped by the complex interplay of communication dynamics and media influence that evoke emotions and shape perceptions. Although media channels are able to develop compelling tourism narratives and attract visitors to destinations, they also have the potential to reinforce stereotypes and simplify complicated realities. The current trend in mainstream media towards 'infotainment' is blurring the lines between news and entertainment, resulting in tourism representations characterized by hyperreal destinations and a range of experiences tailored to postmodern engagement via digital platforms. It can therefore be said that modern media is in some ways helping to shape the human drive for mobility, leading to homogenized tourism experiences and an over-commercialization of tourism in certain contexts.

## Note

[1] Virtual reality is the simulation of reality by technological means using multimedia inputs such as head-mounted displays, data gloves, three-dimensional audio systems and magnetic position trackers. In general, the simulation of reality already exists on the television screen or on the cinema screen and, if we want to go further, in the realistic works of art. Today's virtual reality technology is three-dimensional and interactive (Watson and Hill, 2012, p. 316).

## References


Adler, J. (1989) Travel as performed art. *American Journal of Sociology* 94(6), 1366–1391. DOI: 10.1086/229158.

Anderson, K.J. (1996) Engendering race research. In: Duncan, N. (ed.) *Bodyspace: Destabilizing Geographies of Gender and Sexuality*. Routledge, London, pp. 195–208.

Assael, H. (1981) *Consumer Behavior and Marketing Action*. Kent, Boston, Massachusetts.

Assmann, J. (2011) *Cultural Memory and Early Civilization: Writing, Remembrance, and Political Imagination*. Cambridge University Press, New York. DOI: 10.1017/CBO9780511996306.

Babiniotis, G. (1998) *Dictionary of the Modern Greek Language*. Kentro Lexicologias, Athens. (in Greek).

Batra, R., Myers, J.G. and Aaker, D.A. (1995) Group influence and word-of-mouth advertising. In: Batra, R., Myers, J.G. and Aaker, D.A. (eds) *Advertising Management*, 5th edn. Prentice-Hall, Saddle River, New Jersey, pp. 345–363.

Becker, H.S. (1982) *Art Worlds: Updated and Expanded*. University of California Press, Berkeley, California.

Benjamin, W. (1969) Paris: capital of the nineteenth century. *Perspecta* 12, 165–172. DOI: 10.2307/1566965.

Borden, N.H. (1964) The concept of the marketing mix. *Journal of Advertising Research* 4(2), 2–7.

Bourdieu, P. (2018) Structures, habitus, practices. In: Faubion, J. (ed.) *Rethinking the Subject: An Anthology of Contemporary European Social Thought*. Routledge, New York, pp. 31–45. DOI: 10.4324/9780429497643.

Bowen, H.E. (2002) *Images of Women in Tourism Magazine Advertising: a Content Analysis of Advertising. in 'Travel and Leisure' Magazine from 1969 to 1999*. Texas A & M University, College Station, Texas.

Burton, J. and Burton, L. (1994) *Interpersonal Skills for Travel and Tourism*. Pitman Publishing, London.

Chaudhuri, A. and Buck, R. (1995) Affect, reason, and persuasion: advertising strategies that predict affective and analytic-cognitive responses. *Human Communication Research* 21(3), 422–441. DOI: 10.1111/j.1468-2958.1995.tb00353.x.

Cooley, C.H. (1909) *Social Organization: a Study of the Larger Mind*. Charles Scribner's Sons, New York. DOI: 10.1037/14788-000.

Duncan, N. (1996) Renegotiating sexuality in public and private spaces. In: Duncan, N. (ed.) *Bodyspace: Destabilizing Geographies of Gender and Sexuality*. Routledge, London, pp. 127–144. DOI: 10.4324/9780203974070.

Featherstone, M. (1982) The body in consumer culture. *Theory, Culture & Society* 1(2), 18–33. DOI: 10.1177/026327648200100203.

Goffman, E. (1963) *Behaviour in Public Places*. Free Press, New York.

Huizinga, J. (1955) *Homo Ludens: a Study of the Play-element in Culture*. Beacon Press, Boston, Massachusetts.

Khan, S. (2011) Gendered advertising in tourism: an assessment of tourists' perceptions. *South Asian Journal of Tourism and Heritage* 4(2), 54–63.

Kim, D.Y., Hwang, Y.H. and Fesenmaier, D.R. (2005) Modeling tourism advertising effectiveness. *Journal of Travel Research* 44(1), 42–49. DOI: 10.1177/0047287505276590.

Kinnaird, V. and Hall, D. (eds) (1994) *Tourism: a Gender Analysis*. Wiley, Chichester, UK.

Kirchmajer, L. and Patterson, P. (2004) The role of interpersonal communication in the development of client trust and closeness in a SME professional services context. *Small Enterprise Research* 12(1), 45–56. DOI: 10.5172/ser.12.1.45.

Kotler, P. (1997) *Marketing Management*. Prentice-Hall, Saddle River, New Jersey.

Krippendorff, K. (1994) A recursive theory of communication. In: Crowley, D. and Mitchell, D. (eds) *Communication Theory Today*. Polity Press, Cambridge, UK, pp. 78–104.

Lifshitz, M. (2018) *The Philosophy of Art of Karl Marx*. Topos, Athens. (in Greek).

Lovink, G. and Rasch, M. (2013) *Unlike Us Reader: Social Media Monopolies and Their Alternatives No 8*. Institute of Network Cultures, Amsterdam.

Ludovico, A. and Cirio, P. (2013) Face-to-Facebook, smiling in the eternal party. In: Lovink, G. and Rasch, M. (eds) *Unlike Us Reader: Social Media Monopolies and Their Alternatives*. Institute of Network Cultures, Amsterdam, pp. 254–258.

MacCannell, D. (1973) Staged authenticity: arrangements of social space in tourist settings. *American Journal of Sociology* 79(3), 589–603. DOI: 10.1086/225585.

Masouleh, T. and Morteza, T. (2002) Tools international tourism advertising. *Journal of Marketing Specialist* 13–17.

Morgan, N. and Pritchard, A. (1998) *Tourism Promotion and Power – Creating Images, Creating Identities*. Wiley, Chichester, UK.

Morgan, N. and Pritchard, A. (2018) Gender, advertising and ethics: marketing Cuba. *Tourism Planning & Development* 15(3), 329–346. DOI: 10.1080/21568316.2017.1403372.

Pellegrinis, T. (1998) *The Five Seasons of Philosophy*. Ellinika Grammata, Athens. (in Greek).

Pemayun, A.G.P. and Brata, I.B. (2018) Implementation of tourism facilities expansion to hedonism and community imaging. *International Journal of Social Sciences and Humanities* 2(1), 33–42. DOI: 10.29332/ijssh.v2n1.76.

Pleios, G. (2011) *The Information Society*. Kastaniotis, Athens. (in Greek).

Pritchard, A. (2001) Tourism and representation: a scale for measuring gendered portrayals. *Leisure Studies* 20(2), 79–94. DOI: 10.1080/02614360121679.

Pritchard, A. and Morgan, N. (2011) Tourist bodies, transformation and sensuality. In: Branham, P. and Wagg, S. (eds) *The New Politics of Leisure and Pleasure*. Palgrave Macmillan, London, pp. 153–168. DOI: 10.1057/9780230299979.

Rifkin, J. (2000) *The Age of Access: the New Culture of Hypercapitalism*. Tacher-Putnam Books, New York.

Scruton, R. (1994) Continental philosophy from Fichte to Sartre. In: Kenny, A. (ed.) *The Oxford Illustrated History of Western Philosophy*. Oxford University Press, Oxford, UK, pp. 193–238.

Shannon, C.E. and Weaver, W. (1949) *A Mathematical Theory of Communication*. University of Illinois Press, Champaign, Illinois.

Sirakaya, E. and Sonmez, S. (2000) Gender images in state tourism brochures: an overlooked area in socially responsible tourism marketing. *Journal of Travel Research* 38(4), 353–362. DOI: 10.1177/004728750003800403.

Tsangaris, M. (2016) Sociology of art – Howard S. Becker's approach. In: Nina-Pazarzi, E. and Katsikidis, S. (eds) *From Social Thought to Sociological Theory*. Nomiki Vivliothiki, Athens. (in Greek), pp. 279–290.

Watson, J. and Hill, A. (2012) *Dictionary of Media and Communication Studies*, 8th edn. Bloomsbury Publishing, London.

Watzlawick, P., Bavelas, J.B. and Jackson, D.D. (2011) *Pragmatics of Human Communication: a Study of Interactional Patterns, Pathologies and Paradoxes*. WW Norton and Company, New York.

# 2 Historical Perspective on Human Mobility and Tourism

**Abstract**
This chapter traces human mobility from ancient travel to modern tourism, criticizes the industry-oriented discourse and calls for reflection on what tourism can be. From ancient Greek journeys to medieval pilgrimage routes, the motivations for travelling range from economic, political, cultural and religious factors. The Renaissance promoted trade and culture, the 'Grand Tour' reached its peak in the 18th century, while modern tourism emerged in the 19th century and developed further in the 20th century. In all eras, however, there were people who travelled for adventure, pleasure and a thirst for knowledge. The chapter emphasizes the enduring importance of travelling and offers a broad historical perspective beyond the conventional depiction of tourism.

In the late 1960s, Guy Debord characterized tourism as the packaging of human movement for mass consumption in his seminal work, *The Society of the Spectacle* (1977). He saw tourism as a direct result of the circulation of commodities, where the attraction is to see what has been commodified and trivialized. Debord further argued that the economic organization of travel to different destinations led to a uniformity that robbed these places of their distinctive features. He noted that as modernization progressed, the actual space available for travel was reduced due to shorter travel times, which fundamentally changed the nature and meaning of travel.

Debord's perspective on tourism was an integral part of his broader examination of the modern world, which he saw as completely dominated by the 'realm of economy'. This state was characterized by the domination of money, the objectification of products and human relationships and the manipulation of media representations. According to Debord, all this culminated in the creation of the 'society of the spectacle', a phenomenon that prevents the emergence of an 'authentic life' (Kaufmann, 2003).

In this context, Debord considered the term 'tourism' in its limited meaning as it was coined in the 20th century (i.e. as a purely commercial activity).

As Tribe (2006) points out, the discourses surrounding tourism serve a dual purpose in shaping the interpretation of this concept. They determine not only what is accepted as valid knowledge about this phenomenon, but also what does not fall within this domain. This selective process tends to privilege the perspectives and interests of the tourism industry, while those of other groups involved with the phenomenon are often disregarded. Tribe's position on the overarching conversation about the world of tourism makes it clear that while there are numerous truths on the subject, the complete truth is elusive. This incomplete narrative leads to gaps, silences and misrepresentations. In this context, reflexive academic and tourism research tends to support certain viewpoints and obscure others.

So, in order to develop a more comprehensive concept of tourism, we need to decipher the contradictions that have arisen in the course of its development. To do this, we need to examine how it has been shaped by different influences, including tourism stakeholders and academia, and how the media has portrayed its different manifestations. Most importantly, we need to examine how people's perceptions of tourism have evolved over time. Before we turn to these aspects, however, we should first


DOI: 10.1079/9781800626034.0002

make assumptions about the early existence and development of this phenomenon.

## 2.1 Early Human Mobility

According to some anthropologists, the first human societies were groups of people who moved through space in search of food and temporary shelter in the form of a herd. It is assumed that people in prehistoric times constantly changed their environment, either voluntarily or forced by various external factors such as conflicts, weather conditions or lack of food. They usually moved from place to place in groups, but sometimes individually, in relatively small geographical areas, often crossing the boundaries of that space.

The basic unit of social organization of most primitive groups consisted of 15–50 people, and most of them were linked by kinship ties. It is thought that these groups were probably egalitarian, since leadership was informal and subject to peer pressure, while authority was usually exercised by example rather than command, and finally sharing food and other goods was a basic rule of individual behaviour.

As already indicated, mobility was a central feature of these groups, and an equally important feature was the absence of a fixed size. Individuals or sub-groups within the community had the flexibility to break away from the larger group and form smaller units or join other flocks. Each of these groups went its own way, depending on the internal dynamics of the individual units, external connections and prevailing environmental conditions (Lee, 2007).

When migrating, people either relied on stories from other migrating groups or individuals about rich and fertile areas, or they followed the hordes of other animals, sought a temperate climate, often repeated a predetermined route, or finally followed their intuition.

There is no doubt that travelling back then was a real adventure that went far beyond the boundaries of the ordinary as we understand it today. An adventure that expanded consciousness but could also be deadly, depending on what you encountered along the way, and which of course had nothing to do with travelling today. Another point we should emphasize is that communication between people and groups played a certain role in the choice of destination, giving them an idea of ideal places with plenty of food, water and a good climate.

However, it is believed that population mobility was often due to a positive emotional state that people achieved through movement, in addition to functional reasons (hunting, avoiding anthropogenic or environmental hazards).

The ability to roam ensured that these groups were difficult to track down and gave them a reassuring sense of survival, in contrast to our modern notion of security, which often centres on a fixed abode.

In general, one could argue that mobility was therefore not always a purely rational decision, but rather a compulsion, a manifestation of a certain way of life.

In this context, human mobility can be seen as a complex interplay of rational and spontaneous actions that reflect basic primal needs. These needs include the desire for security, the urge to explore and understand the world, the search for truth and the quest for freedom, all of which contribute to the expansion of one's consciousness (Hupkes, 1982; Bauman, 1998).

The fact that nomadism was an essential part of primitive human life could lead to the hypothesis that moving from place to place was once a natural human condition and thus established itself as an 'archetype' (Jacobi, 2013; Jung, 2014).

Human mobility was thus not just an inherited idea or need, but became a kind of predetermined function of our species, a behaviour that corresponds to the innate way the chick hatches from the egg, the bird builds its nest or the eel finds its way to Bermuda, and so on.

It is obvious that this aspect of the archetype is related to biology. In other words, in human mobility, the social, the biological, the psychological and often the metaphysical parameters are very close together.

## 2.2 Early Travel for Trade, Migration and Pleasure

Over a period of almost two million years, the natural state of humanity was determined by the nomadic way of life of various hunter-gatherer groups. In the following 12,000 years or so, however, a significant change took place, marked by the domestication of 'other animals' and the advent of agriculture.

This transition gradually led to a 'sedentary culture' characterized by the establishment of permanent settlements, forcing people to settle near agricultural centres. Trade quickly became an integral part of human existence and was embraced by all communities. In addition, certain geographical paths, characterized by their natural features, developed into early transport routes that made it easier for people to travel. This series of developments collectively led to what we now call a 'civilized way of life' (Sarigiannis, 1978; Hawkes *et al.*, 2018).[1]

Over time, it became the predominant practice to settle in one place, leading to the development of permanent houses, villages and eventually cities. This transition from nomadism to sedentarization was a crucial step in human history. However, although many people had a deep attachment to the land and chose to live where their ancestors had lived, died and been buried (or cremated), there were cases where the compulsion to migrate arose from conflict, war or natural disasters, among other things. These forced forms of human mobility led to significant population displacements. In each of these scenarios, societies remained connected through travelling and trade.

It is important to keep in mind, however, that the innate desire for movement, ingrained in the hearts of people since the era of nomadic life, remained despite the advent of new ways of life. This urge for mobility not only persisted but evolved, eventually inspiring people to travel in their leisure time and for pleasure, even as they adapted to a more sedentary lifestyle.

In the early days, travellers depended mainly on walking, rafts, canoes, boats and donkeys. In time, horses, carriages and larger ships were added. Later, both land and sea routes were opened up, leading to greater geographical expansion and bringing together different cultures. The sea, when accessible, served as a common route for people living far apart, as land routes were limited, often rudimentary, impassable and fraught with danger.

This is also evident in ancient Greek mythology, for example in the myth that Theseus, a divine hero and king of Athens, encountered and killed several robbers such as Pityokamptis, Skiron, Kerkyonas and Procrustes, on his way from Troizena to Athens. Although this myth reflects the social ideas of the time about the dangers of travel, it ultimately pays tribute to the efforts made in fighting the robbers who attacked the travellers.

The beginnings of harbour facilities date back to the third millennium BC, with remarkable evidence discovered in the Indian and Mesopotamian regions. In this far-reaching epoch, the visualization of distant places took place through stories exchanged among acquaintances, myths, paintings, murals, engraved narratives and similar media. In these representations, the traveller's body often carried significant symbolism, offering insights into the cultural, social and sometimes religious perspectives of the era (Casson, 1994; Mauro, 2019).

Kingdoms soon emerged as organized political entities with territorial boundaries and control mechanisms, although travel was not completely restricted. Individuals and groups could travel beyond their borders for economic, religious and other reasons. During the time of the Egyptian kingdoms, numerous festivals were celebrated every year, such as the Wepet-Renpet festival, which opened the year, or the Beautiful Festival in the Valley, which honoured the souls of the deceased, and these events attracted many people from far-flung areas (Christou, 2022).

It is important to mention here that from around 1 500 BC there is clear evidence of tourism in ancient Egypt as 'travel for curiosity and pleasure' in the sense that the ancient Egyptians began to make special journeys through their country to see the impressive evidence of their powerful past, and they documented their presence through inscriptions on the monuments.

As Casson (1994) notes, the ancient Egyptians were in their own way comparable to the tourists of today. Just like contemporary tourists, they sought tangible evidence to validate their journeys. Carving their names into the surfaces of the places they visited was one of the typical tourist habits of the time. Scholars even go so far as to say that the main reason for the messages engraved by travellers seems to have been the urge to leave one's name as confirmation in a place where one had been. Another characteristic feature of travel behaviour at that time was the urge to take home mementos or souvenirs.

Moreover, there is evidence that the ancient Egyptians, like the Assyrians who created the 'Hanging Gardens of Babylon', created ornamental parks and collected exotic animals to display for their own pleasure, but also for the pleasure of guests; we can therefore say that these cultural practices are the precursors of botanical gardens, zoos and theme parks (Foster, 1999; Christou, 2022).[2]

To return to the essentials: more than 2500 years ago, people in ancient Greece were already undertaking extensive journeys, using various means of transport. Their motives ranged from economic and political goals to participation in religious and cultural events. But there were also people who were not discouraged and travelled simply out of an innate desire. Advances in shipbuilding, in particular, contributed to the fact that more and more people travelled (Casson, 1994).[3]

Archaeological excavations revealed that as early as the Neopalatial period of Crete, (i.e. after 1700 BC) dozens of sanctuaries were founded on the peaks and slopes of the mountains, where the many worshippers from the neighbouring regions gathered. One of the most important was the sanctuary of Juktas, which apparently owed its attraction to its proximity to the palaces of Knossos (Boulotis, 2003).

While visitors in early antiquity had no choice but to rely on private hospitality (i.e. they were hosted by acquaintances, peers or friendly strangers depending on their social status or profession), an infrastructure for hosting travellers gradually developed.

As Casson (1994) suggests, one of the earliest known inns with surviving remains dates from around 1500 BC and was located in Crete. This inn was a modest but elegant building on the road leading from the south coast, right at the entrance to the palace of Knossos.

As the political-economic networks that linked larger communities expanded, so did the infrastructure. Notable examples include the Arkadiko Bridge, one of the oldest surviving bridges in Europe, dating back to the Mycenaean period (Hope Simpson, 1998).

Another example is the Diolkos, a paved road near Corinth in ancient Greece, which was intended for the overland transport of ships across the Isthmus of Corinth, more than 2000 years before the canal was built. It is thought to have been built in the late seventh or early sixth century BC. In addition, various harbours such as Phaistos, Malia, Thira, Pylos and Piraeus played an important role in this period (Mauro, 2019).[4]

### 2.2.1 Events and festivals that attracted travellers

As early as the fifth century BC, many areas where well-known Greek temples were located had their own inns. Gradually, certain events acquired great importance and therefore attracted people from the Greek colonies. Many people travelled from the scattered Greek cities in what is now Spain, France and Italy or from North Africa, Asia Minor and the Black Sea to take part in these major events.

The most famous events were the Olympic Games, the Pythian Games, the Isthmian Games and the Nemean Games, which were sporting events dedicated to the gods and therefore required prayers and animal sacrifices. Sacrificing other animals was an essential and integral part of Greek festivals and religious rituals and attracted people. Therefore, sheep, goats, pigs and cattle were among the most commonly offered sacrifices. It is important to understand that these practices reflected the religious beliefs and cultural norms of ancient Greek society. Although they may seem cruel and unpleasant from today's perspective, they were deeply rooted in the religious practices and worldview of the time.

In this way, people honoured a deity by holding outstanding athletic competitions and other cultural activities that drew crowds of visitors from faraway places. Some ancient scholars, such as Diodorus Siculus (first century BC), have claimed that sometimes 20,000 to 40,000 spectators attended some of these events, but according to modern historians, these figures tend to be exaggerated (Dillon, 2013).

Many have noted that during these games there was a truly festive atmosphere, with not only entertainers such as miracle workers and acrobats, but also a spiritual essence as storytellers and historians presented their works.

The 'Aoidoi' sang epic poems and excerpts from Homer, orators spoke on various topics and sophists imparted the art of rhetoric, thus contributing to this festive atmosphere (Coulet, 1996).

It has been argued that these festivals were a unique combination of elements that captivated tourists across time and cultures. They conveyed the feeling of being part of a great event and offered a special and joyful experience. They exuded a celebratory atmosphere and offered the thrill of participating in high-level artistic and sporting competitions. Most importantly, they provided the opportunity for mobility, allowing travellers to travel around the globe and visit famous landmarks and major cities (Casson, 1994).

In addition, there were numerous secret religious rituals in antiquity, such as the Eleusinian Mysteries organized by the cult of the goddess Demeter and her daughter Persephone, which enjoyed great popularity and brought people together. As already mentioned, temples and sacred sites such as the oracles of Delphi or Dodona attracted devout followers all year round.

It should be made clear at this point that although the travellers were predominantly men, women also often travelled hundreds of miles with their families or alone for religious purposes. Despite rumours about the exclusion of women from these religious events, the finds at Epidaurus or Eleusis show that they often participated. Nevertheless, most of these events were under male control.

In this context, it should be noted that at the Olympic Games, although young girls could participate in certain sporting activities, the event was considered an exclusively male activity to which women were not allowed. The only spectators allowed were the virgins and the priestesses of the goddess Dimitra, while any other woman arrested as a spectator was condemned to be thrown from Mount Typaion. Either way, we can conclude that these religious events were one of the few opportunities for young girls to leave their city and escape the conservatism and narrow-mindedness that characterized patriarchal Greek cities (Dillon, 2013).

It is worth noting that some phenomena of commercial development of the idea of voluntary mass travel for pleasure and cultural purposes already occurred in ancient Greece, but of course not to the same extent as in the age of capitalism. The large numbers of people who gathered at these events needed food and shelter, which boosted trade and the economy in these areas.

The ancient city-state of Elis, for example, where the Olympic Games were held, experienced a great economic and political boom thanks to this event. Particularly interesting is the fact that the start of the Games was announced by heralds to attract a large audience from all over the 'Greek world', which can be compared to a kind of ancient mass media. These emissaries, whose task was to invite the official representatives of the cities, were called *theoroi* or *spondophoroi* and can also be considered as diplomats and 'truce brokers', because during the Olympic Games there was a truce throughout the 'Greek world', but at the same time they were the promoters of the event on a grand scale (Dillon, 2013).

### 2.2.2 Relationship between ancient travel and modern tourism

Given these facts, we can ask ourselves some questions to become aware of the relationship between ancient travel and modern tourism, such as:

- What was the reason that large populations moved at short intervals and changed their daily routines every year?
- Did they do it because they felt the need to form a common social identity?
- Was there perhaps something deeper than the religious basis of the above events that led people to travel?
- Did people associate travel with entertainment?
- Was it the need for a temporary change in their daily routine?
- Was it a cultural need for communication, an urge to discover?
- Or are the causes perhaps deeper and have to do with primitive instincts and the innate human need for mobility in order to expand consciousness?

To add a kind of philosophical dimension to our discussion of human mobility, we can briefly examine how some ancient Greek philosophers viewed movement.

Heraclitus is well known for his philosophy revolving around change and movement. His famous maxim 'Everything flows' underlines his belief that the universe is in constant flux and nothing remains fixed or static. He held that the nature of reality is characterized by continuous and infinite change (Axelos, 1974).

In the context of human mobility, therefore, Heraclitus' overarching philosophy suggests that human movement is a natural and inherent part of an ever-changing world.

Plato attributed great importance to movement in cosmology and associated it with happiness. He believed that the perceptible world had an inherent motion that was different from the divine or soul-controlled motion. Only the soul, in Plato's view, could move voluntarily, while soulless objects moved mechanically. In his dialogue *Theaetetus*, he held that all living organisms emerged from the universal movement that governs the universe and emphasized the value of movement for both the soul and the body (Arvanitakis, 1999).

Other Greek philosophers, such as Aristotle, linked human mobility in geographical space to the 'principle of motion', which states that all living things are subject to a natural, external law that governs their movement (Kalfas, 2015).

## 2.3 Mobility in the Roman Empire

By the time of the Roman Empire, when the Mediterranean had become politically and culturally one world, travel by sea and land could easily spread in all directions thanks to the development of navigation technology and the Roman road network.

In fact, the Romans built a network of roads, admirable for the time, lined with inns and taverns at regular intervals to meet the needs of travellers and to connect the capital with the distant frontier areas, leading to the saying 'All roads lead to Rome'. They travelled all over the then-known world, but especially to Greece, Egypt, Syria and Palestine, to see the places they had heard or read about up close and explore the world.

At the same time, Rome became the most famous city, the centre of the then-known world, which everyone wanted to visit. Moreover, like the Greeks, the Romans held festivals in honour of their gods, such as the 'Veneralia' in honour of Venus, the 'Mercuralia' in honour of Mercury, the 'Saturnalia' in honour of Saturn and the 'Vestalia' in honour of Vesta, which attracted large crowds (Christou, 2022).

The Colosseum, a huge amphitheatre built in the first century BC in the heart of Rome, was a central place of recreation and entertainment during the Roman Empire. It was a centre of attraction, offering a variety of exciting and captivating spectacles that allowed Roman citizens to escape their daily lives and immerse themselves in the excitement of games and competitions. Many people from different parts of the Roman Empire visited the site to witness events and great spectacles, including gladiator fights, animal hunts and other performances that were famous throughout the Empire, so the opportunity to experience these events attracted visitors from different regions.

Against this background, the commercialization of travel for learning and pleasure should not be hastily described as a modern phenomenon, especially since even then, in the cities that the Romans liked to visit, there were guides who, for a small fee, showed visitors around, pointed out and explained the historical background of buildings, monuments, statues and pictorial content, and described local rituals and customs. Even then, people had their own versions of souvenirs depending on the place, such as cheap faience pots, small glass bottles with drawings of the main sights and miniature models of monuments (Casson, 1994).

In addition, the phenomenon of holidays spread at this time, and many upper-class Roman citizens retreated to places other than their winter quarters, usually to villas by the sea or in areas with thermal baths, but sometimes also to the mountains. The most beautiful villas of the wealthy Romans were located on the Gulf of Naples.

Be that as it may, in the summer months the bathing resorts also attracted many people from the lower classes. However, visiting thermal springs was not a unique phenomenon in ancient Rome. The same was done in different historical periods and by many other cultures (Patmore, 1968; Graburn, 1995).

It seems that the mobility of people from place to place, even if it did not reflect everyday life as it might have done in primitive society, nevertheless played a significant role in the economic and cultural spheres within ancient political units. Thus, there were logical reasons for travelling, such as trade, participation in religious events and holidays, while the constant moving of people from one place to another was no longer considered a way of life; as such, it was only attributed to nomadic tribes.

In the fourth and fifth centuries, a great wave of Christian itinerant monks broke out throughout the Roman Empire. These monks followed an apostolic, ascetic lifestyle based on the idea that strict imitation of the life of Jesus and his apostles (i.e. absolute poverty, prayer, spiritual instruction and wandering) was the highest form of Christian life. However, the ascetic itinerant life and the suffering orders were strongly criticized by some representatives of the Church, as they had gained prestige and were seen by the establishment as competitors for social and spiritual power (Caner, 2020).

## 2.4 Travel in the Middle Ages

The Middle Ages was a time when Europe was strongly influenced by a dramatic migration of peoples that began in the third century and was accompanied by the gradual fusion of Greco-Roman with other cultures. Although there was much mobility and various forms of cross-cultural contact, long journeys were very difficult and dangerous, and as Legassie (2017) puts it, travel in the Middle Ages was nasty, brutal and long.

Important political, economic and cultural centres of the Middle Ages, such as Paris, Rome, Constantinople, Florence, Venice, Granada, Seville and Cologne, but also many areas in the East that had to do with religious issues, received numerous visitors every year for fairs and festivals, cultural education, pilgrimages, inner integration and spiritual quests, or simply out of interest and curiosity. All these places represent 'mnemotopes' or places of memory to which people were drawn to strengthen their social identity, their faith or their participation in a culture (Halbwachs, 2020).[5]

At this point it is interesting to mention a repressive policy of the state common in the Middle Ages, which can be placed in the framework of spectacle and tourism in a very tragic way. The punishment of certain crimes was not only a means of preventing deviants, but at the same time an authentic and collective experience, a collective spectacle to which people sometimes travelled to participate in the physical and psychological humiliation of the condemned. In this respect, the religious and feudal values of the time had paradoxically transformed punishment into a public spectacle that attracted many people and often took the form of a festival (Baraz, 2003; Foucault, 2012; Christou, 2022).[6]

Pilgrimage, the journey of the faithful to the holy places as an expression of their worship of God, was for many people an intense inner need, but also a means of social demarcation, a life dream and a goal.

Various small objects, amulets, badges and other symbols that pilgrims acquired at the sacred sites affirmed their religious travel experience and elevated their status. All these objects can in some ways be seen as premature symptoms of the emerging commercial ideology of travel that would establish the modern concept of mass tourism in the 19th and 20th centuries, for they are strongly reminiscent of souvenirs from luxury destinations as part of the 'conspicuous consumption' with which the middle classes enhanced their prestige.

As Bell and Dale (2011) point out, later medieval pilgrimage sites were a kind of franchise operating under the umbrella brand of the Christian church. Additional services to pilgrims, such as the provision of accommodation, food and wine, transport, banking and pilgrim badges, constituted an important economic activity, sometimes involving complex business relationships between the private sector and local church authorities.

During this period, the main pilgrimage destinations of the Western Christian Church were the Vatican in the Italian peninsula, Jerusalem, Bethlehem and Nazareth in the Middle East, St James of Compostela in the Iberian Peninsula, etc., while the main destination of Muslims was

Mecca in the Arabian Peninsula (Swatos and Tomasi, 2002).

### 2.4.1 Expansion of European cultures

The late Middle Ages were marked by a great economic, territorial and cultural expansion of European cultures. The Vikings boarded Greenland, the Crusaders moved east and controlled the Middle East, while Marco Polo, the Vivaldi brothers and many others travelled to China and East Asia to explore new lands and find the best access routes from Europe to distant places already known.

In the 15th and 16th centuries, with the conquest of the seas outside the Mediterranean by the Spanish and Portuguese, the discovery of the New World and the opening of new sea routes linking Europe with Asia, trade developed into a system of exchange which, together with the satisfaction of the material needs and desires of populations and the expansion of knowledge, formed the basis for an extensive network of mobility and communication between people and cultures.

However, it is important to point out that the European voyages of discovery did not only have positive effects. They also brought significant negative consequences, including: (i) the colonization and displacement of indigenous populations; (ii) the spread of diseases that decimated Native American populations; and (iii) the exploitation of human and natural resources in the newly discovered lands.

It can be said that much of the human mobility in mainland Europe during this period, if we disregard the movements caused by the great migrations, wars, etc., was due to the ever-increasing trade that went beyond the narrow confines of local communities. The fairs and international trade favoured the development of various services that later became part of the tourism sector, such as the means of transport and their networks, and the inns and taverns along the roads used by the merchants.

Already in the late Middle Ages, a well-developed network of inns emerged along the main European routes, offering meals, accommodation and stabling to visitors. In many cases, however, they were also secular meeting places that became important centres of communication in public life for all kinds of political and cultural activities.

As Ellias (2000) points out in his book *The Civilising Process* on the origins and historical development of the European habitus, the Dutch philosopher and theologian Erasmus (1466–1536) had already dealt with the different manners in German and French inns towards guests in his work *Diversoria* and commented on what their condition was and should be. In other words, we see that even then there was a change in the way hoteliers and society in general should serve strangers and visitors.

In this context, we can say that the rules and customs for hospitality towards strangers and visitors did not just emerge with the advent of modern tourism in the 19th and 20th centuries, but developed over centuries and across cultures. They were based on a combination of religious beliefs, social norms and practical considerations and have been essential for the comfort, safety and well-being of travellers throughout history.

Meanwhile, the post offices of some European countries began to systematically use advanced horse-drawn carriages, which also frequently carried travellers and became an important means of transporting people from one place to another until the advent of the railways. In addition, shipping continued to develop both within the Old World and between the New and Old Worlds.

In the 17th and 18th centuries, the conquests and colonies of the great European states of the time, such as the Kingdom of Great Britain, the Kingdom of France and the Republic of the Seven United Netherlands, served as a backdrop for the creation of the myth of European cultural superiority over the cultures of the other continents.

Later, when the South Pacific became an adventure, a kind of seafaring hero emerged in the European world, such as James Cook (1728–1779) and his French counterpart Louis Antoine de Bougainville (1729–1811), but also many others. Although these personalities may not have had the direct intention of promoting travel and tourism, their expeditions and the subsequent dissemination of their experiences undoubtedly contributed to a growing interest in discovery and travel. Their reports shaped the cultural mentality of the time and indirectly influenced the development of modern tourism in the following centuries.

Within this framework, a common literary theme emerged in the various European countries, which is of particular interest for our topic and which stimulated the social imagination to travel: the figure of the adventurer who returns from exotic places after having made his fortune there.

By contrast there were also the works in which travel became a suitable metaphor for those who wanted to take a critical look at European societies themselves, starting with *Utopia* (1516) by Thomas More (1478–1535) and continued by Étienne Cabet (1788–1856) with *Travels in Icaria* (1840), and many others (Subrahmanyamn, 2003).

## 2.5 The 'Grand Tour'

The 'Grand Tour', which is one of the most important elements of the historical analysis of tourism, has its roots in the late 16th century. It was first organized by the British aristocracy and later adopted by the bourgeoisie and other Europeans. It involved extensive travelling throughout mainland Europe, encompassing France, Italy, Germany, Switzerland, the Netherlands and later Greece. This cultural expedition usually lasted over 3 years and combined the search for knowledge with the desire for mobility. It was a challenging form of education where participants explored European sites of historical and literary significance and experienced first-hand the monuments of ancient civilizations.

Participating in this adventure provided social prestige and cultivation, familiarized one with political and philosophical concepts, but also offered an experience that broadened one's education and knowledge of the rest of the world and the natural environment, gave the opportunity to discover cultures and valuable remains, including religious texts, and was ultimately an escape (Dritsa, 2019).

Scholars have argued that the emergence of the 'Grand Tour' coincided with the transcendence of Christianity, the Renaissance, an idea of Europe and a newly emerging interest in nations and the differences between them, just as pilgrimage was a ritual through which medieval Christianity constructed its cultural reality (Hay, 1957; Adler, 1989).

Although this adventure was usually for men only, participants were sometimes accompanied by their wives or other female family members (Wilcox, 2017).

However, as Zuelow (2015) notes, women such as the British writer and poet Lady Mary Wortley Montagu (1689–1762), the British noblewoman Lady Mary Coke (1727–1811) and the British writer, philosopher and advocate of women's rights Mary Wollstonecraft (1759–1797) also spent much of their time on the European mainland, where they enjoyed the freedom and received the education normally reserved for men (Dolan, 2001).

Later, the purpose of the 'Grand Tour' excursions expanded to include other topics such as nature experiences, physical exercise, contact with local folk culture, participation in local forms of entertainment and more.

Be that as it may, the 'Grand Tour' gradually changed from a sophisticated educational journey to an adventurous experience in which the middle classes believed they were imitating the lives and habits of the aristocrats (Brendon, 1991; Towner and Wall, 1991).

In the 18th century, the middle classes already formed the majority of participants in the 'Grand Tour'. In those years, however, travel was still understood in a broader sense as a practice that expanded human consciousness and enriched the cultural capital of travellers, rather than as a mere fan, as mass tourism would later become.

On the other hand, the 'Grand Tourists' not only wanted to enrich their spiritual world, but also to have something to show off (i.e. something they could take home to show everyone where they had been and what they had seen). Gradually, a small market developed to satisfy this need, with paintings depicting important landmarks such as the Pantheon and the Colosseum in Rome, the canals in Venice, the Parthenon in Athens and so on. Surprisingly, many of these paintings also show tourists looking at these sights and monuments.

In this context, we can argue that the spirit of 'conspicuous consumption' re-emerged and once again served to enhance the status and social standing of travellers. These early manifestations were a foretaste of the extensive commercialization that was to characterize tourism in the later decades of the 20th century (Zuelow, 2015).

## 2.6 Travel From the End of the 18th Century Onwards

At the end of the 18th century, mobility was relatively improved by the renewal of the historic road network and advances in vehicle technology. At the same time, the creation of new roads, some with turnpikes and others with toll-free passage, combined with amenities such as inns for changing horses and rest stops, led to a significant reduction in travelling time for travellers (Kahk, 2003).

The rise of nation states, however, forced people to stay within certain boundaries. These political entities promoted rationality, unity and a regulated life, and fostered economic, social and cultural cohesion. Historical memory played a crucial role in awakening national consciousness and required the construction of national identities.

The connection between national identity and territorial claims involved defining national geography and constructing national landscapes. At the same time nomadic travel posed a security risk to nation states, as it could erode national identity. Cross-border travel was limited to practical purposes such as trade, diplomacy, leisure and immigration, leading to increased bureaucratic procedures.

As countries consolidated their sovereignty and sought to regulate traffic across their borders, special transport documents and passports became necessary for international travel. The geopolitical dimension of the time thus hindered people's need to move from place to place and completely precluded the consideration of travel as a free, spontaneous and unpretentious act.

On the other hand, nomadic thought, which exists deep within the human mind, cannot be confined to the edifice of a constructed and ordered interiority that depends on external factors. It does not adhere to the artificial division of the land.

Be that as it may, the formation of nation states brought new challenges and regulations to travel and created opportunities for exploration, cultural exchange and the development of tourism in certain countries. Travellers had to deal with changing landscapes, political dynamics and administrative requirements on the developing European continent.

Governments in Europe increasingly tended to take responsibility for numerous institutions that had a significant impact on cultural life. Consequently, various museums, libraries and universities were founded under this ethos. Notable examples include the foundation of the British Museum in 1753, the public accessibility of the Austrian aristocratic collections from 1781, the foundation of the Louvre Museum in 1793, the official designation of the Uffizi as a museum in 1865 and other similar cases (Bled, 2003).

Modern museums emphasized the national consciousness of the population within the state's borders with exhibitions that highlighted the quality of the national culture or the spoils of its conquests, and attracted domestic and foreign tourists dazzled by the state's cultural objects and riches.[7]

As early as 1830, the British Museum possessed two of the richest collections of Greek and Roman antiquities in the world, and during the Victorian years its halls were expanded and enriched with exhibits from Egypt, Carthage and the Levant (Pemble, 1987).

In addition, some cities such as Florence, Rome and Venice began to be considered 'open museums', where visitors could walk and admire the architecture of the buildings, sculptures and fountains, as well as participate in festivals and musical events (Christou, 2022).

Gradually, then, trips to European cities with healing springs became more popular, and many of these places are still popular today, such as Montpelier, Vichy and Aix-les-Bains in France, or Aachen and Baden-Baden in Germany. Some of these towns often offered a range of complementary facilities such as ballrooms, gambling dens and other entertainment venues that brought European cosmopolitans together. Later, the seaside resorts also became attractive for health reasons, as the sun was then considered beneficial to health, and these became increasingly popular at the beginning of the 20th century.

On the other hand, the mountains gradually became fashionable among the upper classes as a means of escaping the cities, and not only for health reasons. Tourist centres such as Davos and St Moritz were known among travellers for the 'cold therapy' that had become fashionable at the time. Towards the end of the 19th century, skiing became an international trend in the European Alps (Bonarou, 2012).

In this context, it is interesting to note that at the end of the 19th century there were some courses on the 'art of travelling' at European universities such as Goettingen and Weimar, in which the traveller was studied as a source of information, as a model for a social way of thinking and knowing, and as a symbol of the relationship of the self to society (West, 1978).

Nevertheless, it is important to stress that after the mid-19th century it was not only members of the royal families, the aristocracy and the wealthy bourgeoisie who undertook Mediterranean travel. Instead, there was a marked shift with a growing influx of middle-class people, not as occasional and foreign visitors like the 'Grand Tourists' of the 18th century, but as a substantial and recurring stream of tourists, or more aptly described as 'mere tourists' (Pemble, 1987).

## 2.7 Summary

To summarize, our study of human mobility throughout history clearly shows that numerous features of modern tourism have always been woven into the fabric of human experience in various shapes and forms. These features include: (i) the search for adventure and pleasure through travel to far-flung destinations; (ii) the notion of hospitality; (iii) the presence of inns and taverns; (iv) the lure of parks with exotic creatures; (v) international gatherings; (vi) the acquisition of souvenirs; (vii) the indispensable role of tour guides; and (viii) the symbolic expression of social status through travel. This revelation challenges us to reconsider the conventional notion that tourism as a social phenomenon only emerged in the 19th century.

## Notes

[1] At this point, I should note that some anthropologists believe that what we call 'civilization' after the advent of agriculture is leading a steady and rapid path to the social and ecological destruction of human life, as well as the extinction of all life on the planet. In short, roughly the last 12 millennia have been a time of decline, a downward course of human development towards self-destruction (Sahlins, 1998).

[2] Ancient cultures such as the Assyrians and Babylonians exhibited exotic animals as symbols of power and wealth. Treatment varied; some respected and cared for these animals, while others mistreated them. The Romans brutally used wild animals for entertainment, and the ancient Greeks had a complex, often abusive relationship with animals. Ethical concerns about animals were less pronounced in the ancient world than in modern societies (Harden, 2013; Kitchell, 2014).

[3] Watercraft, many millennia old, were independently developed by different cultures around the world for water transportation. Early vessels for river navigation probably originated in Mesopotamia and ancient Egypt, but other cultures such as the Phoenicians, Harappians and the ancient Chinese also built sea vessels at various times. The 'Pelasgians', an early seafaring people in the region of Greece, sailed from Ermione to Mylos about ten millennia ago to fetch obsidian. The seafarers of the Aegean also included Cretans and inhabitants of the Cyclades (Lagadas, 1990).

[4] The Diolcus of Corinth was a paved passageway at the narrowest point of the Isthmus of Corinth. It was used to tow boats between the Saronic and Corinthian Gulfs, using cylindrical beams and slave labour, and Corinth levied taxes for this service (Lagadas, 1990). As for harbours, there were many older than those mentioned, such as Wadi el-Jarf on the Red Sea, used by King Snofru and King Cheops, the port of Byblos, etc. (Tufnell and Ward, 1966; Tallet and Marouard, 2014).

[5] Assmann (2011) and Halbwachs (2020) point out that different groups and religions establish sites and monuments to preserve their collective and historical memories. People visit these places at regular

intervals, which, in addition to religious and identity-related motives, also strengthens the human instinct for mobility.

[6] However, there were also many similar situations in antiquity, such as the execution of Christians by wild animals or the duels of gladiators in large stadiums like the Colosseum in Rome.

[7] It is believed that one of the oldest museums in the world was the Museum of Ennigaldi-Nanna, built by Princess Ennigaldi at the end of the Neo-Babylonian Empire around 530 BC, which contained artefacts from earlier Mesopotamian civilizations (Wilkens, 2011). The ancient Greeks and Romans also collected artefacts and cultural elements and displayed them in specific locations. Another important early museum was the Museum of Alexandria, founded by Ptolemy I Soter and his son Ptolemy II Philadelphus (Russo, 2004). However, the establishment of museums similar to modern museums can be traced back to the 15th century, as the first recorded visitor to the Tower of Arms in London dates back to 1498, when entry was by special permission only, while in 1660 the paying public was allowed to enter the Tower to see the exhibits (Royal Armouries, 2018). The 'Chambers of Wonder' in the 16th century, which were collections of remarkable objects, were also a kind of museum (Impey and MacGregor, 2018). The Uffizi, one of the first modern museums, had been open to visitors on request since the 16th century, but was opened to the public by Grand Duke Peter Leopold in 1769 and finally established as a museum in 1865.

## References

Adler, J. (1989) Travel as performed art. *American Journal of Sociology* 94(6), 1366–1391. DOI: 10.1086/229158.

Arvanitakis, T. (1999) *Plato: On Motion*. Zitros Publications, Thessaloniki, Greece. (in Greek).

Assmann, J. (2011) *Cultural Memory and Early Civilization: Writing, Remembrance, and Political Imagination*. Cambridge University Press, New York. DOI: 10.1017/CBO9780511996306.

Axelos, K. (1974) *Heraclitus and Philosophy*. Exantas, Athens. (in Greek).

Baraz, D. (2003) *Medieval Cruelty: Changing Perceptions, Late Antiquity to the Early Modern Period*. Cornell University Press, New York. DOI: 10.7591/9781501723926.

Bauman, Z. (1998) *Globalization: the Human Consequences*. Columbia University Press, New York.

Bell, A.R. and Dale, R.S. (2011) The medieval pilgrimage business. *Enterprise & Society* 12(3), 601–627. DOI: 10.1017/S1467222700010235.

Bled, J.-P. (2003) European cultures or European culture? In: Ahrweiler, H. and Aymard, M. (eds) *The Europeans, Volume I*. Savallas, Athens. (in Greek), pp. 313–343.

Bonarou, C. (2012) *Visual Culture and Tourism. Representations of Greece on Tourist Postcards*. Papazisi Publications, Athens. (in Greek).

Boulotis, C. (2003) The first great civilizations of the Aegean: from prehistory to protohistory. In: Ahrweiler, H. and Aymard, M. (eds) *The Europeans, Volume I*. Savallas, Athens. (in Greek), pp. 24–48.

Brendon, P. (1991) *Thomas Cook: 150 Years of Popular Tourism*. Secker and Warburg, London.

Cabet, E. (1840) *Voyage en Icarie [Travels in Icaria]*. Hippolyte Souverain, Paris. (in French).

Caner, D.F. (2020) *Wandering, Begging Monks: Spiritual Authority and the Promotion of Monasticism in Late Antiquity*. University of California Press, Berkeley, California. DOI: 10.1525/9780520928503.

Casson, L. (1994) *Travel in the Ancient World*. Johns Hopkins University Press, Baltimore, Maryland. DOI: 10.56021/9780801848087.

Christou, P.A. (2022) *The History and Evolution of Tourism*. CAB International, Wallingford, UK. DOI: 10.1079/9781800621282.0000.

Coulet, C. (1996) *Communiquer en Grèce Ancienne: Ecrits, Discours, Information, Voyages*. Les Belles Lettres, Paris. (in French).

Debord, G. (1977) *The Society of the Spectacle*. Black and Red Press, Detroit, Michigan.

Dillon, M. (2013) *Pilgrims and Pilgrimage in Ancient Greece*. Routledge, London. DOI: 10.4324/9780203352441.

Dolan, B. (2001) *Ladies of the Grand Tour: British Women in Pursuit of Enlightenment and Adventure in Eighteenth-century Europe*. HarperCollins, New York.

Dritsa, M. (2019) *Mosaics of Greek Tourism*. Economia publishing, Athens, Athens. (in Greek).

Ellias, N. (2000) *The Civilising Process: the History of Manners*. Blackwell Publishing, Oxford, UK.

Foster, K.P. (1999) The earliest zoos and gardens. *Scientific American* 281(1), 64–71. DOI: 10.1038/scientificamerican0799-64.

Foucault, M. (2012) *Discipline and Punish: the Birth of the Prison*. Vintage, New York.

Graburn, N. (1995) The past in the present in Japan: nostalgia and neo-traditionalism in contemporary Japanese domestic tourism. In: Butler, R.W. and Pearce, D.G. (eds) *Change in Tourism: People, Places Processes*. Routledge, London, pp. 47–70.

Halbwachs, M. (2020) *On Collective Memory*. University of Chicago Press, Chicago, Illinois.

Harden, A. (2013) *Animals in the Classical World: Ethical Perspectives from Greek and Roman Texts*. Palgrave MacMillan, New York. DOI: 10.1057/9781137319319.

Hawkes, K., O'Connell, J. and Blurton Jones, N. (2018) Hunter-gatherer studies and human evolution: a very selective review. *American Journal of Physical Anthropology* 165(4), 777–800. DOI: 10.1002/ajpa.23403.

Hay, D. (1957) *Europe: the Emergence of an Idea*. The University Press, Edinburgh, UK.

Hope Simpson, R. (1998) The Mycenaean highways. *Classical Views* XLII(17), 239–260.

Hupkes, G. (1982) The law of constant travel time and trip-rates. *Futures* 14(1), 38–46. DOI: 10.1016/0016-3287(82)90070-2.

Impey, O. and MacGregor, A. (eds) (2018) *The Origins of Museums: the Cabinet of Curiosities in Sixteenth- and Seventeenth-Century Europe*. Ashmolean Museum, Oxford, UK.

Jacobi, J. (2013) *Complex/Archetype/Symbol in the Psychology of C.G. Jung*. Routledge, London. DOI: 10.4324/9781315009070.

Jung, C.G. (2014) *The Archetypes and the Collective Unconscious*. Routledge, London. DOI: 10.4324/9781315725642.

Kahk, J. (2003) New economic order. In: Ahrweiler, H. and Aymard, M. (eds) *The Europeans, Volume II*. Savallas, Athens. (in Greek), pp. 83–94.

Kalfas, B. (2015) *Aristotle, on Nature*. Kallipos, Athens. Available at: https://repository.kallipos.gr/handle/11419/684 (accessed 2 April 2020).

Kaufmann, V. (2003) *Guy Debord*. Ministère des Affaires étrangères, Paris. (in French).

Kitchell, Jr, K.F. (2014) *Animals in the Ancient World from A to Z*. Routledge, London. DOI: 10.4324/9780203087503.

Lagadas, C. (1990) *The Ancient Greeks in Thalassa*. Chrysafis I. Panezis Publications, Athens. (in Greek).

Lee, R. (2007) Prologue. In: Kouravelos, T. (ed.) *Sharing Societies*. Politiκes Ekdoseis, Athens. (in Greek), pp. 23–30.

Legassie, S.A. (2017) *The Medieval Invention of Travel*. University of Chicago Press, Chicago, Illinois. DOI: 10.7208/chicago/9780226442730.001.0001.

Mauro, C.M. (2019) *Archaic and Classical Harbours of the Greek World: The Aegean and Eastern Ionian Contexts*. Archaeopress Publishing, Oxford, UK. DOI: 10.2307/j.ctvndv7h2.

More, T. (1516) *Utopia*. More, Habsburg, the Netherlands. (in Latin).

Patmore, J. (1968) The spa towns of England and Wales. In: Beckinsale, G. (ed.) *Problems of Urbanisation*. Methuen, York, UK, pp. 47–69.

Pemble, J. (1987) *The Mediterranean Passion: Victorians and Edwardians in the South*. Oxford University Press, Oxford, UK.

Royal Armouries (2018). Available at: https://royalarmouries.org/ (accessed 1 June 2023).

Russo, L. (2004) *The Forgotten Revolution: How Science Was Born in 300 BC and Why It Had to Be Reborn*. Springer, Berlin. DOI: 10.1007/978-3-642-18904-3.

Sahlins, M. (1998) The original affluent society. In: Gowdy, J. (ed.) *Limited Wants, Unlimited Means: A Reader on Hunter-Gatherer Economics and the Environment*. Island Press, Washington, DC, pp. 5–41.

Sarigiannis, G. (1978) The concept, genesis and evolution of the settlement through the etymological evolution and interpretation of the relevant words. (in Greek). *Social Research Review* 33, 327.

Subrahmanyamn, S. (2003) The stages of conquering the seas. In: Ahrweiler, H. and Aymard, M. (eds) *The Europeans, Volume II*. Savallas, Athens. (in Greek), pp. 45–57.

Swatos, W.H. and Tomasi, L. (2002) *From Medieval Pilgrimage to Religious Tourism: the Social and Cultural Economics of Piety*. Praeger, Westport, Connecticut.

Tallet, P. and Marouard, G. (2014) The harbor of Jhufu. *Near Eastern Archaeology* 77(1), 5. DOI: 10.5615/neareastarch.77.1.0004.

Towner, J. and Wall, G. (1991) History and tourism. *Annals of Tourism Research* 18(1), 71–84. DOI: 10.1016/0160-7383(91)90040-I.

Tribe, J. (2006) The truth about tourism. *Annals of Tourism Research* 33(2), 360–381. DOI: 10.1016/j.annals.2005.11.001.

Tufnell, O. and Ward, W.A. (1966) Relations between Byblos, Egypt and Mesopotamia at the end of the third millenium B.C. A study of A Montet Jar. *Syria* 43(Fasc. 3/4), 165–241. DOI: 10.3406/syria.1966.5872.

West, H. (1978) Göttingen and Weimar: the organization of knowledge and social theory in eighteenth-century Germany. *Central European History* 11(2), 150–161. DOI: 10.1017/S0008938900018616.

Wilcox, E.E. (2017) 19th Century Writings on the Grand Tour. Wonders of Nature and Artifice: 7. Available at: https://cupola.gettysburg.edu/wonders_exhibit/7 (accessed 4 October 2021).

Wilkens, A. (2011) The story behind the world's oldest museum, built by a Babylonian princess 2,500 years ago. Gizmodo. Available at: https://gizmodo.com/the-story-behind-the-worlds-oldest-museum-built-by-a-b-5805358 (accessed 1 April 2023).

Zuelow, E. (2015) *A History of Modern Tourism*. Bloomsbury Publishing, London. DOI: 10.1007/978-0-230-36966-5.

# 3 Exploring the Essence of Tourism in its Definitions and Interpretations

**Abstract**

The development of tourism as we know it today is due to changes in social behaviour influenced by historical changes. After the Industrial Revolution, advances in transport and communication and changes in labour dynamics encouraged the growth of tourism. The rise of the middle class, and infrastructure investments in hotels and entertainment facilities further fuelled this trend. At the same time, the mass media began to promote the holiday business, a trend that intensified in the age of mass society. These cumulative changes led to a new understanding of tourism and shaped its own dimension and the behaviour of tourists. The prevailing paradigm, influenced by the mass media and defined by academia, reflects all these far-reaching changes. This chapter looks at the conceptual framework of tourism and examines how historical and societal influences interact to redefine its meaning and perception.

## 3.1 Natural Change of Language

A basic principle of linguistics is the natural change of language. There is no doubt that language is subject to constant evolution, determined by the need to reflect changes in society and to become more effective. According to symbolic interactionism, the meanings of words are not developed by individuals alone, nor are they fixed outside society, but are social products that are the result of social interaction. Therefore, individuals change the 'meanings' with the interpretations they give each time in their transactions (Nina-Pazarzi, 2014).

When a person or a group of people interact repeatedly with other people, objects or situations, some meanings may slowly change as different people may understand things and events in different ways. In this sense, some words acquire additional meanings and are no longer used in their old meaning (Jatowt and Duh, 2014).

Following this train of thought further, it becomes clear that words and language, as fundamental components of human communication, have undergone a profound transformation due to the pervasive influence of evolving commercial practices and ever-expanding consumerism.

It is no exaggeration to say that the way we speak, write and communicate our thoughts has been significantly shaped and moulded in recent centuries by the inexorable progress of commercial exchange and the burgeoning consumer culture. In essence, the dynamic interplay of commerce and consumption has left an indelible imprint on the structure of our linguistic expressions and the way we convey meaning in our interconnected world (Friedman, 1985).

British scholar and writer Raymond Williams (1921–1988) and French literary theorist and philosopher Roland Barthes (1915–1980), as well as many other contemporary scholars, have described in their works how words are tainted with gender, race, class prejudice, etc., and have become aware of how contemporary capitalist society treats and changes words (Barthes, 1957; Williams, 1976).

Recently, in his book *Keywords: the New Language of Capitalism* (2019), John Patrick Leary has described the emergence of a new vocabulary that shapes professional, intellectual and even spiritual life in the 21st century, uncovering the history and common use of popular terms of our time such as 'accountability', 'entrepreneur' or 'innovation', which have their origins in the corporate world and propagate values that are friendly to late capitalism and make it seem natural along with its paradoxes.

Apart from this process of meaning change and polysemy of words, individuals at a given historical moment may perceive the meaning of


DOI: 10.1079/9781800626034.0003

a word somewhat differently depending on their experience.[1]

A journalist may understand the word communication differently from a social psychologist. While the former may place more emphasis on the one-way information aspect of the word, the latter may focus on the two-way understanding between communicators.

Similarly, an entrepreneur may see tourism as a means to make a profit, while a backpacker may see it as a means to adventure and explore the world. Even at a given moment, then, words can be signs that, in order to be decoded and understood by members of a society, they must oscillate between different spectrums of meaning, timbres and perspectives. Not to mention that the meaning of the word also depends on the specific contexts of communicative practice, which change according to the existential orientations of the participants or due to the 'rule' of the 'game' being played (Wittgenstein, 1986).

Thus, words like 'communication', 'tourism', 'tourist', 'traveller', 'wanderer' or 'nomad', are not only somehow fluid, but also meaningful only because we agree on their content; as a linguist would put it, the relationship between their given signifiers and signifieds is conventional.

The importance of these words in describing human experience and social roles becomes even clearer when we consider their historical context.

## 3.2 Historical Context: The Emergence of Tourism

In earlier centuries, in traditional rural societies, it was difficult to distinguish between work and leisure, as many tasks were done at home at all times of the day. The separation introduced by the industrial system with the shift of work to the factory with a specific work schedule was one of the reasons that gradually led to a different perception of the concept of leisure. This was one of the reasons that much later (i.e. at the beginning of the 20th century), after endless and arduous struggles of the working class, led to the institutionalization of seasonal work breaks, while finally leisure became a protected human right in the United Nations Universal Declaration of Human Rights (1948) with Article 24, which states: 'Everyone has the right to rest and leisure, including reasonable limitation of working hours and periodic holidays with pay' (Toti, 1975).

Nevertheless, the absolute separation of work and leisure that industrial society brought about led many workers to want to change their environment and break out of their daily routine in order to experience adventure, travel and have pleasant and unforgettable experiences in their non-working time, namely on holiday, which contributed to the development of the tourism industry.

An important factor in our travel plans is the means of transport we will use, and it goes without saying that transport technology plays a major role in our travel ideas. In early 19th-century Europe, different methods of travel were used by different groups. 'Travellers' and 'tourists' mainly used horse-drawn carriages, ships and railways for their journeys. 'Wanderers' and 'nomads' relied heavily on animals such as horses and reindeer as well as their own mobility to cross different landscapes.

Although we lack precise statistical data on the extent of leisure travel in the early 19th century, it is evident that the practice was relatively rare compared with the present day. Moreover, it was primarily accessible to a select and wealthier part of the population.

It is worth noting, however, that in this historical context only a small part of the European population, about 2%, called themselves nomads, although in some Central European regions the proportion was relatively high, up to 10% (Hauptmeyer, 2003).

Be that as it may, the diversity of origins of the nomadic groups does not allow them to be considered as a homogeneous group. Many of them were extremely poor, bankrupt farmers, unskilled agricultural workers who were no longer in demand, jugglers, entertainers, astrologers, quacks, beggars and so on.

The perpetual wanderers were isolated from the rest of society and even marriage was not enough to reconnect them with it. They usually earned their living in groups, moving around in clans or whole families, and cultivated a special use of language, forming their own dialect codes to protect themselves from the representatives of the state.

Tourism was not actively promoted or regulated by governments at this time, as it was not yet considered an essential part of urban society or labour relations. However, it was very much accepted as a leisure activity, especially by the working bourgeoisie and the aspiring middle class. These people often had the financial means and leisure time to undertake travel for recreation and cultural enrichment.

On the other hand, for most European states, vagabondism and adventurism were the opposite of seriousness, order, stability and rationality, which were necessary for the balance of the developing bourgeois society. Faced with this situation, states therefore introduced laws and policies that classified these lifestyles as marginal or deviant, and took coercive and assimilation measures (Hauptmeyer, 2003).

Be that as it may, what we need to emphasize here is that this century was indeed a time of profound change in which industrial society emerged and became the dominant social and economic structure in many parts of the world.

Transport in Europe underwent profound changes that had a lasting impact on the continent's economy, society and culture. The road network was expanded, canals were built to increase the efficiency of transporting goods, and the rapid spread of railways with steam locomotives drastically reduced the cost and time of transporting goods and people. The advent of steam power also sparked a revolution in maritime transport. Taken together, these transport innovations not only accelerated and streamlined the movement of people and goods, but also played a crucial role in promoting trade, industrialization and urbanization. This transport revolution had a decisive impact on the modern European landscape. It reshaped the continent's economic framework and fundamentally altered its culture by changing the way the world works in general (Lay, 2018).

A main reason for the early emergence of mass leisure travel in Europe was thus the expansion of the railway network in the second half of the 19th century. For the first time, the railway enabled large numbers of low-income travellers to travel safely and cheaply beyond the borders of their region.

The railway companies quickly recognized the emergence of the tourism market and participated in its development. At the same time, the construction of many hotels began and the hotel industry developed around the railway stations, not only in the major population centres such as London and Paris, but in most cities that were connected to the railway network.

Inspired by the new working conditions, the need of a large part of the population to escape from everyday life, the signs of the development of a mass consumer society and the progress of means of transport, some people in Britain began to organize travel systematically and founded institutions that contributed significantly to the conception and development of the tourism industry.

Undoubtedly, pioneers such as Thomas Cook (1808–1892), Henry Gaze (1825–1894), Joseph Crisp, Joseph Dean and John Dawson made a decisive contribution to what we now call the tourism industry through their entrepreneurial activities (Simmons, 1984).

In addition to domestic travel, Thomas Cook gradually began to organize trips from Britain to Europe and America, encouraging the wider middle classes to participate in tourism. After combining various services such as transport, accommodation, catering, entertainment and sightseeing into a single independent entity with defined functions, Cook was the first to systematically sell and advertise package holidays to attract customers. This made his office one of the first international tourism companies to crystallize and promote the idea of a tourism industry. In short, he was one of the first to establish and promote tourism as an exclusively commercial activity, as a series of combined economic transactions, and finally as an industry (Simmons, 1984).

During this period, various tourist attractions were created under state auspices, for example more and more museums were founded – the Prado Museum in Madrid was officially opened in 1819, the National Gallery in London in 1824, the Reale Galleria in Turin in 1832, the Munich Gallery in 1834 and the State Hermitage Museum in St Petersburg in 1852 – and many national natural parks were demarcated. The National Archaeological Museum of Greece was founded in 1829 with its original seat in Aegina, the first capital of Greece, and moved to Athens in 1834, while a small Acropolis Museum was founded in 1865 at a site south-east of the Parthenon.

It is worth noting here that museums are important public spaces that can be seen as rhetorical, visual and communicative media for the dissemination of a culture's 'heritage'. As institutionalized points of attraction, they determine what of a culture should and should not be shown. This can inform how a society sees itself and how it wants to be seen by others, such as tourists who perceive that society and its heritage. Therefore, museums are key factors in shaping a country's image (Ashley, 2010).

Transport and communication technology continued to develop. The automobile came on the market in the 1890s; a few years later it was mass produced by the Ford industry in the USA and by other European companies in Germany, France, the UK, Italy, etc. While automobiles were powered by petrol, ships began to use oil and diesel engines. In addition, Alexander Graham Bell invented the electric telephone in 1876, which initially switched calls through a switchboard. In 1885, George Eastman invented the plastic film roll, which replaced bulky photographic plates and introduced a camera that was both cheaper and easier to transport and use (Davis, 1996).

All these technical developments and inventions, together with the spectacular development of the mass media (cheap press, cinema and, a little later, radio) which completely changed the dimensions of space and time in the Western world, created the basic conditions for the conception, organization and promotion of the tourism industry. They provided the opportunity for the formation of an international tourism circuit that enabled all parties involved to coordinate their actions between geographically distant areas and to combine the companies involved in leisure travel into a quantifiable business category.

## 3.3 Transformation of Language and Creation of New Concepts

The profound changes in technology, society and the economy are not only changing the way people think, act and communicate, but are also leading to a transformation of language and the creation of new concepts.

The terms 'tourist' and 'tourism' were already widely used at the beginning of the 19th century. However, their meaning had changed considerably. Eventually, the terms became closely associated with leisure travel undertaken by individuals in relatively large numbers, although their colloquial use in the past was more associated with participants in the 'Grand Tour'. We must therefore stress that the term 'tourist', which was associated with the concept of the 'Grand Tour' as early as the 17th century, has since moved away from this aristocratic adventure (Fuster, 1967; Leiper, 1983).[2]

Buzard (1993), however, argues that the pejorative connotation of the term tourist, as opposed to traveller, had already developed in late 18th-century English literature.[3]

In any case, Larousse's (1876) French *Grande Dictionaire Universel DU XIXe SIECLE* notes the following next to the term tourism: 'Passion, tourist habits: these happy country people suddenly feel bitten by the tourist fly, an inevitable contagion of the elegant mountain', and defines tourists as people who travel out of curiosity and idleness (Larousse, 1876, p. 360).

*The Century Dictionary* (Century Company, 1889–1891, p. 6403), published in the last decades of the 19th century, defines tourism only as 'travelling for pleasure' and the tourist as 'one who makes a tour: one who makes a journey for pleasure, stopping at a number of places for the purpose of seeing the sights, scenery, etc.'.

The 1933 *Oxford English Dictionary* defines the word tourism as 'the theory and practice of touring; travelling for pleasure', while tourist is defined as 'one who makes a tour or tours; one who does this for recreation; one who travels for pleasure or culture visiting a number of places for their objects of interest, scenery or the like' (Oxford University Press, 1933). However, the dictionary also contains examples of the same word as an adjective, such as 'tourist agency', 'tourist country', 'tourist rendezvous', 'tourist ticket', 'tourist crammed', 'tourist haunted', 'tourist loaded' and 'tourist car', which are associated with the notion of mass tourism and the tourism industry or have a kind of pejorative meaning (Oxford University Press, 1933, p. 190).[4]

Thus, in the 20th century, critical opinions about tourism were included in the dictionary and the word was obviously also associated with the business world.

In any case, when the term 'tourism' was first used, there was no comprehensive, systematized sector of services, products and labour large enough to be understood as a tourism industry as it is defined today by academics, economists and international organizations.

It can thus be said that the concept of the modern tourism phenomenon gradually gained content and substance during this period, even if it was not mentioned in the dictionary in the purely economic meaning that was later attributed to it. In fact, the old meaning of the word was a backward remnant of the 'Grand Tour' from the time before mass society, so it had to be modernized in line with the developments of an increasingly commercialized world.

With the development of the modern travel infrastructure and all the commercial enterprises that together make up the tourism industry, the term began to differ in part from the past and became more and more an element of the commercial world.

### 3.3.1 Early academic research

There is no doubt that the research and contributions of the academic community in the early 20th century also played a crucial role in defining and shaping the meaning of 'tourism' as it is understood today. This academic work formed the basis for the development of the field of tourism studies and helped define the terminology and concepts associated with tourism.

Of course, there were already some important studies dealing with the economics of tourism at this time, such as those of the Austrian Stradner (1890), the Swiss Guyer-Freuler (1905) or the work of the Italian Mariotti (1928) *Lezioni di Economia Turistica* (Dann and Parrinello, 2009).

However, in this early phase, tourism also attracted the attention of geographers and historians (Parkes, 1925), while some other academics and researchers, such as the Swiss Walter Hunziker (1899–1974) and Kurt Krapf (1907–1963) or the German Leopold von Wiese (1876–1969), studied tourism not only from its economic but also from its sociological side, trying to understand its deeper meaning in relation to human society (von Wiese, 1930; Hunziker and Krapf, 1942).

The fact is that most of the systematic forms of academic discourse that have emerged within the framework of the 'spirit of capitalism' have sought to define the legitimate way of representing the various and diverse aspects of the tourist phenomenon, especially through the economic lens; that is, to transform tourism into a purely economic act that can be identified with the 'tourism industry'.

In addition to the forms of academic discourse that have emerged, a number of other discourses have been developed in the mass media about the optimal way to consume the 'tourism commodity' and about the appropriate forms of participation in the social practices that are part of the tourist activity.

Be that as it may, our basic position in this book is that the systemic manipulation of basic human attitudes by the capitalist system to fit them as commodities into the circuit of production and consumption can be powerful if it presupposes the existence of genuine desires, genuine primal instincts, to which industries are directed; and in the present case, the tourism industry presupposes the concept of tourism as a consciousness-expanding human quality.

### 3.3.2 Tourism as a source of national income

In the course of the 20th century, leisure and tourism, through their continuous commercialization, undoubtedly became very important sources of national income and of the 'supposed prosperity' of Western countries at the macroeconomic level.[5]

The ever-growing number of travel agencies, hotels and all kinds of businesses related to tourism, combined with economic prosperity and increasing leisure opportunities for the middle class, as well as technological developments in transport and communications, have contributed to a remarkable development of the tourism industry in many Western European countries.

Until the middle of the 20th century, however, long trips abroad were a privilege for those with sufficient leisure time and money.

The character of wealthy people who undertook long exotic journeys was usually associated with vitality, individuality, freedom, culture, but often also with adventurism, which brought great recognition and attracted the admiration of people from all other social classes.

Basically, this was the time when the idea of an integrated tourism industry with all its plans and programmes for economic development was properly formed. This included not only the pure tourism enterprises such as travel agencies, hotels and transport companies, but also others involved in their procurement, promotion, accounting, insurance, etc., and thus tourism was understood as a developing industry; in short, it was believed that tourism could stand on its own and be considered a new industry. In this respect, the term 'tourism' was reintroduced from a purely economic point of view.

Gradually, in countries such as Switzerland and New Zealand, the first national tourism organizations were created to promote tourism (Capuzzo, 2020; Hay, 2021; Christou, 2022).

The Swiss National Tourist Board has existed since 1889, the New Zealand Tourist Board was founded in 1901, the Italian Tourist Board in 1919, the French Tourist Office in 1920, the British Travel Association in 1929 and the Japanese Tourist Board in 1930.

In 1934, the International Union of Official Tourist Propaganda Organizations (IUOTPO) was founded, and after a long journey, the United Nations World Tourism Organization (UNWTO) emerged in the early 1970s, defining and defusing the new commercialized meaning of tourism, alongside all its support for the development of the tourism industry and in cooperation with the World Travel and Tourism Council.

After the Second World War, a remarkable change occurred. This change was characterized by a reduction in working hours and an increase in disposable income. At the same time, the standard of living of the growing middle class in the advanced Western countries improved considerably. This confluence of factors led to an increased appetite for leisure items, with a clear emphasis on long-distance travel, reflecting a growing desire for discovery and adventure in far-flung places.

The post-war period is characterized by the progress of audio-visual entertainment and leisure technologies, but also more generally by the important role of the mass media in influencing the trends and ideas of society. The broad development of communication, which at least in the Western world was able to transcend the borders of states, contributed to the interaction of national cultures and to a kind of unification of ways of thinking.

### 3.3.3 Tourism becomes a field of study in its own right

As we have already noted, due to the increasing importance of tourism in the economies of countries and its direct relationship with the hospitality industry, many academics studying the subject were initially economists, while the first courses in tourism were mainly focused on vocational training in the leisure sector (Airey, 2004).

However, with the steady growth of the tourism and travel industry in the 1960s, many other scholars from various disciplines such as anthropology, geography, sociology, political science and psychology also became involved with the related issues. As some contemporary researchers in the field claim, it was during this period that the concept of tourism emerged as a field of study in its own right (Airey, 2004; Ateljević, 2014).

These years can be considered the time of the important 'first generation' tourism researchers, including Dennison Nash, Jeremy Boissevain, Daniel Boorstin, Erik Cohen, Malcolm Crick, Graham Dann, Nelson Graburn, Jafar Jafari, Marie-Françoise Lanfant, Dean MacCannell, Michel Picard, Valene Smith, Margaret Byrne Swain, Pierre L. van den Berghe and many others (Jamal and Kim, 2005; Nash, 2007).

Gradually, the aviation industry experienced a remarkable boom. As more and more consumers were able to take advantage of air travel, new, far-flung destinations in exotic lands appeared in travel agents' catalogues. The advertising of airlines in the tourism industry gradually increased and at this point we can note that as early as 1953 the Greek National Tourism Organisation cooperated with joint advertising campaigns with the company TWA and a little later with other companies such as the British European Airways

Corporation, Olympic Airlines and Air France (Katsigiannis, 2017).

The development of charter flights in the 1970s and the emergence of American-style low-cost airlines in Europe a few decades later have significantly changed the data on tourist air travel.

Amusement parks, tourist attractions and resorts of all kinds also experienced a worldwide boom during these years. Disneyland, which opened as early as 1955 in Anaheim, California, was one of the major players, followed by others such as: (i) Europa-Park in Rust, Germany, in 1975; (ii) Tokyo Disneyland in Tokyo, Japan, in 1983; (iii) Astérix Park in Plailly, France, in 1989; (iv) Euro Disney Resort in Marne-la-Vallée, France, in 1992; and (v) PortAventura Park in Salou, Spain, in 1995.[6]

In the media, tourism was increasingly presented as an indispensable and universal leisure activity, a kind of basic social right and consumable pleasure available to all. It was presented as a service that everyone can enjoy in their free time, as an ordinary activity that is integrated into the life of every citizen and that is not questioned or discussed.

As a result, the study of tourism became strongly oriented towards the applied sciences, business administration, marketing, but also economics in general; tourism became an industry that had to be studied scientifically. At the same time, the field of vocational and technical education was also growing steadily. Research focused on supporting and promoting the tourism market, with important representatives such as Stephen J. Page, J. Christopher Holloway, Luiz Moutinho, William C. Gartner, Chuck Y. Gee, James C. Makens, Dexter J.L. Choy, Michael Hall, Chris Ryan and others.

The researchers' aim was to select typologies that can methodically support the management needs of the tourism industry and tourism policy. Issues of consumer satisfaction, opportunities for the development of new tourism products, image of destinations, perception of tourism services, segmentation of the tourism market, marketing, advertising, sustainable development of tourism, etc. were basic attractions for the researchers' attention (Ateljević, 2014).

In addition, the geopolitical changes at the international level that accompanied the fall of the Berlin Wall, the dissolution of the Soviet Union and the collapse of the international bipolar political system led to major changes in the tourism industry and significantly increased international tourist flows. As a result, growing tourism from countries in the Western world as well as from the countries of the former Eastern bloc was an important source of revenue to support and sustain the economies of many developing countries.

The post-industrial era, characterized by the dominance of a global economic system that favours oversupply and overconsumption, has accelerated the development of the tourism industry.

Since the 1980s, when academic research focused heavily on economics and business administration, tourism research has developed into a multifaceted academic discipline that has spawned a large number of journals. It is primarily concerned with business and economic development, but also encompasses a wide range of sociological, psychological, anthropological, cultural, geographical and environmental perspectives (Pritchard and Morgan, 2007; Ateljević, 2014).

### 3.3.4 Change in the concept of tourism

It is important to point out a crucial change in the 20th century concerning the concept of tourism. The global development of political ideologies and the pervasive influence of strictly utilitarian economic principles in various spheres of life have led to a transformation of the concept. It has moved away from elements of paramount importance, such as the intrinsic desire to travel in order to expand one's consciousness, which can be considered the core of this phenomenon. Over time, this fundamental aspect has been expanded to include a variety of other dimensions that go beyond the simple pursuit of 'pleasure'.

The fundamental question that arises is: without the innate desire to travel, the urge to discover, the instinctive curiosity about the unknown, the thrill, the adventure, the relaxation, the unlimited possibilities and the lack of predetermined itineraries, can we still legitimately call these experiences tourism?

In order to organize and improve tourism statistics at the international level, the UNWTO, in cooperation with the World Travel and Tourism Council, has developed a basic system of concepts, classifications, definitions, tables and formulae that enable national tourism organizations to collect, classify and compare quantitative data from the domestic tourism industry.[7]

As a result, the term 'tourism' acquired an instrumental character that made it an 'economic indicator' that also included other activities related to travel and accommodation abroad or even within the country. However, it can be argued that many of the services and practices subsumed under the term tourism today were far removed from the meaning of the term in the 18th century.[8]

At this point, however, it is worth noting that the dominant ideology and the current global order are surrounded by many different and intersecting regimes of mobility and travel, in which the movements of certain groups of people are 'normalized' and others 'criminalized' (Schiller and Salazar, 2013).

For example, the mobilization of tourists or students travelling to other countries is constantly presented by the Western media as revitalizing societies, while at the same time the situations created by other travellers such as nomads and wanderers, vagabonds, immigrants from poor countries and refugees are presented as social problems.

However, many researchers believe that the widely used terms nomadism, vagabondism, migration and tourism and the meanings attributed to them are mainly Western-oriented and need to be examined from a more complex perspective and in different cultural and social contexts (Adams, 2020; Choe and Lugosi, 2022).

While the second half of the 20th century was characterized by mass tourism and standardized package tours, the organization of the tourism industry has changed dramatically at the beginning of the 21st century.

Tourists who had experienced mass tourism began to reject the package tours offered by travel agencies and preferred individual trips. The new information technologies and networks of recent years have enabled consumers of tourism products and services to obtain information instantly, to plan their trips individually and to carry out direct and secure financial transactions at any time from home with the help of new forms of online electronic agents.

All the developments in the media landscape have undoubtedly led to a change in the image of tourism. In recent years, the demands, role and identity of the modern tourist have become more complex, as evidenced by the replacement of mass tourism by alternative tourism and the specializations initiated by the tourism industry such as agrotourism, gastronomic tourism, voluntourism, festival tourism, creative tourism and lesbian, gay, bisexual and transgender (LGBT) tourism.

These new forms of tourism fragment the tourism market into countless consumer groups that are ideal target groups for businesses. As a result, tourists are no longer addressed as a mass, but individually, although the number of tourists has reached enormous proportions compared with the past.

## 3.4 Summary

In summary, today's understanding of tourism is primarily a product of various developments that took place in the second half of the 20th century. These include: (i) the progressive commercialization of services traditionally associated with tourism; (ii) the consolidation of these services into a single industry; (iii) the waning influence of earlier understandings of the term; and (iv) more recently, the division and specialization within the tourism market.

Academic publications in journals, scientific conferences and international forums also played a crucial role in shaping the concept of tourism. This development did not occur in isolation, but was closely interwoven with advances in transport, communications and information technology, as well as changes in the global political landscape that drove the process of globalization.

When we talk about tourism today, we usually focus on the tourism industry and the associated amenities it provides. In this contemporary perspective, we may inadvertently neglect the subtleties and nuances that defy quantification but add depth to this phenomenon. We overlook the intangible qualities of tourism, which is fundamentally a uniquely human practice rooted in mobility and the pursuit of greater awareness.

## Notes

[1] The term polysemy refers to the presence of more than one meaning within a word. Polysemous words contrast with monosemous words, which have only one meaning. These are characteristic of technical discourse: the word 'Internet', for example, is basically monosemous (Martin and Ringham, 2000).

[2] At this point it would be interesting to take a brief look at the etymology of the word tourism. Indeed, there are many versions for the origin of this word, but we will mention only some of them here. The most common version is that it comes from the French word 'tourisme', which in turn is derived from the French word 'tour'. The word 'tour' has a rich etymological history, as it can refer to various things, such as a tower, a turn, a tour or journey and even a trick. However, it probably derives from the Latin word *tornus* which referred to a lathe or tool used to shape objects by turning them. The Latin term was derived from the Greek word *tornos*, which also referred to a lathe or circular instrument for drawing or shaping. Over time, the meaning of the word 'tour' evolved from its original association with tools and shaping to a journey undertaken by a person who often visits several places or sites and then returns home. Korstanje (2007), who has done a very interesting etymological analysis of the word 'tourism', presents some other versions regarding the origin of this word. Thus, it can also be argued that the roots of the word 'tourism' come from the Old English verb 'tyrnan', which means 'to turn around'. The root of this word probably goes back to a common Germanic antecedent and reflects the basic idea of changing direction or turning around. However, it is thought that in practice it often referred to leisure travel in the 12th century as well (Fuster, 1967). In the mid-18th century, English aristocrats used the term 'turn' mainly to refer to travel for the purpose of seeking, educating and exploring culture. However, the English term 'turn' was soon abandoned as the term 'tour' prevailed after it was coined in French in connection with the expansion of aristocratic travel to France with the 'Grand Tour'. Another version for the origin of the word tourism is put forward by the Belgian journalist and poet Arthur Houlot (1913–2005), who claims that it does not originate from Latin, Greek or Germanic, but from ancient Aramaic. According to Houlot (1961), the term 'tur' was used in ancient Aramaic for travelling, exploring and wandering. This word was first used when Moses began his campaign in the land of Canaan. Houlot (1961) claims that the term 'tur' was adopted almost simultaneously from European languages. Finally, another version associates the word with the surname of the French nobleman 'de la Tour'. According to this version, after the signing of the Franco-Spanish Treaty between the Roman-German Emperor Charles V (1500–1558) and Francis I (1494–1547) in 1516, the king granted the 'de la Tour' family the right to trade on behalf of English merchants and provided them with travel and accommodation facilities. Soon English merchants spoke of 'a tour' when they had to travel to France on business. Over time, the new terms 'tourist' and 'tourism' emerged. If this version is correct, tourism theoretically refers to an activity that was directly linked to the mediation of trade (Korstanje, 2007).

[3] Buzard (1993) notes that according to the *Oxford English Dictionary*, the word tourist appeared in the late 18th century as a synonym for traveller, but in his study of European tourism literature from 1800 to 1918, he emphasizes that the term 'tourism' already had a pejorative connotation when it appeared in poems and other literary texts from this period. The dichotomy between tourists and travellers already existed as travellers sought to distinguish themselves from the 'mere tourists' they saw or 'imagined' around them. Tourists were seen as people who travelled but looked at their surroundings superficially and recklessly, moving along the marked paths of the railways or major sea routes as part of an economic transaction; they could be compared to parcels programmed to follow a relative trajectory. On the other hand, the traveller was seen as a person who travels, always eager to see things and learn about new cultures, and who enjoys every moment of a journey. Perhaps the 'supposed' difference between tourists and travellers can be better viewed from a perspective closer to social exchange theory. Thus, it can be said that the tourist's relations are purely economic, like commercial transactions (i.e. the tourist has paid a certain amount of money for the contributions made by the various factors involved in the tourism process – employees of tourism enterprises, local people, local services, even nature, etc. – and this demands from these contributors a certain behaviour, quality, quantity, etc., which the tourist believes corresponds to what he/she has offered to pay). The traveller, on the other hand, perceives the whole

system completely differently, because he/she believes that what he/she offers is his time and attention and that what he/she expects is generally the extension of his/her consciousness.

[4] Incidentally, it's quite intriguing and amusing that the *Oxford English Dictionary* (1933) includes an example sentence where the word 'tourism' is used, stating, 'Tourism was born in the 17th century, and Englishmen were the first to engage in it'.

[5] 'Supposed prosperity' in the sense that we doubt, both from a philosophical and a political point of view, that the development of a country's economic indicators alone certifies the well-being of all its inhabitants (i.e. personal and social 'eudaimonia').

[6] It is fair to say that amusement and theme parks have a long history when associated with similar amusement venues such as trolley parks and Ferris wheels, not to mention pleasure gardens, circuses and regular fairs dating back to ancient times.

[7] According to UNWTO (2010, p. 10):

> tourism refers to the activity of visitors [and] a visitor is a traveller taking a trip to a main destination outside his or her usual environment for less than a year, for any main purpose (business, leisure or other personal purpose) other than to be employed by a resident entity in the country or place visited. These trips taken by visitors qualify as tourism trips.

As can be seen from the book by Soulantigas (1950), similar definitions have been given several times in the past by different organizations such as the League of Nations or by economists.

[8] Mobilities such as sending workers to business conferences, meetings and congresses, attending professional or activist forums in remote areas, short-term professional training in the parent company, or travelling to a country with advanced health technology to undergo surgery are probably not what was understood by tourism in the 19th century. We should emphasize here that the verb travel is not synonymous with tourism. On the other hand, we believe that people who travel constantly for pleasure or to get to know the world (e.g. through Europe, America or around the world, with stopovers where they work temporarily in the local economy to secure their resources for onward travel), or those who travel as a way of life, are closer to the word tourist but are excluded from the UNWTO definition because they work in the countries they visit. Finally, one can only laugh at the fact that someone who has spent 11 months in one place is considered a tourist, while someone who has spent 13 months in the same place is not.

## References


Adams, K.M. (2020) What western tourism concepts obscure: intersections of migration and tourism in Indonesia. *Tourism Geographies* 23(4), 678–703. DOI: 10.1080/14616688.2020.1765010.

Airey, D. (2004) From here to uncertainty. In: *Proceedings of the 2004 Conference 'Critical Issues in Tourism Education', Missenden Abbey, Buckinghamshire, UK, 1–3 December 2004*, Association for Tourism in Higher Education (ATHE), Buckinghamshire, UK, pp. 9–15.

Ashley, S.L. (2010) Vision, translation, rhetoric: constructing heritage in museum exhibitions. In: Burns, P., Lester, J.-A. and Bibbings, L. (eds) *Tourism and Visual Culture, Volume 2: Methods and Cases*. CAB International, Wallingford, UK, pp. 24–34. DOI: 10.1079/9781845936112.0000.

Ateljević, I. (2014) Mapping a history and development of tourism studies field. Turizam: Eđunarodni Znanstveno-Stručni Casopis 62(1), 75–101.

Barthes, R. (1957) *Mythologies*. Editions du Seuil, Paris. (in French).

Buzard, J. (1993) *The Beaten Track. European Tourism, Literature, and the Ways to Culture, 1800–1918*. Clarendon Press, Oxford, UK. DOI: 10.1093/acprof:oso/9780198122760.001.0001.

Capuzzo, E. (2020) War tourism in Italy (1919–1939). In: Martínez, C.P. and Aranda, M.L. (eds) *Inter and Post-War Tourism in Western Europe, 1916–1960*. Springer, Cham, Switzerland, pp. 35–63. DOI: 10.1007/978-3-030-39597-1.

Century Company (1889–1891) *The Century Dictionary*. An encyclopedic lexicon of the English language. Prepared under the superintendence of William Dwight Witney. Century Company, New York.

Choe, J. and Lugosi, P. (2022) Migration, tourism and social sustainability. *Tourism Geographies* 24(1), 1–8. DOI: 10.1080/14616688.2021.1965203.

Christou, P.A. (2022) *The History and Evolution of Tourism*. CAB International, Wallingford, UK. DOI: 10.1079/9781800621282.0000.
Dann, G. and Parrinello, G. (2009) Setting the scene. In: Sharpley, R.A. (ed.) *The Sociology of Tourism: European Origins and Developments*. Emerald, Bingley, UK, pp. 1–63.
Davis, E. (1996) *Inventions*. Pataki Publications, Athens. (in Greek).
Friedman, M. (1985) The changing language of a consumer society: brand name usage in popular American novels in the postwar era. *Journal of Consumer Research* 11(4), 927–938. DOI: 10.1086/209027.
Fuster, L.F. (1967) *Teoría y Técnica del Turismo*. Nacional, Madrid. (in Spanish).
Guyer-Freuler, E. (1905) Fremdenverkehr und hotelwesen. In: Reichesberg, N. (ed.) *Handworterbuch der Schweizerischen Volkswirtschaft, Sozialpolitik und Verwaltung*. Verlag Enzyklopadie, Bern, Switzerland.
Hauptmeyer, C.H. (2003) New marginal groups and lower classes. In: Ahrweiler, H. and Aymard, M. (eds) *The Europeans, Volume I*. Savallas, Athens. (in Greek), pp. 158–170.
Hay, B. (2021) The future of national tourism organisations marketing functions – there is no future? *Journal of Tourism Futures* 7(2), 179–183. DOI: 10.1108/JTF-08-2019-0075.
Houlot, A. (1961) *Le Turisme et La Biblie*. Revue l'Académie Internationale du Turisme, Monaco. (in French).
Hunziker, W. and Krapf, K. (1942) *Algemeine Frendenverkehrslehre*. Berna University, Zurique, Switzerland.
Jamal, T. and Kim, H. (2005) Bridging the Interdisciplinary divide: towards an integrated framework for heritage tourism research. *Tourist Studies* 5(1), 55–85. DOI: 10.1177/1468797605062715.
Jatowt, A. and Duh, K. (2014) A framework for analyzing semantic change of words across time. In: *IEEE/ACM Joint Conference on Digital Libraries (JCDL),London, UK, 8–12 September 2014*, Institute of Electrical and Electronics Engineers (IEEE), London, pp. 229–238. DOI: 10.1109/JCDL.2014.6970173.
Katsigiannis, K.X. (2017) *Hellenic Tourism Organization: Travel Through Time*. Private Edition, Athens. (in Greek).
Korstanje, M. (2007) The origin and meaning of tourism: etymological study. *E-Review of Tourism Research* 5(5), 100–108.
Larousse (1876) *Grande Dictionaire Universel DU XIXe SIECLE*. Larousse, Paris. (in French).
Lay, M.G. (2018) *The Harnessing of Power: How 19th Century Transport Innovators Transformed the Way the World Operates*. Cambridge Scholars Publishing, Newcastle upon Tyne, UK.
Leary, J.P. (2019) *Keywords: the New Language of Capitalism*. Haymarket Books, Chicago, Illinois.
Leiper, N. (1983) An etymology of tourism. *Annals of Tourism Research* 10(2), 277–280. DOI: 10.1016/0160-7383(83)90033-6.
Mariotti, A. (1928) *Lezioni di Economia Turistica* Tiber, Rome. (in Italian).
Martin, B. and Ringham, F. (2000) *Dictionary of Semiotics*. Bloomsbury Publishing, New York.
Nash, D. (2007) The Study of Tourism: Anthropological and Sociological Beginnings. Elsevier, Oxford, UK.
Nina-Pazarzi, E. (2014) *Sociology: the Study and Understanding of Human Societies*. Da Vinci, Athens. (in Greek).
Oxford University Press (1933) *The Oxford English Dictionary*. Oxford University Press, Oxford, UK.
Parkes, J. (1925) *Travel in England in the Seventeenth Century*. Oxford University Press, Oxford, UK.
Pritchard, A. and Morgan, N. (2007) De-centring tourism's intellectual universe, or traversing the dialogue between change and tradition. In: Ateljevic, I., Pritchard, A. and Morgan, N. (eds) *The Critical Turn in Tourism Studies: Innovative Research Methodologies*. Elsevier, Oxford, UK, pp. 11–28.
Schiller, N.G. and Salazar, N.B. (2013) Regimes of mobility across the globe. *Journal of Ethnic and Migration Studies* 39(2), 183–200. DOI: 10.1080/1369183X.2013.723253.
Simmons, J. (1984) Railways, Hotels, and Tourism in Great Britain 1839-1914. *Journal of Contemporary History* 19(2), 201–222. DOI: 10.1177/002200948401900203.
Soulantigas, P. (1950) *Tourism*. Dialismas SA, Athens. (in Greek).
Stradner, J. (1890) Die forderung des fremdenverkehrs. In: Wilhelm, G. (ed.) *Kulturbilder Aus Steiermark*. Leykam, Graz, Austria, pp. 257–288.
Toti, G. (1975) *Il Tempo Libero*. Editori Riuniti, Rome. (in Italian).
United Nations World Tourism Organization (UNWTO) (2010) *International Recommendations for Tourism Statistics 2008*. United Nations Publications, New York.
von Wiese, L. (1930) Fremdenverkehr als zwischenmenschliche Beziehung. *Archiy Fur Fremdenverkehr* 1.
Williams, R. (1976) *Keywords: a Vocabulary of Culture and Society*. Oxford University Press, Oxford, UK.
Wittgenstein, L. (1986) *Philosophical Investigations*. Basil Blackwell, Oxford, UK.

# 4 Culture and Tourism

**Abstract**
Culture is a fluid, evolving phenomenon that is influenced by the collective needs of individuals, group dynamics and engagement with the natural and social environment. It manifests itself in specific actions, ideas and customs within communities. This chapter addresses the complex interaction between culture and tourism and emphasizes the importance of refraining from hierarchical judgements about cultures based on personal prejudices. Culture is presented as a universal human experience with unique differences between societies, which encourages comparative studies. Nomadic and sedentary cultures, multicultural societies and concepts such as consumer and mass culture are analysed. The interaction between globalization and tourism is examined in detail, as are specific topics in tourism studies such as cultural preservation, cultural exchange, cultural heritage, sustainable and creative tourism and cultural routes. The central role of the mass media in shaping tourism trends and facilitating cultural exchange is consistently recognized.

Throughout history, culture has played a crucial role in the mobility of people. Travellers rooted in their own cultural background go on journeys driven by curiosity to explore faraway places and learn about other cultures, thus expanding their consciousness.

To begin our exploration of the term 'culture', we want to pose some definitional questions and establish a clear framework for its use in this book. We will look at issues such as: (i) the distinction between 'civilization' and 'culture'; (ii) whether only humans among all animals have culture and civilization; (iii) the concept of superior and inferior cultures; and (iv) how we can define culture to relate it to tourism. By answering these questions, we aim to provide a basis for understanding the intrinsic connection between culture and tourism – a theme that runs like a common thread through the development of this book.

## 4.1 Distinction Between Civilization and Culture

The word 'civilization' has its origins in the Latin word *civis*, which means citizen, but is also a root for the word 'city', so that 'civil' can also refer to various aspects related to the city. In its colloquial meaning, the term usually refers to an advanced intellectual, cultural and material phase in the development of human society, characterized by advances in architecture, technology, more generally in the arts and sciences, the spread of writing and the emergence of complex political and social institutions. In the context of its etymology, the term can be more closely associated with the foundation of the city or a society consisting of cities, and in any case with civil law and governance (i.e. with an advanced state of human society characterized by the development of complex social, political and cultural institutions).

On the other hand, the word 'culture' derives from the Latin word *cultus*, which means to inhabit, cultivate, honour, etc. So we can assume that, apart from its first and colloquial meaning, which generally refers to the patterns of human activities and the symbolic structures that give meaning to these activities, the term has more to do with something that is perhaps related to a process, that has an immediate, developmental and not so cumulative meaning; and that is dynamic, or we could say growing or changing.

Culture can be seen as a constantly evolving phenomenon, shaped by human needs arising



DOI: 10.1079/9781800626034.0004

from the natural environment and manifested in a set of customs and circumstances within a social community. There is a fundamental functional relationship between culture, human needs and the ability of people to organize and institutionalize practices and issues (Malinowski, 2002; Uscatescu, 1986).

Be that as it may, the fact is that civilization and culture are two very fluid and complex concepts about which there have been many long but fruitful discussions, and it would probably be better to think of them as mobile signifiers that allow us to talk about human activity in different ways and through different optical prisms depending on the objective, and as such they can always be associated with tourism.

So, if we look at earlier discussions, we can assume that the concept of civilization is sometimes seen as inseparable from the idea of historical social progress (i.e. the triumph of rationality over the transcendent, the decline of local special customs and the rise of the natural sciences). The word civilization has often been used to refer to any complex human society, usually consisting of advanced cities and exhibiting certain features of cultural and technological development; but also as a comparative and evaluative term for classification, sometimes considering great civilizations as morally good and culturally advanced, while others were considered backward. Be that as it may, the term has also been applied generally to the process of raising a society to a higher level of humanity and submission to the law.

In the same context, the term culture was sometimes considered to refer to a particular way of life of a larger community or social group, a method of deciphering the world, or a form of expression of the members of a society (or social group). The word culture has also been used as a tool for evaluating a society (or social group), as a standard of comparison or as a means of categorization. In particular, culture may be considered as the everyday practical life, the art and artistic development of a society (or social group), the organizational structure, the high or low level of education, a phase of forms or behaviour of a society (or social group), etc.

Sometimes civilization is seen as a progressive, measurable, cumulative human achievement, while culture is seen as a current form of a society's self-awareness. In other cases, culture and civilization are distinguished by an idealistic view based on the distinction between 'matter and mind' (Vassiliou and Stamatakis, 2000).

In many cases, the terms civilization and culture were even regarded as identical. At the end of the 19th century, the British anthropologist Edward Burnett Tylor (1832–1917) defined both terms using a broad ethnographic concept that still causes some confusion in distinguishing the terms today. He regarded civilization and culture as the complex totality of 'knowledge, beliefs, arts, morals, law, customs and all other faculties and habits' that people have acquired as members of a society in the various societies of mankind. He argued that civilization and culture can be studied in general principles and lend themselves to the study of the laws of human thought and action (Tylor, 1871).

We see, then, that because of the multiplicity of dozens of understandings of civilization and culture, there can be no objectively correct or objectively incorrect definitions, for ultimately none of these definitions is incorrect in the sense that it fails to describe an object correctly. In short, the meanings and the way these terms are used change according to historical epoch, according to national language, but also because thinkers coming from different scientific fields or ideological positions use them in different ways each time and try to develop different themes. In this framework, Norbert Elias (1897–1990), in his work *The Civilizing Process* (1939), in which he attempts a formal analysis and theory of civilization, presented many different definitions of the concepts of civilization and culture in 19th-century France and Germany (Barker, 2004).

Be that as it may, civilization and culture permeate all aspects of our lives, be it in terms of our personal and social relationships or our lifestyle, consumption, technology, art, politics, environment, mobility and so on.

As far as our civilization, but also our culture, are concerned, we usually think of them as facets that we simply absorb into our society, we accept them as they appear to us, they come into our possession without asking how we got to this point, we never ask ourselves whether they are processes or part of a process in which we find ourselves. All the individual elements we include in them (i.e. traditions, art, scientific discoveries, machines, political forms, habits, etc. and everything else), are testimonies to a certain structure of human relations, our society, and thus to a certain kind of human behaviour (Elias, 1939).

In his book *Culture and Society, 1780–1950* (Williams, 1960), the British writer, academic and novelist Raymond Williams (1921–1988) systematically analysed the development of the concept of culture, observing how the use of this word has been influenced and changed over time by various economic, political and social changes. For Williams, the history of the concept of culture is the record of our intellectual and emotional responses to the changing conditions of our daily lives (Smith and Riley, 2008).

According to Williams, there are three main categories in which scholars commonly understand and use the term culture. These are, first, a general process of intellectual, spiritual and aesthetic development; secondly, the particular way of life of a group or epoch; and thirdly, the works and practices of intellectuals and artistic activities (Buzard, 1993).

In using the terms 'civilization' and 'culture', then, certain associations must be taken into account, since they depend on the discussion and the point of view from which we are looking at something; but we should in some general way determine how we shall use them in our development.

So we can say that the term culture in this book generally encompasses values, lifestyles and institutional products, as well as the practices of shaping and reshaping them over time, whereas when we speak of civilization we are referring to a particular and accumulated form rather than to the analysis of that form and the practices that produce and simultaneously change it (Smith and Riley, 2008).

## 4.2 Do Only Humans Among Animals Have Culture and Civilization?

Since ancient times, most philosophers and theorists who have reflected on human civilization and human culture, from Aristotle to Thomas Aquinas (1225–1274), René Descartes (1596–1650), Immanuel Kant (1724–1804) and so on, have maintained that there is an essential difference between humans and other animals.[1]

Indeed, it has often been said that what distinguishes man from other animals is culture. In general, it can be said that in the Middle Ages the soul was considered the main distinguishing feature between non-human beings and humans, but in recent years rationality seems to have taken over this role. In the last century, the view has gained ground that we are distinguished from other animals by our capacity for complex thought, our use of complex language, our ability to solve difficult problems, and our self-awareness and introspection (i.e. our ability to describe our own thoughts and feelings).

However, a growing number of contemporary thinkers argue that all these differences are hypothetical, exaggerated or at least not essential. New evidence is being systematically presented in both the popular media and the scientific literature that the boundaries between humans and other animals are not as impermeable as we once thought. According to John Zerzan, the human species is certainly unique, but so is every other species, and in this sense we need to remember that we are no more different from others than other species are from each other (Zerzan, 2014).

It is believed that a large part of culture has to do with the issue of social meaning-making (i.e. the different ways in which we give meaning to the world). However, the different meanings we give to things and ideas do not exist independently 'out there' but are created through social conventions that we form through signs when we communicate with each other.

The study of culture is therefore closely linked to the study of communication, the analysis of the meaning of signs, the study of codes and language.

However, as we have seen, for a long time the view prevailed that the systematic use of symbolic codes and language is practised only by humans and not by other animals, although today this view is rejected by various researchers.[2]

So, after all this discussion, we can say that it only depends on how we define these terms, and depending on which definition we accept, we can judge whether other animals can have a civilization and culture like humans.

For example, if we assume that civilization is a complex idea that includes urban development, advanced architecture and technology, symbolic communication (e.g. language, writing and mediated communication), social stratification and the perception of separation from nature, then we can say that civilization distinguishes humans from other animals. In the same way we can say culture distinguishes humans from other animals, if we assume that culture is art, morality, law, education, crafts and customs, and all other skills.

In reality, we usually define civilization and culture in anthropocentric terms, encompassing elements such as organized settlements, language, science, art, education, customs and a refined way of life. Consequently, it is not possible to label other species as civilized according to these human-centric criteria.

## 4.3 The Concept of Superior and Inferior Cultures

In our effort to be as neutral as possible, we will do our best not to use culture as a means of distinguishing between the best and the worst human practices (i.e. we will not use it as a yardstick to determine which works of art and human practices are good and which are bad, which represent a superior and which an inferior civilization, etc.).

In other words, we will not consider culture as a tool for hierarchical judgement of societies, but we will also try not to get hung up on the question of whether there are always improvements and advances in the culture of a particular society over time.

The concept of superior and inferior cultures is highly subjective and even contested over time for the same society. Cultural superiority or inferiority implies the assumption that the way of life and general customs and beliefs of a particular group of people at a particular time are inherently better or worse than those of another. However, such assumptions can only be based on personal prejudices, ethnocentrism and biased or limited views.

We will approach our topic, shall we say, in the spirit of an anthropological understanding that describes culture as a universal concept but also as a particular way of life, which means that we are all cultured. In other words, we will accept culture as a comparative rather than a hierarchical means of evaluating human societies; we will treat it as a differentiating concept that provides recognition factors for the internal cohesion and external distinction of a group or society.

## 4.4 How to Define Culture to Relate It to Tourism

There are several approaches to culture that are of particular interest for the development of the topic of tourism. Before discussing these approaches, however, we must first very briefly introduce some key terms used in this study in relation to the concept of culture (i.e. what we mean when we speak of 'nomadic culture' and 'sedentary culture', 'mass culture' and 'consumer culture', 'multicultural societies' and 'cultural globalization').

### 4.4.1 Nomadic culture and sedentary culture

Over the millennia, human life on our planet has gone through different phases characterized by certain cultural differences. One of the most important changes came with the discovery of agriculture, when more and more groups of people abandoned their nomadic way of life and settled in settlements. This led to sedentary societies developing a different culture.

'Nomadic culture' is based on the constant movement of groups and therefore allows them little intervention in the environment, while it is characterized by adaptation, contingency, variability and flexibility in order to survive. In our time, the imaginary relationship between movement and freedom, created by the reflections of postmodern philosophers, social anthropologists, primitivists, etc., has given nomadic culture an attractive aura of utopian society (Deleuze and Guattari, 1986; Zerzan, 1994; Peters, 2013; Engebrigtsen, 2017).

'Sedentary culture' developed when people began to domesticate plants and animals about 10,000 years ago and settled permanently in one place instead of moving from one place to another. This allowed them to farm better and develop different methods and techniques for production and organization, which educated them to values such as classification and order and made them more rigid and unadaptable to change. In a stable environment, life became more predictable and secure, but perhaps also more boring.

The sedentary culture gave the terms travelling, touring, hiking, exploring, etc. a different social meaning, as these activities were no longer part of everyday life.

In the context of human history, this transition represents only a small fraction of the time that *Homo sapiens* has existed. So the time when humans settled permanently in one place is a relatively recent development compared with the time they spent as nomadic groups.

What captivates us about the use of the terms 'nomadic culture' and 'sedentary culture' is the transformative effect they signify for human existence, especially in the context of mobility. This shift underscores our deep recognition that humans have an innate propensity to move. The tourism industry has recognized and harnessed precisely this facet of human nature. Thanks to this innate human trait, the tourism industry is poised for even greater success in the future.

### 4.4.2 Mass culture and consumer culture

'Mass culture' is a term that has been used since the 1930s, especially by literary critics, sociologists and philosophers (e.g. the Frankfurt School), and usually in a negative sense. Although it can simply refer to culture created in the age of mass society, it had evolved into a judgemental position about the inferiority of culture in the age of capitalism as well as the detrimental quality of life of mass society. According to this view, 'mass culture' was considered inauthentic, manipulative and inferior because it was based on the mass production of identical products. High culture was the crowning glory of the cultures of earlier eras, but has almost perished with the development of capitalism and the mass production of cultural products.

In this context, Theodor Adorno (1903–1969) and Max Horkheimer (1895–1973) used the term 'culture industry'. They argued that contemporary culture is primarily shaped by capitalist enterprises that offer cultural products to the public but ultimately promote homogeneity and conformity (Adorno, 1991).

In this view, 'mass culture' was a manipulative tool that simultaneously deceived and undermined the intellectual capacities of the masses.

Today, the term 'cultural industry' encompasses various areas of cultural production, including galleries, theatre, cinema, music and mass media, as well as tourism, without necessarily having a negative connotation.

As mentioned earlier, we believe that the evaluation of aesthetic quality is inherently debatable. Each of us has the freedom to form a judgement on any subject and to judge its inferiority or superiority subjectively and according to our personal taste. This underlines that universal judgements about cultural artefacts are essentially social constructs or conventions that are subject to evolution. We argue that concepts such as beauty, harmony of form and quality can be applied to a motorbike in the same way as to a novel, a film or a painting, which underlines their cultural significance (Barker, 2004).

Today's tourism is closely interwoven with the effects of the Industrial Revolution and all the cultural and social changes that followed. These changes include: (i) technological advances; (ii) expanded leisure opportunities; (iii) improved transportation; (iv) increased disposable income; (v) the influence of the media; (vi) globalization; and (vii) the rise of consumerism. Taken together, these elements contribute to the phenomenon known as 'mass culture' that characterizes the landscape of modern tourism.

However, it is important to understand that tourism is a multi-layered phenomenon characterized by a variety of elements. These also include the inherent human 'seeking instinct', individual preferences, the richness of cultural diversity, historical heritage, economic dynamics, ecological considerations and much more.

The term 'mass culture' is directly related to the term consumer culture, which refers to the concept of consumerism in the age of mass production. 'Consumer culture' states that everything in modern societies revolves around constructed consumption, from products and services to ideas, human relationships, values, ethics, social behaviour and so on.

Apart from that, the purchase and consumption of goods and services are directly related to the construction of individual and social identities and the formation of lifestyles for the purpose of social categorization.

Nowadays, anyone can visit a holiday destination that is considered 'exclusive' in order to fit into a certain social category or lifestyle that values that destination. The mass media and advertising play an important role in this. They offer a repertoire of images and representations associated with certain goods and services, and inform about what constitutes a good life or what one has to do to belong to a certain social group that follows a 'lifestyle'.

In addition, 'consumer culture' is driven by the desire to expand the range of goods and meanings through the constant search for new products and services, ceaseless technological innovation and the pursuit of the exotic in order to increase corporate sales and profits.

Thus, our daily lives are increasingly dominated by the constant consumption of goods, services and experiences that take place in a wide variety of places, such as shopping centres, department stores, historic city centres, theme parks, tourist attractions, spas and cultural monuments, which offer a range of constructed pleasures, simulations and spectacular spaces of consumption (Featherstone, 1982).

### 4.4.3 Multicultural societies

Various historical factors related to colonialism and the rapid growth of cities and urban areas, as well as current geopolitical and economic conditions on the international stage and the development of transport and communication technologies, have gradually led to the emergence of 'multicultural societies'.

In the past, intercultural relations and the nature of interaction and integration of different groups in larger cities of imperial or global political entities were referred to by the Greek word 'cosmopolitanism'. This word could sometimes refer to an elite and denoted the sophistication and cognitive development resulting from familiarity with the foreign. The same word was also used metaphorically to refer to sensitivity and tolerance of otherness, contact with and enrichment by other cultures, independence from particular local authorities, intercultural and transnational realities and demands, and mobility and migration. Opposing concepts included xenophobia, stability, regionalism, limited sovereignty and fanatical loyalty to a homeland or nation state.

From ancient to modern times, great cities such as Alexandria, Rome, Constantinople, London, Paris, Amsterdam and New York were considered cosmopolitan, admired and attracted visitors from all over the world. It could be argued that in some ways they reflect the concept of 'multicultural society' (Ribeiro, 2001).

Multiculturalism is often an attraction for tourism because it generally offers different authentic experiences, even if artificially concentrated in one place, and can even simulate tourist experiences for the permanent residents of a city.

Nowadays, tourists enjoy visiting international festivals that celebrate and showcase the different cultures, traditions, art forms, music, food and other aspects of different countries from around the world, or visiting ethnic enclaves such as Chinatowns, Italian or Greek quarters and other ethnic neighbourhoods that have sprung up in all major cities around the world. They enjoy strolling through the colourful specialty shops, tasting exotic foods, smelling dozens of unusual perfumes, trying on ethnic clothing, listening to world music, etc.

Be that as it may, today multicultural societies are considered societies in which people from different cultural backgrounds live side by side and interact with each other. In short, they are mosaic societies composed of a mixture of different social groups or individuals from different cultures (i.e. with differences in ethnicity, religion, language, traditions, etc.).

Multiculturalism as a value recognizes cultural diversity within a society and has as its stated aim the promotion of understanding, respect and equality between different social groups and individuals.

In a multicultural society, groups and individuals are encouraged to maintain their own cultural identity while remaining connected to the wider society. This may include maintaining their customs and religious ceremonies, celebrating cultural practices and using their languages. In this regard, multicultural states often support policies and initiatives to promote and preserve diversity (e.g. anti-discrimination laws, teaching of different languages, cultural integration programmes and support for undervalued minority identities).

The benefits of multicultural societies include: (i) broadening the consciousness of the population through the enrichment of cultural experiences; (ii) fruitful innovation through the exchange of very different ideas; (iii) increased tolerance and acceptance of the 'other'; and (iv) economic benefits through diverse perspectives and talents and the easy creation of trade and tourism networks between the host country of cultural groups and their places of origin. However, such situations can also lead to misunderstandings and conflicts if the groups are not open to peaceful coexistence. Therefore, there is a great need to reduce cultural tensions, promote social cohesion and ensure equal treatment and opportunities for all people, regardless of their cultural background.

### 4.4.4 Interaction between globalization and tourism

In the second half of the 20th century, the pace of 'globalization' of consumer culture began to increase dynamically. This has become particularly clear since the 1980s with the dismantling of tariffs through World Trade Organization agreements, state deregulation, the development of the international market, the collapse of the Eastern Bloc and the development of transport and communication technology (Featherstone, 1982).

However, 'globalization' does not only mean the internationalization of consumer culture, but also encompasses many other things. In essence, globalization theory reflects the emergence of a 'global village', meaning that the entire world is increasingly interconnected (McLuhan, 1962).

According to Beck (1999), the term globalization refers to a complex historical process that, reinforced by the new technological possibilities developed in the neo-liberal context, shapes the idea of a global society that brings the planet's populations into direct economic, political and cultural contact, making the world a single community. Many researchers believe that globalization has empowered citizens around the world by broadening their horizons and giving them more opportunities to act in all areas (Giulianotti and Robertson, 2004; Shin, 2009).

The phenomenon of tourism is indeed directly linked to globalization. International tourism, which includes both inbound and outbound tourism (i.e. all movements and tourism activities by visitors who are not permanent residents of a country), is a large part of this phenomenon.

The United Nations World Tourism Organization (UNWTO) estimated the number of international tourist arrivals in 1950 at almost 25 million . At the beginning of the 21st century, the growth of international tourism has continued to increase due to various global developments that more or less coincide with the causes that led to the extensive globalization mentioned earlier. These include, namely, the cultural and political democratization of many developing countries, the easier and more extensive movement of people and capital due to the development of new transport and communication technologies and the emergence of alternative lifestyles and consumption patterns. In this context, the number of international arrivals has increased to 2.4 billion tourists in 2019 (World Bank Group, n.d.).

An important role in international coordination and in the process of globalization of tourism has undoubtedly been played by the development of international organizations and professional associations in this field. These include the UNWTO, the World Travel and Tourism Council (WTTC), the International Air Transport Association (IATA), the International Civil Aviation Organization (ICAO), the Air Transport Action Group (ATAG) and the International Hotel and Restaurant Association among others.

The conditions of globalization have led to new patterns of travel, highlighted alternative forms of tourism consumption and introduced many new destinations. Examples include the rapid development of medical tourism, which has been interpreted as a direct and positive expression of globalization to better meet the needs of medical care, and conference tourism, which supports international meetings (Hall, 2005, 2011; Hannam *et al.*, 2006).

It should also be noted that developments in technology and operational organization in the field of aviation have enabled many remote areas of the world to be added to the list of attractive destinations. Globalization has drawn the attention of the Western world to remote areas rich in ecological or cultural aspects, which have been featured in documentaries in the media for many years and represent a dream destination for the public.

In this context, many remote countries that did not have sufficient tourism resources have succeeded in becoming centres of attraction for mass tourism offers, as well as for certain alternative forms of tourism through the development of technical infrastructure (hotel facilities, transport networks, etc.) and the development of a tourism economy with investments from international tourism companies or with state support.

Thus, globalization has put great competitive pressure on states that already had a high economic dependence on tourism, forcing them in many cases to organize central support for the planning of a unified tourism development policy. However, the role of the state in tourism planning always depends on the objectives and political philosophy of the respective government.

On the other hand, many believe that globalization has benefitted some but not all of the world's citizens, creating new conditions with advantages and disadvantages. They argue that globalization does not lead to the development of small local economies, but is a force that serves the economically powerful and leads to the transformation and homogenization of cultures, ideas, behaviours and lifestyles and reinforces economic-cultural expansionism, a view reminiscent of the position of cultural imperialism (Wood, 1984; Golding and Harris, 1996).

The dominance of Western states in the development of the tourism industry and their preferential control have led many to believe that tourism is ultimately a form of imperialism.[3]

The American sociologist Ritzer (2004) uses the term 'McDonaldization' to argue that the principles of fast food restaurants, especially the possibility of alternative composition, efficiency, precise calculation, predictability and intense control at every point of production, are now very dominant not only in the business world but in all human activities in modern society. It is believed that this culture extends across the world and reflects the universal dominance of modern Western business ideology.

Ritzer believes that tourism is a clear example of this inevitable process, as most tourists: (i) seek alternative but common tourist experiences at fixed prices; (ii) visit identical fragmented cityscapes and similar multicultural environments; (iii) have the ability to assemble their tourist experiences; (iv) have full control over the holiday schedule; and (v) can predict the experiences of their trip with great consistency in advance.

It can thus be said that the tourism industry has transformed places unknown to most tourists into familiar and convenient destinations with the same tourism infrastructure. Although freedom of choice and diversity of destinations have increased, people enjoy standardized forms of tourism consumption in various combinations.

The dialectic of global and local can also be seen as a form of globalization, which can be found in foreign literature under the term 'glocalization' and concerns the simultaneous occurrence of universal and local trends in modern social, political and economic systems (Blatter, 2007).

This hybrid term (globalization-localization) was introduced into social science discourse by Robertson (1994), who assumes that the 'local', the 'global' and 'the dialectic of the global with the local' do not lie outside the global but converge within it. However, the result of the convergence of the global with the local may ultimately lead to global elements returning to the international stage, but under the influence of the local. Therefore, the concept of 'glocalization' can be defined as the refraction of the global through the local (Roudometof, 2016).

So instead of offering classic internationalized products, a local tourism enterprise can try to adapt them to its own local conditions, and this glocalization can then be exported and be successful abroad. In this context, a Greek restaurant in Athens could include an internationalized dish, such as an Italian pizza, in its menu. However, the chef of the restaurant could invent an original version with elements of Greek cuisine (e.g. with the ingredients of the traditional Greek salad) in addition to all the other classic pizzas he prepares. So it would not be at all unlikely that this hybrid pizza, which the chef even calls 'Greek Lovers', would become popular in the pizzerias of the tourists' countries of origin when they return home.

## 4.5 Culture-Related Topics of Relevance to Tourism Studies

Having talked about the culture-related concepts that we will refer to frequently in this book, we can also present some issues and approaches to tourism in the cultural field that are in any case closely related and of particular interest to tourism scholars.

The 'cultural preservation' approach, for example, focuses on protecting and preserving the cultural heritage of a destination. The cultural, historical and natural resources of a community are valuable and often irreplaceable, but tourism is often perceived as a threat to the preservation of cultural heritage. With the threat of change to local culture in general, great efforts are now being made to protect historic sites and old towns, traditional areas, customs, local art forms and other intangible cultural practices. The protection of local culture helps to preserve the authenticity of a destination and offer tourists unique experiences. The relationship between tourism, mass media and cultural preservation is essentially multidimensional. The power of mass media to shape perceptions, promote destinations and inform travellers makes it a crucial tool in balancing the demands of tourism with the need to preserve cultural heritage for present and future generations (Reinfeld, 2003; Chong and Balasingam, 2019).

The concept of 'cultural exchange', on the other hand, focuses on promoting cultural interaction between tourists and local communities. Tourism that promotes bidirectional acculturation or meaningful cultural exchange between tourists and locals equalizes the relationship between hosts and guests. This includes initiatives such as guesthouses, community-based tourism and cultural immersion programmes that allow tourists to engage with local people, their traditions and their way of life. These interactions promote mutual understanding and appreciation of different cultures. The mass media acts as a bridge that facilitates cultural exchange, enhances tourists' experiences and promotes a deeper understanding and appreciation of the world's diverse cultures (Jack and Phipps, 2005).

The interpretation of 'cultural heritage' is about deciphering the cultural heritage associated with a place and communicating it to tourists. The focus is on meaningful and informative experiences through guided tours, exhibitions, talks in museums, theatre walks, storytelling and other forms of outreach. The goal of heritage outreach is to enhance and enrich visitors' experiences by helping them understand the significance of the place they are visiting and connecting that meaning to their own lives.

In short, heritage education helps visitors understand the historical, social and cultural significance of a place, enhancing the overall cultural experience. The ability of mass media to communicate and appeal to a wide audience increases the impact of heritage communication, enhances the tourist experience, raises awareness and contributes to the preservation of cultural and historical treasures (Nowacki, 2021).

While sustainability is not itself a culture, it is deeply embedded in and influenced by the cultures in which it is practised. Different cultures have unique approaches to sustainability and prioritize environmental protection, social justice and economic development differently depending on their values and worldviews. The idea of sustainability in tourism as a desired direction of change is now widespread. It has become an explicit structuring concept for tourism worldwide, although there are differing views on what exactly it entails and research shows that its practical application can be limited and superficial.

'Sustainable tourism' aims to minimize the negative impacts of tourism on cultural heritage as well as on the environment of the place that is part of its culture and to promote sustainable practices. However, sustainable tourism does not aim to revitalize disappearing ecosystems or revive extinct cultures. Rather, it is a form that contributes to optimal socio-economic change by preventing the degradation of social, cultural and ecological systems. This includes responsible tourism management with strategies to limit overpopulation, conserve natural resources, respect local communities and support the local economy. Sustainable tourism ensures the long-term preservation and enjoyment of cultural assets for both tourists and future generations of local communities. The ability of mass media to influence travellers' behaviour, shape the image of destinations and raise awareness makes it a powerful tool for promoting and expanding sustainable tourism practices, supporting responsible travel and protecting the environment and local cultures (Bramwell and Lane, 1993; Clarke, 1997; Choi, 2016).

Culture serves as a source for the promotion of creative experiences, a concept that is emphasized by the emergence of 'creative tourism'. According to the United Nations Educational, Scientific and Cultural Organization (2006), creative tourism is a particular form of travel that focuses on immersive and authentic encounters and emphasizes participatory learning linked to art, heritage or the special character of a place. This approach creates meaningful connections with the local community and contributes to the vibrancy of culture. In essence, creative tourism is about involving tourists in hands-on, culturally rooted activities such as art workshops, agricultural activities, cookery classes, traditional crafts and active participation in cultural festivals.

By actively involving visitors in the cultural practices of a destination, creative tourism promotes not only creativity but also the learning and appreciation of local traditions. The mass media play a central role in spotlighting these creative activities. It serves as a channel to inform travellers and promote cultural exchange. Through its influence, the mass media plays a crucial role in promoting distinctive and enriching travel experiences centred on cultural and artistic engagement (Guerreiro, 2016; Galvagno and Giaccone, 2019).

The concept of 'cultural routes' has also attracted a great deal of attention in the tourism industry. In 1987, the Council of Europe launched the 'Cultural Routes' programme to show how the cultural heritage of different European countries contributes to a common cultural lineage by travelling through space and time. These programmes are collaborations between cultural heritage and tourism that aim to develop and promote routes based on historical routes, cultural concepts, influential personalities or transnational phenomena of cultural significance and importance for intercultural understanding and respect. Essentially, 'cultural routes' are thematic routes that emphasize the cultural, historical and natural heritage of major regions or specific themes. These routes link different destinations under a common cultural theme, such as ancient trade routes, pilgrimage routes or literary trails.

'Cultural routes' offer tourists a framework to explore different cultural landscapes one after the other and promote regional cooperation in tourism development. The power of mass media to inform, inspire and support these programmes can significantly influence their visibility, success and positive contribution to cultural exchange, heritage preservation and sustainable tourism development (Majdoub, 2010; Richards, 2011; Berti, 2013).

## 4.6 Summary

In summary, the study of the intricate relationship between communication, culture, civilization and tourism enriches our understanding of human behaviour and our society. This comprehensive study debunks the fallacy of perceived cultural hierarchies and promotes a nuanced understanding of different cultures. Themes and practices such as 'cultural preservation', 'cultural exchange', 'cultural heritage', 'sustainable tourism' and 'cultural routes' emphasize the intrinsic value of each culture and its central role in tourism, a narrative that is increasingly reinforced by the media. These initiatives aim to provide enriching experiences while protecting cultural and natural heritage and promoting travellers' appreciation of all cultures. In any case, it is important to recognize the downsides of tourism, particularly the phenomenon of overtourism, which poses a significant threat to cultural and environmental sustainability.

## Notes

[1] Aristotle believed that plants have life, animals have life and perception, and humans have both qualities together with rationality which makes the difference. Thomas Aquinas believed that God is pure intelligence and below him hierarchically were the beings with more developed intelligence (i.e. the angels, then the humans), then the beings with lower intellectual capacities, the animals, and finally the plants without intelligence. Descartes considered bodies as kind of machines, but believed that the animals are non-intelligent machines that cannot express linguistically pure thoughts anyway. Kant believed that humans exist as ends in themselves, while animals have no self-consciousness, exist for the satisfaction of humans, cannot use symbols and have no symbolically structured language, and so on (Georgopoulos, 2022).

[2] The debate about whether other species besides humans can deal with symbolic references has occupied communication research for many years (Gallistel, 1990). For a long time, symbolic communication in other animals was thought to be completely impossible. However, this view was challenged by a series of studies that began when a study of monkeys in Kenya showed that they had the ability to perceive symbols (Slobodchikoff *et al.*, 1991; Greene and Meagher, 1998). All these studies on different animals eventually led to the hypothesis that symbolic and referential meaning is not an exclusive element that distinguishes human from non-human communication (Håkansson and Westander, 2013).

[3] Before the concept of globalization took hold, the term 'cultural imperialism' was popular, and many social scientists resorted to it in the 1960s when commenting on the world situation. This term was directly related to the 'dependency theory' and the 'world system theory', which said that the international order that prevailed and reproduced itself between rich and poor states was due to the historical coincidence of the imperialist expansion of the West and the functioning of the world system. The most important proponents of this theory for analysing world systems were Andre Gunder Frank and Immanuel Wallerstein. However, numerous thinkers have taken positions that follow the same logic, such as Rosa Luxemburg, Oliver Cox, Samir Amin, Giovanni Arrighi, Christopher Chase-Dunn, Beverly Silver, Volker Bornschier, Janet Abu Lughod, John W. Meyer and others. The position of cultural imperialism has largely been abandoned today, as many of the issues it raised are now addressed with a milder critique through the concept of globalization.

## References


Adorno, T.W. (1991) *The Culture Industry: Selected Essays on Mass Culture*. Routledge, London.

Barker, C. (2004) *The Sage Dictionary of Cultural Studies*. Sage, London. DOI: 10.4135/9781446221280.

Beck, U. (1999) *What is Globalization?* Polity Press, Cambridge, UK.

Berti, E. (2013) Cultural routes of the Council of Europe: new paradigms for the territorial project and landscape. *Almatourism – Journal of Tourism, Culture and Territorial Development* 4(7), 1–12.

Blatter, J. (2007) Glocalization. *Encyclopædia Britannica*. Available at: https://www.britannica.com/topic/glocalization (accessed 16 December 2020).

Bramwell, B. and Lane, B. (1993) Sustainable tourism: An evolving global approach. *Journal of Sustainable Tourism* 1(1), 1–5. DOI: 10.1080/09669589309514792.

Buzard, J. (1993) *The Beaten Track: European Tourism, Literature, and the Ways to Culture, 1800–1918*. Clarendon Press, Oxford, UK. DOI: 10.1093/acprof:oso/9780198122760.001.0001.

Choi, H.S. (2016) Sustainable tourism. In: Jafari, J. and Xiao, H. (eds) *Encyclopedia of Tourism*. Springer, Cham, Switzerland, pp. 916–917. DOI: 10.1007/978-3-319-01384-8.

Chong, K.Y. and Balasingam, A.S. (2019) Tourism sustainability: economic benefits and strategies for preservation and conservation of heritage sites in Southeast Asia. *Tourism Review* 74(2), 268–279. DOI: 10.1108/TR-11-2017-0182.

Clarke, J. (1997) A framework of approaches to sustainable tourism. *Journal of Sustainable Tourism* 5(3), 224–233. DOI: 10.1080/09669589708667287.

Deleuze, G. and Guattari, F. (1986) *Nomadology: the War Machine*. Semiotext(e), New York.

Elias, N. (1939) *The Civilizing Process: the History of Manners*. Urizen Books, New York.

Engebrigtsen, A.I. (2017) Key figure of mobility: the nomad. *Social Anthropology/Anthropologie Sociale* 25(1), 42–54. DOI: 10.1111/1469-8676.12379.

Featherstone, M. (1982) The body in consumer culture. *Theory, Culture & Society* 1(2), 18–33. DOI: 10.1177/026327648200100203.
Gallistel, C.R. (1990) Representations in animal cognition: an introduction. *Cognition* 37(1–2), 1–22. DOI: 10.1016/0010-0277(90)90016-d.
Galvagno, M. and Giaccone, S.C. (2019) Mapping creative tourism research: reviewing the field and outlining future directions. *Journal of Hospitality & Tourism Research* 43(8), 1256–1280. DOI: 10.1177/1096348019862030.
Georgopoulos, A. (2022) *Environmental Ethics*. Gutenberg, Athens. (in Greek).
Giulianotti, R. and Robertson, R. (2004) The globalization of football: a study in the glocalization of the serious life. *The British Journal of Sociology* 55(4), 545–568. DOI: 10.1111/j.1468-4446.2004.00037.x.
Golding, P. and Harris, P. (eds) (1996) *Beyond Cultural Imperialism: Globalization, Communication and the New International Order*. Sage, London.
Greene, E. and Meagher, T. (1998) Red squirrels, *Tamiasciurus hudsonicus*, produce predator-class specific alarm calls. *Animal Behaviour* 55(3), 511–518. DOI: 10.1006/anbe.1997.0620.
Guerreiro, M. (2016) Creative tourism. In: Jafari, J. and Xiao, H. (eds) *Encyclopedia of Tourism*. Springer, Cham, Switzerland, pp. 198–199. DOI: 10.1007/978-3-319-01384-8.
Håkansson, G. and Westander, J. (2013) Communication in humans and other animals. John Benjamins Publishing, Amsterdam. DOI: 10.1075/ais.4.
Hall, C. (2005) *Tourism: Rethinking the Social Science of Mobility*. Prentice Hall, Harlow, UK.
Hall, C. (ed.) (2011) *Medical Tourism: the Ethics, Regulation and Marketing of Health Mobility*. Routledge, London.
Hannam, K., Sheller, M. and Urry, J. (2006) Editorial: mobilities, immobilities and moorings. *Mobilities* 1(1), 1–22. DOI: 10.1080/17450100500489189.
Jack, G. and Phipps, A. (2005) *Tourism and Intercultural Exchange: Why Tourism Matters*, Vol. 4. Channel view Publications, Bristol, UK. DOI: 10.21832/9781845410193.
Majdoub, W. (2010) Analyzing cultural routes from a multidimensional perspective. *Almatourism – Journal of Tourism, Culture and Territorial Development* 1(2), 29–37.
Malinowski, B. (2002) *Argonauts of the Western Pacific: an Account of Native Enterprise and Adventure in the Archipelagoes of Melanesian New Guinea*. Routledge, London. DOI: 10.4324/9780203421260.
McLuhan, M. (1962) *The Gutenberg Galaxy: the Making of Typographic Man*. University of Toronto Press, Toronto, Canada.
Nowacki, M. (2021) Heritage interpretation and sustainable development: a systematic literature review. *Sustainability* 13(8), 4383. DOI: 10.3390/su13084383.
Peters, J.D. (2013) Exile, nomadism, and diaspora: the stakes of mobility in the western canon. In: Naficy, H. (ed.) *Home, Exile, Homeland: Film, Media, and the Politics of Place*. Routledge, New York, pp. 17–41.
Reinfeld, M.A. (2003) Tourism and the politics of cultural preservation: a case study of Bhutan. *Journal of Public and International Affairs* 14, 125–143.
Ribeiro, G.L. (2001) Cosmopolitanism. In: Smelser, N.J. and Baltes, P.B. (eds) *International Encyclopedia of the Social and Behavioral Sciences*. Elsevier, Amsterdam, pp. 2842–2845.
Richards, G. (2011) Tourism trends: tourism, culture and cultural routes. Cultural tourism trends in Europe: a context for the development of cultural routes. In: Khovanova-Rubicondo, K. (ed.) *Impact of European Cultural Routes on SMEs' Innovation and Competitiveness*. Council of Europe Publishing, Strasbourg, France, pp. 21–39.
Ritzer, G. (2004) *The McDonaldization of Society*. Thousand Oaks, Newbury Park, California.
Robertson, R. (1994) Globalisation or glocalisation? *Journal of International Communication* 1(1), 33–52. DOI: 10.1080/13216597.1994.9751780.
Roudometof, V. (2016) Theorizing glocalization: three interpretations. *European Journal of Social Theory* 19(3), 391–408. DOI: 10.1177/1368431015605443.
Shin, S.H. (2009) A study on the economic benefits of globalization: focusing on the poverty and inequality between the rich and the poor. *International Area Review* 12(2), 191–214. DOI: 10.1177/223386590901200210.
Slobodchikoff, C.N., Kiriazis, J., Fischer, C. and Creef, E. (1991) Semantic information distinguishing individual predators in the alarm calls of Gunnison's prairie dogs. *Animal Behaviour* 42(5), 713–719. DOI: 10.1016/S0003-3472(05)80117-4.
Smith, P. and Riley, A. (2008) *Cultural Theory: an Introduction*. Wiley, Oxford, UK.

Tylor, E.B. (1871) *Primitive Culture: Researches into the Development of Mythology, Philosophy, Religion, Art and Custom*, Vol. 1. John Murray, London.

United Nations Educational, Scientific and Cultural Organization (2006) In: *Towards Sustainable Strategies for Creative Tourism: Discussion Report of the Planning Meeting for the 2008 International Conference on Creative Tourism,Santa Fe, New Mexico, October 25–27 2006*, UNESCO and Creative Cities Network, Paris.

United Nations World Tourism Organization (2012) *Global Report on Aviation. Responding to the needs of new tourism markets and destinations*. AM Reports: Volume five. Available at: https://www.e-unwto.org/doi/pdf/10.18111/9789284414925 (accessed 28 October 2023).

Uscatescu, G. (1986) *Brief Theory and History of Culture*. Ekdosis Grigori, Athens. (in Greek).

Vassiliou, T. and Stamatakis, N. (2000) *Dictionary of Human Sciences*. Gutenberg, Athens. (in Greek).

Williams, R. (1960) *Culture and Society, 1780–1950*. Anchor Books, New York.

Wood, R.E. (1984) Ethnic tourism, the state, and cultural change in Southeast Asia. *Annals of Tourism Research* 11(3), 353–374. DOI: 10.1016/0160-7383(84)90027-6.

World Bank Group (n.d.) International tourism, number of arrivals. Available at: https://data.worldbank.org/indicator/ST.INT.ARVL (accessed 28 October 2023).

Zerzan, J. (1994) *Future Primitive and Other Essays*. Autonomedia, New York.

Zerzan, J. (2014) Animal Dreams. *Black Seed Issue #1*. Available at: https://theanarchistlibrary.org/library/john-zerzan-animal-dreams-black-seed-issue-one (accessed 12 October 2023).

# 5 Theories of Mass Media Effects and Mass Tourism

**Abstract**

The mass media play a crucial role in disseminating information about destinations, shaping public perceptions of travel and tourism and influencing people's behaviour. This chapter examines 20th-century theories in the field of mass media and tourism studies, emphasizing their parallel development and possible connections. Theories discussed include the strong influence of the media, the 'two-phase flow model', 'cultivation theory', the 'Palo Alto group' perspective, structuralist and post-structuralist theories, 'frame analysis' and 'uses and gratifications' theory. With parallels to the established tourism theories of Daniel Boorstin, Dean MacCannell, Erik Cohen, the 'tourist gaze' of John Urry, and Maxine Feifer, the intersection of media and tourism offers fertile ground for interdisciplinary research. This research is becoming increasingly important in view of the changing media landscape and ongoing developments in the tourism industry.

In the previous chapters, we have argued that words such as 'communication' and 'tourism' can be interpreted differently depending on the conversational dynamics between the interlocutors, their social position, their cultural background and the historical context in which these terms are discussed. In this chapter, we will extend this investigation to the realm of theories. We generally assume that the prevailing paradigms, methods, ideologies, the different social positions, gender perspectives, ethnic backgrounds of the scholars and the historical background can contribute to the emergence of different theories that either clash, coincide or represent different points of view on a common topic.

Scholars of an epistemological current that emerged in the early 1960s, whose most prominent representatives were Thomas Kuhn (1922–1996), Paul Feyerabend (1924–1994), Russell Hanson (1924–1967), Stephen Toulmin (1922–2009) and others, held that scientific theories are historical entities whose lifeline follows the stages of birth, growth and decline, and that they not only lead to truth but often prove to be false. The basic idea was that the development of science is a discontinuous process with many surprises and constant upheavals (Kuhn, 1970).

Feyerabend even went so far as to argue that the only principle that does not impede scientific progress is that everything is permitted. Essentially, he was concerned with removing the obstacles that intellectuals and specialized scientists face when they deviate from established knowledge and conventional scientific methods (Feyerabend, 2010).

Before turning to the theories of communication and tourism that follow the theme of this book, it is important to emphasize at the outset that most psychological and neuropsychiatric theories of instinct (as well as motivation and needs) generally recognize the fact that humans and other species spend a considerable part of their lives exploring and seeking sensory data, although the extent to which this is emphasized may vary according to theory and perspective. It should be noted that some of these theories accept this behaviour as the result of some primary motivators they have established, such as hunger or sex, while others regard this behaviour as fundamentally independent and autonomous.

Moreover, the 'seeking system', as it is called by neuropsychologists and neuropsychiatrists, can be considered as the system responsible for the species' motivated actions related to survival

 *Communication and Tourism: Reflecting on the construction of the tourist image of Greece* (Michael Tsangaris)
DOI: 10.1079/9781800626034.0005

and desires. We can argue that the term 'seeking instinct' refers generally to a basic and innate process or urge that can be observed in many living things. While the specific behaviours associated with the 'seeking instinct' may vary greatly from species to species, its basic purpose is to help organisms meet their basic needs and adapt to their environment.

The 'seeking instinct' and 'sensation seeking' are related concepts, both involving the pursuit of novelty, excitement and stimulation, but they are not exactly the same thing.

In particular, 'sensation seeking' is a characteristic that refers to the need for 'varied', 'novel' and 'complex' sensory impressions and experiences and the willingness to take physical and social 'risks' for such experiences. In this formulation, 'varied' reflects the need for variety, 'novel' indicates the need for something that is different in some way from previous experiences, 'complexity' refers to the number of stimulus elements and their complex arrangement, and finally, 'risk' can be defined as the degree of uncertainty, the estimated probability of a negative outcome in the process (Wright and Panksepp, 2012; Zuckerman, 2014).

The 'seeking instinct' can be seen as a broader, more general concept that encompasses different behaviours and motivations, including the search for feelings. It encompasses a wide range of behaviours that go beyond the pursuit of excitement, such as exploring new ideas, social interactions and learning in general. 'Sensation seeking', on the other hand, focuses specifically on the pursuit of sensory and experiential stimuli.

Anyway, before we look at the theories of 'communication' and 'tourism', and without going into the different views of psychological and neuropsychiatric theories, we are in any case interested in the fact that there are several theses of important scientists about the 'seeking instinct' (i.e. about what we consider to be the basis of human mobility and tourism).

The 'seeking instinct' is also found in Bergson's *élan vital* (vital impulse or life force), which implies an innate drive or instinct for self-preservation and growth. Bergson believed that living organisms have an innate tendency to seek out new opportunities and adapt to their environment. This instinct to seek is not only driven by external factors or a predetermined plan, but is an essential part of the *élan* that guides organisms on their evolutionary path. We therefore honestly state from the outset that our basic attitude suggests a deep connection between the development of tourism and the innate human drive to discover and expand consciousness.

In the following, we will present some of the basic theories developed in the 20th century in the fields of mass media studies and tourism studies, and try to show some connections and frictions between them, their continuity and relativity.

## 5.1 Theories of Mass Media Studies and Tourism Studies

The truth is that systematic scientific thinking in these two areas developed only in the course of the 20th century. We can therefore say that the hypotheses and theories are more recent and in some ways have a more fluid form than those of the major social sciences.

Of course, much earlier there were various social studies, economic reports or philosophical works that dealt in one way or another with the 'idea' of tourism or the concept of communication, but in this book we will mainly focus on some of the basic theories that emerged in the 20th century and led to them being considered as separate branches of the social sciences. We start with the theories on the power of media effects at the beginning of the century, while in the field of tourism we start with some historical, sociological and anthropological theories on mass tourism that emerged in the 1960s, without considering the economic or other theories.

We believe that the brief presentation of some basic theories from these two areas is a prerequisite for understanding the relationship between media and tourism, which will be discussed more practically in the following chapters.

### 5.1.1 The strong influence of the media

In the first half of the 20th century, following the dynamic development of the press, radio and cinema, many scholars and experts in the field of communication, such as Walter Lippmann (1889–1974), Harold Lasswell (1902–1978) and Edward Bernays (1891–1995), saw the mass media as instruments of great power to influence people.

It should be noted that this view was also supported by the prevailing opinion on the decisive contribution of propaganda to the developments of the two great wars. At the same time, however, other scholars were surprised by the enormous cultural impact of cinema, especially Hollywood, on people all over the world (Thorp, 1939; Powdermaker, 1950).

In this respect, and without forgetting that behaviourism was developing at this time, which paid great research attention to the 'stimulus-response' model, social scientists set the starting point of their research in the field of communication in the fact that mass media messages can produce powerful effects. All these considerations led to the general view that the media can have an enormous impact on individuals as well as on society as a whole. This paradigm is often referred to by the terms 'hypodermic needle model' or 'magic bullet theory' (Thibault, 2016).

In the 1930s, a series of organized surveys on mass media communication began, based on the above views and initially aimed at determining the exact impact of the press and radio on people. The surveys were primarily concerned with political communication (e.g. how and to what extent the media influenced the final vote in election campaigns). However, some other studies conducted at the same time also looked at commercial advertising and the extent to which an advertisement for a product in the press or on the radio can influence consumer buying behaviour.

The main arguments of these studies were the following: (i) the media were able to convey messages equally to all people; (ii) the messages were stimuli that could exert a strong influence on the thinking, attitudes and behaviour of the public; and (iii) the public perceived the messages uniformly (i.e. in the same way), reacted to them and acted accordingly (De Fleur and Ball-Rokeach, 1983; Demertzis, 2002).

According to this logic, then, the images of exotic, faraway places shown in the cinema, the advertising posters of the major travel agencies of the time, such as Thomas Cook & Son, or described in the announcement of an excursion on the radio, could have an immediate and uniform effect on the public which, if positive, would certainly promote tourism.

One of the first simple definitions of mass communication proposed by Lasswell (1927) was based on a simple linear form of message transmission as follows – 'Communication is: Who? Says what? In what channel? To whom? and With what effect?'. This model of communication followed a process whereby a sender sends a message to a receiver and produces an effect, which inadvertently supported the assumption that mass communication is a one-way, mechanical process of influence.

A few years later, another, more technical model was proposed by Shannon and Weaver (1949), based on the same logic and considered by many researchers to be the starting point for all communication models. This model also envisaged communication taking place with a sender relaying a message to a receiver in the form of a linear, directed trajectory, but also introduced other concepts such as feedback and noise. This basic concept for mass media was adopted and served as a guide for practitioners and researchers in the field for many years.

According to this model, a tour operator can disseminate direct messages to inform the public about an organized trip, for example through the press or radio. The messages should motivate all people to participate without taking into account other social or personal characteristics of the recipients. If the message is cleverly and positively worded and not marred by various 'noise', the end result would be a large public participation in the trip, higher profits for the travel agency and mass tourism. It should be noted that this kind of logic still has a great impact on many people's views on the concept of communication today (McQuail and Windahl, 2015).

It can be said that all these early media theories were developed in the spirit of the times (i.e. in the context of the basic sociological and social psychological theories about the development of urbanization, industrialization and the modernization of society and mass behaviour), such as the theories of Max Weber, Emile Durkheim and Gustave Le Bon. However, they were also under the influence of behaviourism and structuralism, and of course they emphasize the great technological development of mass communication and the rise of journalism, within a general assumption that great changes have taken place in Western society.

Media effects research was intensively conducted until the 1960s, but it is worth noting that similar studies are still being conducted today (Lazarsfeld *et al.*, 1944; Dewenter *et al.*, 2019). In summary, the first organized studies of the media focused primarily on their potential use for planned persuasion (or information), political campaigns and advertising of consumer goods. Later, however, interest expanded to include the potential negative consequences of the media in the form of increased crime, the spread of prejudice and other similar issues.

Surprisingly, studies on the effects of political advertising a few years later found that citizens' voting decisions were not significantly influenced by the media. This led to the conclusion that the media ultimately did not have that much power to influence public opinion. Gradually, researchers began to differentiate their views on the impact of the press, cinema and radio, taking into account other variables such as interpersonal relationships, the social environment, and social and psychological characteristics of the recipients (Samaras, 2008).

At that time, the majority of researchers were sceptical about the influence of the media, and this trend was sealed in the mid-20th century by a publication by Joseph Klapper (1917–1984). Klapper concluded that mass communication does not necessarily and directly influence the public, but indirectly through a combination of mediating factors that influence individuals' decisions. In short, various groups such as friends, relatives or even the closest social environment were identified as key factors in influencing individuals (Klapper, 1949).

At this point, however, it must be stressed that all these new claims did not prove conclusively and absolutely that newspapers, cinema or radio had no influence on the public, but only that they worked within a certain atmosphere or pre-existing structure of social relations that determined their impact. Nor did they rule out the possibility that information about a subject, such as the proposed political programme of a candidate or the advertised benefits of a tourism offer, might not change the recipient's attitude.

### 5.1.2 The 'two-phase flow model' and 'cultivation theory' and return of the idea that media have a major influence

Various sociologists and researchers in the field of media have argued that influencing a person can be gradual and that more than one factor contributes. In this context, they have presented the 'two-phase flow model', which is structured as follows: first, the media influence certain individuals, the so-called 'opinion leaders', and then these individuals influence the majority of the population. According to this model, the catalyst for influencing the public's beliefs is interpersonal communication rather than mass communication (Katz and Lazarsfeld, 1955).

In this case, a newspaper article or a radio broadcast about a place may positively influence some people who know how to evaluate and appreciate tourist destinations and who may be highly regarded in their circle of friends. These people would then start a process of idealizing that place, which would probably also influence their friends and wider social network, eventually leading to a large wave of tourists to that destination.

But then, in the 1960s, after a new intense media research, significant doubts began to emerge that again challenged views about the media. Lang and Lang (1981) argued in a study that the data examined in earlier research was insufficient to justify a firm position that the media was powerless or had no influence on public opinion. In any case, many researchers felt that too much attention was paid to the short-term functioning of the media, leading to discouraging studies on the power of the media, such as that of Klapper (1949).

At this point, however, it should be noted that the main role in the return of views advocating the decisive influence of the media in society was played by the great development of television, which by this time had begun to be excessively attractive to the public.

More or less during these years, Daniel Boorstin (1914–2004) devoted himself to a comprehensive critique of the phenomenon of mass tourism, which was undergoing a dynamic development. At that time, there was an unprecedented trend of mass participation in tourism by the populations of developed Western countries.

In his work, Boorstin (1964) pointed out that tourists, who were isolated from the real environment of the place they visited and had almost no contact with the inhabitants, were led in guided groups that were offered packages of predetermined attractions by various players in the tourism industry. He therefore considered that through this massively planned process, tourists participated in fabricated 'pseudo-events' and were completely indifferent to the real life of local communities (Boorstin, 1964).

Rojek (2000) referred to Boorstin's view as the fundamental 'structuralist' position in tourism research. It can be argued that Boorstin shares some common ground with the work of numerous structuralist thinkers, such as the early work of Barthes (1957), and Althusser (2014). This position therefore places particular emphasis on the structural forces of the tourism industry that lead to preconceived arrangements of tourism experiences. In other words, the tourism industry has created a specific tourist reality that is simultaneously modelled and promoted by the media.[1]

In Boorstin's theory, tourists are defined as non-participating social actors who have no control over their decisions and cannot influence either the tourism market or the tourism industry, but are passive and instructed by the tourism industry, a defenceless mass influenced by the media (Rojek, 2000).

George Gerbner (1919–2005) introduced the cultivation theory, which states that long-term, intensive exposure to mass media shapes the way media consumers perceive the world and behave (Gerbner and Gross, 1976).

People who watch television frequently are more likely to have beliefs and ideas about the world that match what they see on TV (i.e. are closer to the reality shown on the screen). In this way, the media 'cultivate' the attitudes and values of a culture, for example viewers are educated to reject certain acts of violence while being educated to accept or tolerate others. Similarly, audiences may be educated to reject certain forms of travel that are very similar to the concept of tourism, such as aimless wandering, hitchhiking or vagabonding, while being educated to accept others, such as mass tourism and summer holidays in the sun.

In this sense, the distorted and incomplete reality as presented by the mass media ultimately represents for many people what the world actually looks like.

An important idea that emerged from these theories was that of 'mainstreaming', (i.e. the phenomenon that groups that otherwise have different opinions and beliefs on certain issues become more similar to each other when they use television intensively) (Signorielli and Morgan, 1990; Shanahan and Morgan, 1999).

The return of the idea that the media have a great influence on individuals has also been reinforced by the new directions of psychological and cultural research in the field, based on the social characteristics and physical state of recipients when receiving messages. As a result, the relationships between exposure to images, sounds and media texts in combination with other intervening variables, such as personal or social influences, have again been studied more systematically.

In addition, research has focused more on the long-term impact of newspapers, film, radio and television, on the codes and content of their messages, but also on the assumption that the structure of the media's organization and the interests of business owners can be a key factor influencing and controlling the structure of the messages.

The work of Noelle-Neumann (1984) made an important contribution to this trend, as did earlier research by the Frankfurt School and the New Left in the 1960s.[2]

Within this framework, it can be assumed that tourist representations in the media provide people with the conformist boundaries within which they must operate, so that they are reluctant to turn to tourist events that are rejected by the social environment. Or, critically, tourism advertising as part of the mass media industry can be accused of deceiving people and creating false needs in them so that they feel compelled to visit certain destinations it promotes. Critics of tourism advertising might argue that it serves the interests of the tourism industry and capitalist corporations, sometimes at the expense of authenticity and genuine cultural experiences. However, it must be noted here that these manufactured false needs for commercial tourism, which may have been referred to by a socially critical commentator, are, in our opinion, based on the basic human need for mobility.

In the 1960s, however, structuralism was challenged by various post-structuralist theories that rejected the possibility of stable communication structures and emphasized the polysemy and fluidity of any structure, including science and human knowledge (Coward and Ellis, 1977).

The postmodern theories of tourism, which were developed much later and which we will discuss in more detail in Section 5.2 'Other Perspectives on Tourism Communication', can also be counted among these theories. Nevertheless, the structuralist models and methods have been further developed by scholars and are still used as analytical tools for various communication practices and institutions.

Be that as it may, regardless of the angle and point of view from which we look at the issue, it can be noted that opinions about the power of the media were constantly changing and divided. As we will see below, views on tourism were also relatively contradictory.

### 5.1.3 The 'Palo Alto group' perspective, the 'Encoding-Decoding Model', and MacCannell's, Cohen's and Plog's theories for tourism

In contrast to linear or mechanistic theories, some scholars such as Gregory Bateson (Bateson and Jackson, 1964), Paul Watzlawick (Watzlawick *et al.*, 1967), Don D. Jackson (Jackson, 1968) and John Weakland (Weakland, 1967) considered communication as a dynamic whole, taking into account both the interactive relationships between people, their connections within particular communication frames and the overall context. This type of interactionist theory, which views communication as a circular process within a social and cultural milieu, is also known as the 'invisible university' or the so-called School of Palo Alto (Porcar and Hainic, 2011; Kaitatzi-Whitlock, 2012).

Although not directly associated with the Mental Research Institute (MRI) in Palo Alto, Birdwhistell (1955), Hall (1963), Goffman (1955), Mead (1964) and many others have contributed to developing ideas for the general framework of interactionist communication theory.

In this case, the study of the communication process becomes more holistic, because a message to the media about a place can have a positive effect on a group of people after a long and complex process. A tourist TV advertisement or the projection of Paris in a film have an effect on people that depends not only on the verbal content of the messages, but also on their character, temperament, cultural and social capital, profession, friends, etc. They can also be influenced by the time they receive the message (i.e. their specific financial situation, their tiredness, their mood and the stage they are at in their life cycle). Most importantly, however, is the feedback they give to travel agents about their preferences after being exposed to media messages, which can trigger a new phase of communication. In short, message recipients can initiate a broader interaction process regarding the 'image of a destination' and its attractiveness by expressing their like or dislike for that place.

In the 1970s, Stuart Hall (1932–2014) and his colleagues at the Centre for Contemporary Cultural Studies (CCCS) at the University of Birmingham proposed a theory of reception developed primarily in the field of cultural studies. This model was essentially based on critical theory, on the study of the receiver's reception of messages and on the analysis of the construction of communicative texts. It proposed four stages of communication, namely: (i) 'production'; (ii) 'circulation'; (iii) 'use'; and (iv) 'reproduction'. This process involves the encoding of the message by the producer and its decoding by the audience (Hall, 1980).

One of the most important elements of this view was the recognition of the intellectual independence of the receiver and thus his or her contribution to the construction of the meaning of the messages transmitted by the mass media.

Hall (1980) therefore argued that there are three ways of decoding the messages received from the media. One is the 'dominant', which leads to acceptance of the messages; the second is the 'negotiated', where the receiver negotiates the truth of the message; and the third is the 'oppositional', where the receiver reacts negatively to the message (McQuail and Windahl, 2015).

In this case, a tour operator might codify a message about travel to traditional Spain according to its idea of what would be an attractive image for the public to travel to that destination country by presenting images of a bullfight in a crowded stadium. This message can be conveyed with large posters that can be placed in the centre of a northern European city. Many people will take the message as an invitation to travel to Spain to explore its customs, others who love animals may have doubts about the poster's appeal but be attracted to this foreign culture, and finally there will be some supporters of the animal rights movement who will take the same message as an insult, (i.e. in an antithetical way because it contradicts their world view).

At the same time, MacCannell (1973) developed his theory of tourism from a different perspective, namely in relation to the search for 'authenticity'. In his view, the alienated bourgeoisie of the second half of the 20th century hoped to discover forgotten places through tourism where they could regain the experience of a whole that was missing in the normal everyday life of modern cities.

In this case, Rojek (2000) referred to MacCannell's view as the 'voluntarist' position of tourism. Other researchers also moved within this view, emphasizing mainly the subjective elements and personal views of tourists, such as Krippendorf (2010). In this interpretation, tourists were presented as conscious social actors who have control over their decisions and can influence the tourism market and industry to a certain extent.

Finally, however, MacCannell acknowledged that the search for undeniably authentic places is sometimes weakened by the 'staged authenticity' offered in destinations. The question of authenticity has remained an important aspect of social tourism research to this day (MacCannell, 1973).

Erik Cohen argued that it is indeed possible to bridge this relative opposition between Boorstin's and MacCannell's positions on tourism (Uriely, 1997).

Cohen (1972) presented a multi-layered typology based on the degree of emotional attachment people have to the environment of the places they visit and are named after. He distinguished: (i) the 'organized mass tourists', who undertake planned trips (package tours) without coming into contact with the communities of the host places; (ii) the 'independent mass tourists', who undertake planned trips but also seek other activities outside the programme; (iii) the 'explorers', who organize their trips themselves and come into direct contact with the inhabitants of the host country and their way of life; and (iv) the 'wandering tourists', who usually do not come into contact with other tourists or with pure tourism enterprises, but only with local people.

In this context, Plog (1974, 2001) presented two main psychographic positions regarding the character of tourists and the internal psychological factors that influence their choice of destination: (i) the reliable or 'psychocentric'; and (ii) the adventurer or 'allocentric'. He described dependable or psychocentric tourists as those who prefer familiar situations, choose destinations with a fixed schedule and often a ready-made environment, and always avoid surprises because they want their holiday to go smoothly. On the other hand, he described adventurer or allocentric tourists as those independent people who prefer adventure and always try to tread unknown paths. Between these two extreme categories lies the vast majority of tourists who are likely to choose one destination or the other depending on their personality traits.[3]

We can assume that the categories of tourists identified by Cohen (1972) and Plog (1974) cover a spectrum within which each one has a particular relationship with the media. 'Organized mass tourists' and 'psychocentric' tourists tend to follow mainstream media narratives in interpreting tourism industry messages, while at the other end of the spectrum 'wandering tourists' and 'allocentrics' resist these media-driven messages about destinations. It is important to note that the second case is not an absolute rejection of the media's influence on tourism as a whole. If we consider alternative media that present unconventional travel experiences, people exploring unique forms of tourism may also be directly influenced by media perspectives that challenge dominant narratives.[4]

### 5.1.4 'The agenda-setting model' of communication and 'frame analysis'

In the last three decades of the 20th century, interest has developed in the 'agenda-setting' model of communication. According to this model, the media determine the issues and the ranking of discussion topics presented to the public depending on their political beliefs or interests and not necessarily on the basis of the 'objective' importance or journalistic significance of the facts (McCombs and Shaw, 1972; Funkhouser and Shaw, 1990).

For this reason, it has been pointed out that newspapers, radio, television and cinema do not determine the precise content of the public's thoughts, but rather the contours of the issues they engage with and think about (i.e. the media ultimately determine which issues people attach more importance to in their daily discussions) (Demertzis, 2002).

So, a travel poster, a radio advertisement or a TV documentary about a seaside resort do not force people to travel to that particular place, but they give them a direction that influences them. They construct general 'tourist orientations' like 'summer holidays in the sun' that prescribe what tourism must include, and they present them with ratings, but of course they do not talk about other forms of travel, like hitchhiking, constant walking to explore moors in winter, and so on.

In this brief overview of media and tourism theories, it would be a serious omission not to mention the model of 'frame analysis', which was widely used around the same time as the above models in a number of studies on the influence of the media on the public, highlighting the so-called framing effect.[5]

'Framing' can be seen as another process by which the media delineate reality in a certain 'context' (i.e. they colour reality in a certain way). In other words, 'frame analysis' deals with factors such as the situation that determines the communication event and the mental interpretation of the actors. It analyses under which conditions and from which view of the situation something is said at a certain point in time, which is the key parameter, the basic frame in which a discussion is formatted, and which other parameters are used to shape the communicative reality.

For example, the decision to impose a temporary ban on tourism businesses because of the coronavirus disease 2019 (COVID-19) pandemic may be presented on television from different angles, confirmed by different 'frames' (common interpretative conventions). When the issue is presented by a TV channel in an economic context, this decision is disastrous. There is no doubt that the closure of tourism businesses is a disaster for the economy of a place. However, another TV channel may take up the same issue in the context of public health, and in this case the decision appears sensible as the only way to reduce the spread of the disease and survive as a society is to temporarily stop all kinds of business and stay at home at all costs. The way journalists frame issues may differ from one television station to another, from one newspaper to another or from one website to another, but it may also vary over time within one medium (Cappella and Jamieson, 1997; Scheufele, 1999).

### 5.1.5 'Uses and gratifications' model and 'social constructions of reality'

Another, more general approach to understanding mass communication, which encompasses many variants, is the 'uses and gratifications' model. In general, this model assumes that members of the public voluntarily and actively choose both the means and the content to inform and entertain themselves according to their needs and concerns. Individuals are thus not at the mercy of media outcomes in their daily lives, but can determine for themselves which communication channels to use and how to obtain information or engage in some other communication activity (Blumler and Katz, 1974; Palmgreen *et al.*, 1985).

It could be said that with the development of new digital media and the expansion of information sources in our time, the value of this model has been widely recognized.

Potential tourists thus know what kind of tourism and which areas they are interested in, and therefore use the media and certain channels that very often report on their specific areas of interest. 'Mass tourists' use the mainstream media to inform themselves about tour packages, 'wandering tourists' and 'allocentric tourists' use alternative forms of media, etc. This model thus fits well with today's reality of tourism communication and practice, where tourists themselves search for their preferences and plan their trips on the online platforms of travel agencies.

Although various epistemological issues related to objectivity, empiricism, rationalism, etc. have preoccupied scholars since ancient times and have led to endless philosophical debates about reality, some theories have been associated with constructivist models, as mentioned above, while the second half of the 20th century saw a growing interest in semiotic and phenomenological theories about the 'social construction of reality'. Most of these views recognize both the power of the media to influence people's reality in relation to the issues they raise, and the power of individuals to perceive news critically and act in a field of constant negotiation to construct reality (Berger and Luckmann, 1967; Gamson and Modigliani, 1989).

Linking tourists' motivations to theories of the social construction of reality and phenomenological theories can shed light on how individuals perceive and experience their travel experiences. In essence, we can achieve interesting results by examining how social constructions shape their motivations and how tourists subjectively interpret and give meaning to their travel experiences according to their personal perspectives and desires.

John Urry (1946–2016) developed the well-known theory of the 'tourist gaze', which assumes the dominance of the visual dimension among the senses and suggests that the tourist experience involves a particular 'way of seeing' whose codes are mainly structured, controlled and organized by the tourism industry (Urry, 1990).

However, it can always be argued that nowadays individuals, depending on their habitus and cultural capital, contribute significantly to the concept of the 'tourist gaze' through social media when they post images and their views of tourist sites.

Another perspective for interpreting tourist phenomena is based on the work of the great British social anthropologist Victor Turner (1920–1983). Turner's work on pilgrims was an important starting point for social research on tourism (Turner, 1973; Turner and Turner, 1978).

According to this view, tourism has replaced the reasons and goals of religious rites in traditional societies. Travel can be perceived as a periodic, ritual event outside everyday life that has a significant impact on people's social and personal lives.

In this context, Graburn (1977), in an article entitled 'Tourism: the Sacred Journey', argued that tourism, although a secular event, can be equated with religious rituals. There is no doubt that travelling is out of the routine and ordinary activities of daily life. He claimed that tourists who escape their usual environment go to the place of the extraordinary and solemn, only to find the banality of everyday life upon their return. That is, they experience situations and follow sequences similar to the phases that characterize sacred rituals such as baptism and marriage (Dann and Parrinello, 2009).

At this point we must also note the relative similarity of Graburn's position to the dipole of everyday life and religious ritual developed by Durkheim (1995) in his book *The Elementary Forms of Religious Life*, in which he analyses religion as a social phenomenon.

## 5.2 Other Perspectives on Tourism Communication

In addition to the aforementioned approaches from history, sociology, culture, psychology and anthropology, which together form a comprehensive interdisciplinary stance on tourism and media, numerous other perspectives from various academic disciplines have been integrated into the discussion on tourism communication over time. These include areas such as business administration and management, marketing, advertising, public relations and rhetoric (argumentation).

The approaches of these academic fields, as well as many manuals for business people, have mainly focused on the methodical development of suggestions for the use of communication tools to achieve immediate practical goals (e.g. increasing market share, building a positive reputation, changing consumer preferences, manipulating the public) through practical guidance. Undoubtedly, all these researchers and professionals in the field of business research have made important contributions to the development of mass communication studies and tourism (Coolsen, 1947; Khang *et al.*, 2016).

As early as the late 1980s, Feifer (1985) saw a change in tourism characterized by post-tourists (i.e. the tendency of people to voluntarily switch between different types of tourist characters and psychographic locations depending on their mood). This view emphasizes the diversity and flexibility of tourist experiences in postmodernity and contrasts with older theories that view societies, social classes and social groups as compact and inelastic units.

At this point, we should refer to the view of the famous sociologist and philosopher Zygmunt Bauman (1925–2017). He argued that the problem in modernity was how individuals constructed an identity and how they could keep it compact and stable, while the problem in the current phase is the compact identity they have created and how they can avoid being obsessed with it in order to keep their choices open for the many different roles that modernity has imposed on them (Bauman, 2016).

In other words, whereas the main concern of tourist identity in modernity was the worry about its resilience, today it is the fear of being permanently fixed to a particular tourist category. For example, people from the upper or middle classes no longer try to identify exclusively with luxury tourism to demonstrate a higher social status. On the contrary, it is very likely that they often want to flaunt an alternative lifestyle by regularly posing as 'backpackers', 'adventure travellers', and so on.

Indeed, tourism research has long been concerned with the diversity of tourist categories, tourist experiences and tourist environments, as well as many other specific issues that underline the complexity of this phenomenon.

According to Munt (1994), after the initial opposition of Boorstin and MacCannell, these two major theoretical currents have also prevailed in the postmodern discussion. In this case, the two currents developed in creative and complementary ways, forming complex and hybrid perspectives of enquiry without contradicting each other (Uriely, 1997).

The first postmodern research tendency focused on analysing 'hyperreal' experiences (e.g. 'simulated' theme parks or other fancy and unusual attractions as typical postmodern environments) largely following Boorstin's notion of pseudo-events in relation to postmodern literature (Baudrillard, 1983; Eco, 1986; Featherstone, 1991; Lash and Urry, 1994; Gottdiener, 1995; Kulhanek, 2013).

The second tendency is oriented towards another postmodern tourism based on the search for the 'real', the 'primitive' and the 'genuine'. Studies belonging to this model aimed to consider both the attractiveness and the purity of tradition, heritage, landscape, ecology and nature in general as postmodern elements. These studies generally followed MacCannell's original argument of the search for 'authenticity', albeit a relative, changeable and kaleidoscopic authenticity (Barrett, 1989; Poon, 1989; Urry, 1990; Munt, 1994).

Later studies on communication and tourism retained elements of the original powerful media theories, but emphasized the influence of the social environment, globalization, the Internet and the role of the recipient's cognitive and psychological processes in shaping individual reality.

Research approaches varied. Some focused on theoretical analyses, others used quantitative or qualitative methods, depending on the objectives of the research.

For example, marketing research on the impact of tourism advertising usually involves questionnaires and statistical analysis, while qualitative and semiotic methods are often used to analyse how posters of Internet campaigns or videos shape the image of a destination. Some studies have shown that media, including cinema, radio, television and new media, construct social reality and even reshape historical narratives. Conversely, other studies have examined how individuals can critically construct their own worldview.

Since the early 2000s, with the advent of new media and Web 2.0 technologies, the impact of tourism on society has changed significantly in theory. Changes include:

- the interplay between the diversification of daily life and the digital tourism experience;
- new interdisciplinary research in the field of mobility studies;
- the emergence of new tourism platforms;
- the ability of tourists to create their own tourism image using social media;

- the use of smart tourism and wireless networks;
- an increase in environmental concerns, debates about tourism in social media, and concern about discrimination and human rights issues; and
- an increased focus on the subjectivity of tourism experiences in certain social groups and subcultures.

In response, theoretical frameworks have adapted to modern, postmodern and flexible paradigms that are in line with contemporary directions in the social sciences.

## 5.3 Summary

To summarize, an in-depth understanding of the connections between media and tourism can emerge through the lens of communication and tourism theories that illuminate their nuanced interpretations and intersections within the evolving landscape of social dynamics and technological advancement. As some of these theories demonstrate, the media has an impressive ability to shape perceptions, construct realities and guide decision-making processes. However, in addition to the influence of the media, a multitude of economic, sociocultural, psychological and historical variables interact to orchestrate the complex relationship between tourism and individual preferences and experiences. Taking all these aspects into account, communication and media theories can be critically linked to industrialization, post-industrial developments and the new network society. In this way, the emergence of today's tourism overload is highlighted as a pressing phenomenon driven by mass society, hyper-commercialization and the mass consumption of travel experiences.

## Notes

[1] Structuralism within media theory serves as a theoretical lens with the aim of uncovering the underlying frameworks and systems that exert influence on human culture and communication. When we apply this perspective to the analysis of tourism advertising, for example, it is fascinating to observe how it reveals the hidden structures, languages and symbols that advertisers use to shape people's perceptions and desires for travel experiences. This approach proves indispensable in gaining insight into the cultural and symbolic dimensions of tourism advertising that go beyond the superficial content.

[2] It is not the purpose of this book to develop the theories of communication in great detail, but for the better understanding of the reader we can make a small description of these two theories. The 'spiral of silence' theory by Elisabeth Noelle-Neumann states that people, as conformists, do not want to be isolated; they therefore constantly observe their social environment, and if one of them notices that his/her opinion is different, he/she becomes afraid and falls silent. Thus, the social group that has the power of speech and is present in public appears much stronger than it actually is (Noelle-Neumann, 1984). The scholars of the 'Frankfurt School', better known as the Critical Theorists, were concerned, among other things, with the deception of people through commercial propaganda and the associated creation of false needs. Mass culture and consumerism, which enslave and alienate people, were seen as central problems of capitalist societies. In general, they believed that consumer society, mass culture and the culture industry contribute to the commercialization of human relations, while the media promote the dominant ideology (Wiggershaus, 1994).

[3] At this point, of course, it is interesting to mention Plato again, who made his categorization of the types of visitors to Athens long before Eric Cohen or Stanley Plog's categorization of tourists. In his book *Laws*, the great ancient Greek philosopher attempted to divide Athens' foreign visitors into four main categories: (i) official foreigners who were members of international missions; (ii) those who came for return visits; (iii) those who participated in various cultural events; and (iv) traders who came by sea, especially in the summer. This is most likely the first categorization of travellers worldwide according to the purpose of their trip (Katsigiannis, 2017).

[4] Alternative media can be defined as any form of media used to convey messages that are not mainstream or subject to more liberal controls, content rules, forms, etc. that sometimes challenge traditional controls over what is published.

[5] Frame analysis examines how individuals perceive the nuances of the world in particular situations and is often associated with interactionist theory and the Palo Alto group. It has its origins in Goffman's (1974) idea of 'frames', shared interpretive conventions that are crucial to understanding cultural belief systems. This approach has evolved in several directions, from political analysis to media research and marketing, what Entman (2007) calls a fragmented paradigm. Media researchers use frame analysis to assess how the media use visual cues such as economic, environmental, altruistic, patriotic, anthropocentric, geocentric and feminist interpretations to shape and convey messages about reality.

## References


Althusser, L. (2014) *On the Reproduction of Capitalism: Ideology and Ideological State Apparatuses*. Verso, London.

Barrett, F. (1989) *The Independent Guide to Real Holidays Abroad*. The Independent, London.

Barthes, R. (1957) *Mythologies*. Editions du Seuil, Paris.

Bateson, G. and Jackson, D.D. (1964) Social factors and disorders of communication. Some varieties of pathogenic organization. *Research Publications – Association for Research in Nervous and Mental Disease* 42, 270–290.

Baudrillard, J. (1983) *Simulations*. Semiotext(e), New York.

Bauman, Z. (2016) Zygmund Bauman From Pilgrim to Tourist – or a Brief History of Identity. ResPublica. Available at: http://www.respublica.gr/2016/04/post/bauman3/ (accessed 1 September 2019).

Berger, P. and Luckmann, T. (1967) *The Social Construction of Reality*. Doubleday, New York.

Birdwhistell, R.L. (1955) Background to kinesics. *Review of General Semantics* 13, 10–18.

Blumler, J. and Katz, E. (1974) *The Uses of Mass Communication: Current Perspectives on Gratification Research*. Sage, London.

Boorstin, D. (1964) *The Image: a Guide to Pseudo-events in America*. Harper and Row, New York.

Cappella, J.N. and Jamieson, K.H. (1997) *Spiral of Cynicism: the Press and the Public Good*. Oxford University Press, Oxford, UK. DOI: 10.1093/oso/9780195090635.001.0001.

Cohen, E. (1972) Toward a sociology of international tourism. *Social Research* 39(1), 164–182.

Coolsen, F.G. (1947) Pioneers in the development of advertising. *Journal of Marketing* 12(1), 80–86. DOI: 10.1177/002224294701200110.

Coward, R. and Ellis, J. (1977) *Language and Materialism*. Routledge Kegan Paul, London.

Dann, G. and Parrinello, G. (2009) Setting the scene. In: Sharpley, R.A. (ed.) *The Sociology of Tourism: European Origins and Developments*. Emerald, Leeds, UK, pp. 1–63.

De Fleur, M.L. and Ball-Rokeach, S.J. (1983) *Teorías de la Comunicación de Masas*. Paidós, Barcelona, Spain. (in Spanish).

Demertzis, N. (2002) *Political Communication: Risk, Publicity, Internet*. Papazisis, Athens. (in Greek).

Dewenter, R., Linder, M. and Thomas, T. (2019) Can media drive the electorate? The impact of media coverage on voting intentions. *European Journal of Political Economy* 58, 245–261. DOI: 10.1016/j.ejpoleco.2018.12.003.

Durkheim, E. (1995) *The Elementary Forms of Religious Life*. The Free Press, New York.

Eco, U. (1986) *Travels in Hyper-Reality*. Picador, London.

Entman, R.M. (2007) Framing bias: media in the distribution of power. *Journal of Communication* 57(1), 163–173. DOI: 10.1111/j.1460-2466.2006.00336.x.

Featherstone, M. (1991) *Consumer Culture and Postmodernism*. Sage, London.

Feifer, M. (1985) *Going Places: the Ways of the Tourist from Imperial Rome to the Present Day*. MacMillan, London.

Feyerabend, P. (2010) *Against Method: Outline of an Anarchistic Theory of Knowledge*. Verso Books, New York.

Funkhouser, G. and Shaw, E.F. (1990) How synthetic experience shapes social reality. *Journal of Communication* 40(2), 75–87. DOI: 10.1111/j.1460-2466.1990.tb02263.x.

Gamson, W.A. and Modigliani, A. (1989) Media discourse and public opinion on nuclear power: a constructionist approach. *American Journal of Sociology* 95(1), 1–37. DOI: 10.1086/229213.

Gerbner, G. and Gross, L. (1976) Living with television: the violence profile. *Journal of Communication* 26(2), 173–199. DOI: 10.1111/j.1460-2466.1976.tb01397.x.

Goffman, E. (1955) On face-work; an analysis of ritual elements in social interaction. *Psychiatry* 18(3), 213–231. DOI: 10.1080/00332747.1955.11023008.

Goffman, E. (1974) *Frame Analysis: an Essay on the Organization of Experience*. Harper and Row, New York.

Gottdiener, M. (1995) *Postmodern Semiotics: Material Culture and the Forms of Postmodern Life*. Blackwell, Cambridge, UK.

Graburn, N. (1977) Tourism: the sacred journey. In: Smith, V. (ed.) *Hosts and Guests: The Anthropology of Tourism*. University of Pennsylvania Press, Philadelphia, Pennsylvania, pp. 21–36.

Hall, E.T. (1963) A system for the notation of proxemic behavior. *American Anthropologist* 65(5), 1003–1026. DOI: 10.1525/aa.1963.65.5.02a00020.

Hall, S. (1980) Encoding/decoding. In: Hall, S., Hobson, D., Love, A. and Willis, P. (eds) *Culture, Media, Language*. Hutchinson, London, pp. 117–127.

Jackson, D. (ed.) (1968) *Communication, Family and Marriage: Human Communication,* Volume 1. Science and Behavior Books, Palo Alto, California.

Kaitatzi-Whitlock, S. (2012) *Communication: Theory in Practice*. A.A. Livani, Athens. (in Greek).

Katsigiannis, K.X. (2017) *Hellenic Tourism Organization: Travel Through Time*. Private Edition, Athens. (in Greek).

Katz, E. and Lazarsfeld, P.F. (1955) *Personal Influence: the Part Played by People in the Flow of Mass Communications*. Free Press, Glencoe, UK.

Khang, H., Han, S., Shin, S., Jung, A.R. and Kim, M.J. (2016) A retrospective on the state of international advertising research in advertising, communication, and marketing journals: 1963–2014. *International Journal of Advertising* 35(3), 540–568. DOI: 10.1080/02650487.2015.1066477.

Klapper, J.T. (1949) *Effects of Mass Media: a Report to the Director of the Public Library Inquiry*. Bureau of Applied Social Research, New York.

Krippendorf, J. (2010) *Holiday Makers*. Routledge, London. DOI: 10.4324/9780080939032.

Kuhn, T.S. (1970) *The Structure of Scientific Revolutions*. Routledge and Kegan Paul, London.

Kulhanek, I. (2013) *Hyperreality in Post-Tourism: an Analysis of Fun, Space and the Experience of Hyperreality Aboard Carnival's Cruise Ships*. AV Akademikerverlag, Berlin.

Lang, G.E. and Lang, K. (1981) Mass communication and public opinion: strategies for research. In: Rosenberg, M. and Turner, R.H. (eds) *Social Psychology: Sociological Perspectives*. Basic Books, New York, pp. 653–682.

Lash, S. and Urry, J. (1994) *Economies of Signs and Space*. Sage, London.

Lasswell, H.D. (1927) *Propaganda Technique in the World War*. Paul Kegan, London.

Lazarsfeld, P.F., Berelson, B. and Gaudet, H. (1944) *The People's Choice*. Sloan and Pearce, Duell, New York.

MacCannell, D. (1973) Staged authenticity: arrangements of social space in tourist settings. *American Journal of Sociology* 79(3), 589–603. DOI: 10.1086/225585.

McCombs, M.E. and Shaw, D.L. (1972) The agenda-setting function of mass media. *Public Opinion Quarterly* 36(2), 176–187. DOI: 10.1086/267990.

McQuail, D. and Windahl, S. (2015) *Communication Models for the Study of Mass Communications*. Routledge, London. DOI: 10.4324/9781315846378.

Mead, M. (1964) *Continuities in Cultural Evolution*. Yale University Press, New Haven, Connecticut.

Munt, I. (1994) The other postmodern tourism: culture, travel and the new middle classes. *Theory, Culture & Society* 11(3), 101–123. DOI: 10.1177/026327694011003005.

Noelle-Neumann, E. (1984) *The Spiral of Silence: Public Opinion—Our Social Skin*. University of Chicago Press, Chicago, Illinois.

Palmgreen, P., Wenner, L.A. and Rosengren, K.E. (1985) Uses and gratifications research: the past ten years. In: Rosengren, K.E., Wenner, L.A. and Palmgreen, P. (eds) *Media Gratifications Research: Current Perspectives*. Sage, Beverly Hills, California, pp. 11–37.

Plog, S.C. (1974) Why destination areas rise and fall in popularity. *Cornell Hotel and Restaurant Administration Quarterly* 14(4), 55–58. DOI: 10.1177/001088047401400409.

Plog, S.C. (2001) Why destination areas rise and fall in popularity: an update of a *Cornell Quarterly* classic. *The Cornell Hotel and Restaurant Administration Quarterly* 42(3), 13–24. DOI: 10.1016/S0010-8804(01)81020-X.

Poon, A. (1989) Competitive strategies for a new tourism. In: Cooper, C. (ed.) *Progress in Tourism, Recreation and Hospitality Management*. Belhaven, London, pp. 130–142. DOI: 10.21832/9781873150467.

Porcar, C. and Hainic, C. (2011) The interactive dimension of communication: the pragmatics of Palo Alto group. *Journal for Communication & Culture* 1(2), 4–19.

Powdermaker, H. (1950) *Hollywood, the Dream Factory*. Little Brown, Boston, Massachusetts.

Rojek, C. (2000) Mass tourism or the re-enchantment of the world? Issues and contradictions in the study of travel. In: Arditi, J., Cook, D., Bramlett, M.D. and Cerulo, K.A. (eds) *New Forms of Consumption: Consumers, Culture, and Commodification*. Rowman and Littlefield, Lanham, Maryland, pp. 51–70.

Samaras, A. (2008) *Television Political Advertising in Greece*. Kastaniotis Publications, Athens. (in Greek).

Scheufele, D.A. (1999) Framing as a theory of media effects. *Journal of Communication* 49(1), 103–122. DOI: 10.1093/joc/49.1.103.

Shanahan, J. and Morgan, M. (1999) *Television and Its Viewers: Cultivation Theory and Research*. Cambridge University Press, Cambridge, UK. DOI: 10.1017/CBO9780511488924.

Shannon, C.E. and Weaver, W. (1949) *The Mathematical Theory of Communication*. University of Illinois Press, Champaign, Illinois.

Signorielli, N. and Morgan, M. (eds) (1990) *Cultivation Analysis: New Directions in Media Effects Research*. Sage, Newbury Park, California.

Thibault, G. (2016) Needles and bullets: media theory, medicine, and propaganda, 1910–1940. In: Nixon, K. and Servitje, L. (eds) *Endemic: Essays in Contagion Theory*. Palgrave Macmillan, Basingstoke, UK, pp. 67–92. DOI: 10.1057/978-1-137-52141-5.

Thorp, M.F. (1939) *America at the Movies*. Yale University Press, New Haven, Connecticut.

Turner, V. (1973) The center out there: pilgrim's goal. *History of Religions* 12(3), 191–230. DOI: 10.1086/462677.

Turner, V. and Turner, E. (1978) *Image and Pilgrimage in Christian Culture*. Columbia University Press, New York.

Uriely, N. (1997) Theories of modern and postmodern tourism. *Annals of Tourism Research* 24(4), 982–985. DOI: 10.1016/S0160-7383(97)00029-7.

Urry, J. (1990) *Tourist Gaze: Travel, Leisure and Society*. Sage, London.

Watzlawick, P., Beavin-Bavelas, J. and Jackson, D. (1967) Some tentative axioms of communication. In: Watzlawick, P., Beavin-Bavelas, J. and Jackson, D. (eds) *Pragmatics of Human Communication – a Study of Interactional Patterns, Pathologies and Paradoxes*. W.W. Norton, New York, pp. 74–80. DOI: 10.4324/9781315080918.

Weakland, J. (1967) Communication and behavior. *American Behavioral Scientist* 10(8), 1–3. DOI: 10.1177/000276420100801.

Wiggershaus, R. (1994) *The Frankfurt School: its History, Theories, and Political Significance*. MIT Press, Cambridge, Massachusetts.

Wright, J.S. and Panksepp, J. (2012) An evolutionary framework to understand foraging, wanting, and desire: the neuropsychology of the SEEKING system. *Neuropsychoanalysis* 14(1), 5–39. DOI: 10.1080/15294145.2012.10773683.

Zuckerman, M. (2014) *Sensation Seeking (Psychology Revivals): Beyond the Optimal Level of Arousal*. Psychology Press, London. DOI: 10.4324/9781315755496.

# 6 Revealing the Complex World of Tourism Imagery

**Abstract**

Tourist images are created in a multi-layered process that is characterized by various influences. They can originate from different media and art forms and reflect the attractiveness and authenticity of a destination without this being the intention of the creator. Furthermore, the tourism industry strategically constructs images by incorporating elements such as collective memory, mass media, art and cultural diplomacy. This chapter examines the complex creation of tourism images and emphasizes the central role of media coverage, artistic expression and marketing. Simple examples illustrate how these factors shape perceptions and highlight the transformative impact of the social media era. In today's landscape, travellers are also actively contributing to the representation of destinations, fostering a dynamic and diverse narrative through shared experiences. This shift marks a profound change in the way tourism images are created and disseminated.

## 6.1 Creation of the Tourist Image of a Destination

It is perhaps no coincidence that the French sociologist and philosopher Maurice Halbwachs (1877–1945) begins the first chapter of his book on collective memory (*Mémoire Collective* (1950) published posthumously) with an example from a trip to London, with which he tries to introduce the reader to the subject in order to develop his point of view. Some of the streets, alleys and landmarks the visitor sees, together with the descriptions of monuments, architectural buildings and historical events made by friends, architects, artists, historians, etc., as well as things he has read in books and guidebooks, help him to slowly recall the whole 'image of the city' of London in his memory. From this he concludes that other people have also formed his memories along with him.

So it is true that these images, imposed on us by memories and narratives from our social environment and the media, together with our personal experiences, can determine our perception of a destination. We can therefore say that the concept of the tourist 'image of a destination' is also influenced by and directly linked to collective memory.[1]

Indeed, empiricists claim that human knowledge comes exclusively from sensory experiences and perceptions. People perceive different destinations and form mental images or conceptions based on sensory impressions they receive from the outside world. This concept is in line with the viewpoint of the Anglo-Irish philosopher George Berkeley, who argued that our perception defines reality, since human existence is limited to what we can perceive with our senses (Winkler, 2005).

In terms of destinations and their tourist image, this means that logic alone is not able to invent a correct mental image of a particular place *ex nihilo*. Instead, our perception of a destination is formed by absorbing information, personal encounters, external influences from different media sources and the collective memory.

In order for a destination to integrate into our global tourism landscape, it is essentially necessary to obtain comprehensive information. The more convincing and thorough this information is, the more positive and complete our perception of the place will be. In addition to our direct experiences on the ground and conversations with family members and friends who have visited the place, written texts, images and audio-visual material we come across are also


DOI: 10.1079/9781800626034.0006

important sources that shape our impression of any destination.

In this context, the 'image of a destination' draws on a range of sources such as newspapers, books, photographs, television programmes, films and more. Because of their widespread distribution, these media have the power not only to influence our individual perceptions, but also to shape our social representations about places. The media's influences are of course complemented by our first-hand experiences when we actually visit the place, and by conversations with our interpersonal relations.

However, the attraction that a place presented in the media exerts on people can in many cases be an inherent feature of the media narratives themselves (i.e. it can occur without the intention of the producers to promote tourism). Simply put, when people listen to stories, read novels, watch films or see documentaries about a particular place, they may be influenced and this can lead them to develop a desire to visit it, even if this was not the original intention of the storyteller or producer.

When it comes to perception and the idea we form of a place, people often use the terms 'image of a country', 'image of a nation', 'image of a place', 'image of a city', etc. synonymously and sometimes without regard to their respective meanings. Be that as it may, in this book we will mainly refer to the tourist dimension of image, whether we are talking about a state, a country, a place, a city, etc., without taking into account the specificities of the political or sociolinguistic use of each term. What interests us most when we refer to related concepts is the tourist aspect of image (i.e. the 'image of a destination') and the way it is created and perceived.[2]

### 6.1.1 Media influences – 'organic' and 'induced' tourism images

To investigate how different sources of information influence our perception of a destination, Gunn (1972) developed a seven-stage model for the travel experience. It starts with the formation of mental images of a place and ends with how these images change after the actual visit and personal experiences. Using this model, it is easier to see that in addition to the first-hand experiences we usually have when we finally visit the place (and which sometimes change the original image of the destination), two main types of media influences are distinguished.

First, there are the images created by factors not directly related to the tourism industry (novels, paintings, news, films, etc.), which are called 'organic tourism images'. Indeed, images of places and written and oral narratives surround us every day and form an essential part of our world. All these messages have something to say about a place, something to suggest, whether they are informal like family snapshots and discussions on social media or imposing like sermons in churches and paintings in galleries.

These images usually come from non-advertising media texts provided by ordinary people, writers, photographers, film directors, artists, etc., expressing the authentic essence of a place and its inhabitants as they see it. They are not intended to encourage us to visit and often reflect the perspective of the creators, who have nothing to do with the tourism industry. 'Organic tourism images' contrast with overly staged or manipulated photos for tourism advertising, which can present a destination in a glamorous or artificial way.

In all these cases of organic tourism imagery, then, the media are autonomous mediators who suggest destinations, as their narratives essentially offer images that emotionally connect the audience to the particular places where the action takes place, thus forming tourist representations (Gunn, 1972; Gartner, 1994).

For greater clarity, let us look at the specific factors that can unintentionally influence how tourists perceive a country and the representations associated with it. Artists such as poets, painters, film directors, singers, but also historians, geographers or anthropologists, through their work as well as friends who share their travel experiences, may not have the explicit intention of motivating us to travel by shaping a 'tourist image' or creating a 'tourist imaginary'.[3]

Nonetheless, their contributions influence these perceptions, which may or may not be consistent with the images promoted by official bodies such as national tourism organizations. Importantly, the unique characteristics and cultural background of the audience, whether individuals or social groups, play a crucial role in shaping these impressions and should not be underestimated.

'Organic tourism images' are thus based on unintentional representations of all kinds, on real experiences and personal stories, which are often more credible and trustworthy than traditional tourism advertising. It is important to note that while 'organic tourism images' are not the result of the tourism industry, destination stakeholders and national tourism promotion organizations have recognized their importance and are beginning to support them indirectly by offering free facilities for their production, tax exemptions, etc.

On the other hand, Gunn (1972) argued that there are a considerable number of tourist images that are deliberately constructed by the various actors in the tourism industry to promote certain destinations, calling them 'induced tourism images' (Echtner and Ritchie, 1993; Bolan and Williams, 2008).

Thus, in terms of the dimension of induced tourism promotion, a country's tourism image is indeed constructed through a combination of factors such as marketing strategies, public relations, cultural diplomacy and so on. Apart from the unintended effects of media messages, creating a positive and appealing tourism image therefore requires coordinated efforts by government agencies, tourism associations, businesses and local communities.

Any paid form of positive media coverage, including travel articles, television programmes, sponsored feature films and documentaries, can certainly have a significant impact on the perception of a country as a destination. Tourist boards and businesses also use social media platforms to showcase stunning images, tell travel stories and engage with potential visitors. In addition, the public relations (PR) strategies of national tourism organizations and ministries of tourism, which take a multi-layered approach that uses media, content, partnerships, events and online platforms, play a crucial role in promoting a country.

Inviting celebrities, diplomats, singers, actors, etc. or hosting international events, festivals, conferences and sports tournaments can also attract the attention of the global public and contribute significantly to the country's image as a vibrant and exciting destination.

In this context, it can be said that all these induced media narratives mentioned above are structured on the basis of the language of tourism, which we can consider as the sophisticated code that coordinates the 'tourist gaze' (Dann, 1996; Urry, 2005; Fox, 2008; Thurlow and Jaworski, 2010).

The language of tourism is now an everyday aspect of our lives, as it is used by the modern tourism industry as a powerful tool to shape and promote certain consumer behaviours in relation to tourism products.

However, a closer look reveals that the tourism industry prescribes certain rules of conduct for tourists. Despite the seemingly liberating concept of tourism, the various media representations of destinations, hotels and other tourist products created by the industry declare certain places to be must-visit tourist hotspots and determine tourists' behaviour within the framework of consumer society. These representations institutionalize certain tourist activities and exclude others, ultimately limiting the freedom of travellers, which is the essence of this phenomenon of human mobility (Irimiea, 2018).

Be that as it may, most modern states use the mass media to shape their image through 'cultural diplomacy'. In its effort to promote the 'image of the nation', cultural diplomacy aims to protect, promote and utilize cultural heritage. More precisely, cultural diplomacy is the methodical and hierarchical use of a country's culture in its foreign policy. It aims, on the one hand, to promote the country's cultural values and achievements abroad and to improve the country's image, and, on the other hand, to strengthen relations with other nations and consolidate a climate of trust and mutual understanding (Tzumaka, 2005).

Thus, based on the concepts of intrinsic and cultivated tourism narratives that we have already discussed, we can conclude that in addition to today's tourism industry that promotes tourism with marketing strategies, there was also an earlier conception of tourism that consisted of purely organic images of tourism, and that today it is somewhat difficult to distinguish the elements that each of these channels has given to today's image of tourism.

In short, the organic representation hypothesis states that the media, especially through the presentation of artistic or quasi-artistic narratives, can inspire people to conjure up mental images that stimulate a longing for imaginary journeys or awaken an innate wanderlust.

In this sense, various media promoted the idea of tourism long before the arrival of modernity and the industrialized, urbanized society that gave rise to the tourism industry, as we know it today.

It can be argued then that the contrast between unintentional and intentional efforts to create tourist images and fantasies is related to the dual nature of the term 'tourism', namely as a human urge to expand consciousness through travel, and as an industry. In both cases, the media play a central role in shaping the attractiveness of destinations.

In light of what has been said so far, it is important to realize that the term 'image of a destination' encompasses two different aspects: (i) how we perceive an image; and (ii) how an image is intentionally created and disseminated by the tourism industry.

In other words, in the first option, this term can refer to the development of a destination's image as a mental process that takes place in the minds of individuals or groups who form personal or social representations shaped by literature, art and, of course, the promotional practices of the tourism industry.

### 6.1.2 Tourist image of a destination in our minds

So, if we consider destination images as personal or social representations, we can say that they consist of four elements. These are: (i) the cognitive element (i.e. what we know about the destination in question); (ii) the emotional element (i.e. how we think about the destination and what feelings it triggers in us when we think about it); (iii) the evaluative element (i.e. the way we evaluate the destination); and (iv) the behavioural element (i.e. the tendencies we have regarding our behaviour towards the destination and everything related to it) (Boulding, 1959; Avraham and Ketter, 2008; Samaras, 2014).

The tourist image of a destination in our minds can thus be unintentionally shaped by a variety of positive and negative factors influenced by word of mouth, history, literature, art, media coverage and social media, but also intentionally by tourism marketing.

It can be deduced from this that the perceived destination image is quite complicated as it relates to specific recipients or groups, their characteristics, their education, their social capital in general and the specific way they decode the projected images, and is usually studied through interviews, ethnomethodological analysis, etc.

In this way, an image can be positive in one way or another if it leads to positive opinions in its final form, or negative if it leads to negative opinions. It can also be rich if it is formed from many sources, or weak if it is almost invisible and comes from limited sources.

On the other hand, the term 'image of a destination' can also be seen as a controlled process of construction and dissemination promoted by the tourism industry and the mass media. This includes a series of narratives that form a rhetoric aimed at deliberately constructing and promoting a certain destination image.

### 6.1.3 Developing a good tourist image for a destination

The development of a good tourist image for a destination by the tourism industry or, more precisely, by national tourist boards involves several phases, which are often interlinked and may overlap. These include: (i) researching and analysing the destination and the target market; (ii) creating a unique positioning of the destination in the minds of travellers; (iii) developing a visual and verbal identity for the destination; (iv) researching appropriate media; (v) building relationships with the media; (vi) developing PR campaigns; and (vii) monitoring and measuring the effectiveness of image building (Molina *et al.*, 2010; Ispas *et al.*, 2016).

The tourism image of a country projected by the modern mass media and the practices of tourism industry shareholders can be studied by analysing marketing and cultural diplomacy strategies, and the narratives that make up the media's promotional messages (case studies, comparative analysis, quantitative or qualitative textual analysis, semiotic analysis, geosemiotics, etc.).

The arguments we have put forward suggest that various media narratives about the natural beauty and wealth of a place or about the civilization and history of certain peoples have most likely inspired travel fantasies since antiquity. In a broader context, then, we can say that in addition to interpersonal communication, all kinds of media that convey information, stories and, above all, representations of places can form internal images in people's minds of the geography of those places and their physical and constructed surroundings. In this sense, there have always been crucial factors that influence the imagination and mobility of individuals and populations towards these places.

It is not difficult to assume that the 'image of a destination' also includes the typical landscapes of a country. In many cases, however, a country's typical landscapes have undergone complex processes of redefinition and enrichment over time, depending on the influences of culture, aesthetics, tourism strategies, international tourism preferences of the time, and so on. Sometimes the meaning of the landscapes themselves also changes depending on the economic, social and political objectives of the state. Be that as it may, much has been written on the subject and in general we can accept that perceptions and 'ways of seeing' landscape vary from time to time and also from culture to culture. Ultimately, we should accept that the tourist landscape of a country is largely one of the fundamental features in shaping the destination image and is generally reproduced through art and mass media.

The spoken language, the written word, drawings and, more generally, the representations included in any form of communication (i.e. discussions and lectures, ancient stone tablets, ceramic vases, newspapers, drawings, photographs, moving images) as symbolic systems, are key elements in the construction of the 'image of a destination'. In this respect, they have always been unconscious factors that encouraged people to visit that place; and if we want to use the terminology of tourism studies, they have influenced tourist demand.

## 6.2 How We Understand Media Messages

Without relying on any particular theory of cognitive psychology or accepting unquestioningly the theory of powerful media, or in any case without endorsing the force of a strictly structuralist position, we will try at this point to break down into simpler parts the way we normally proceed to understand the meaning of media messages. Our aim is merely to provide a simplified description of the process of image formation in tourists, not a theoretical cognitive model for message perception.

Starting from the simple assertion that listening to a speech or reading a text is the extraction of meaning from verbally or visually represented symbols, while looking at a picture or a moving image is generally also the extraction of meaning from visually represented signs, we can arrive at the following considerations.

The processes involved in listening or reading are based on auditory or visual perception, recognition, memory and the code (language) we use (Cohen, 1972).

Auditory or visual perception of texts of any kind (written works or images) leads the individual to first take in some messages. With the help of his/her memory, the individual recognizes and processes these messages, recalling his/her personal experience and knowledge.[4] Here, however, we must note that cognition is a process of construction rather than a process of perception. Through the development of his/her imagination and creativity, the individual makes various evaluations, comparisons and linkages of incoming information with his/her experiences on the basis of a structural code, and acquires or changes knowledge, skills, beliefs, attitudes and behaviour.

Greatly simplified, it can be said that the difference between words and pictures (i.e. between verbal and pictorial representation), is that the codes of spoken and written language are always based on a social contract, whereas pictures can be represented by their appearance.

As Berger (1972) states, seeing comes before words (i.e. a child understands images before it learns to speak). In this sense, images can give us a more direct testimony about the world, and therefore images can be more universally understandable than literature. But things are not so simple: images can also be symbolic and based on a code that follows certain conventions created by culture (Berger, 1972; Cohen, 1972; Lopes, 2004).

Sketches, paintings, photographs, films and any other form of visual representation are usually linked to their meaning through a relationship of similarity-identity or through the basis of an existential bond, a relationship. Sometimes images can be metaphorical for various reasons, so their meaning in this case can be based on completely arbitrary cultural conventions established by society.

So, images can also be part of a code, like the written language mentioned above (e.g. the language of cinema or the system of street signs). In all these cases, individuals observe and interpret virtual representations in a process similar to reading (i.e. they follow the procedures of perceiving, recognizing, retrieving and decoding described above).

It should also be noted that the decoding of the message can very often be interpreted quite differently from what the encoder was trying to convey, if distortions or misunderstandings arise from a lack of cultural equivalence between the two sides of the communicative exchange (Hall, 1980).

It is often argued that in addition to information, the aesthetics and artistic value of a text or image (or moving image) are of particular importance in shaping the 'image of a destination' and in easily creating 'tourist imaginaries'.

This view is related to the attraction of art, which was discussed in Chapter 1, this volume (section '1.4 Art as a Form of Communication'). Thus, when listening to a poem, reading a literary work, looking at a painting or a film, one can experience unique moments, be moved, fascinated and attracted by the places described or the characters in the narrative, and without being aware of it, the desire can arise to actually visit the place where the action takes place one day.

Using this logic, we can say, for example, that films such as *Amélie* (2001) by Jean-Pierre Jeunet, *Moulin Rouge!* (2001) by Baz Luhrmann or *Paris, je t'aime* (2006) by Gurinder Chadha and other directors are perhaps as important for the tourist image of Paris as the original advertising campaigns of the 'Bureau des Congrès et des Salons de Paris', an organization promoting tourism in Paris.

There is no denying that virtual media representations are very important for the tourism industry and its reproduction, but also for the tourists themselves when they return to their everyday lives.

When we remember something we experienced in the past at a certain time and place, our thoughts, in combination with the images we ourselves have retained in our memory, often draw on visual information derived from photographs or videos presented and created by the media about that place. From these photographic images of the media, we often form the core images (i.e. the background in which we place the events of our lives), and it is in these photographed spaces that our memories eventually come to life, especially if they are travel memories (Antoniadis, 1995).

When we think of a trip to Paris, for example, we often add to our imagination and personal memories typical images of this city and its sights, such as already familiar views of the city from old films, newsreels we have seen, books we have read, stories our teachers have told us, but also other views as presented in the tourist media, on posters, T-shirts, in souvenir shops and so on. And here we can argue that the tourist image of the 'City of Light' has been strongly influenced by the media since the end of the 19th century and after.

Although Paris existed in the collective memory of the Western world much earlier as an attractive cultural destination, shaped over many centuries and passing through different historical eras, it was not until the 19th century that many called it the 'capital of the century' (Benjamin, 1969).

This overwhelming impression had to do with the new building materials and techniques, with iron and glass, which were used nowhere else as generously and artfully as they were then to construct the Eiffel Tower, the underground tunnels, the arcades, the exhibition halls and railway stations. All these buildings of iron and glass represented a renewal of architecture. And then there were the new boulevards designed by Baron Haussmann (1809–1891), who had been commissioned by Emperor Napoleon III to carry out a major urban renewal in Paris. The new boulevards not only provided space for the army to put down each new revolution, but also functioned as modern transport axes between the city's end points and extended sight lines between monuments and large public buildings that citizens and visitors could admire as national symbols (Mak, 2008).

In the late 19th and early 20th centuries, the composition of this image was widely disseminated through the press, cinema, painting and literature, and it seems that some of the basic elements of the tourist image of Paris were created during this period and are still valid today.

Nowadays, it can be said that the image of Paris as a tourist destination contains elements that are very often used by all media as signifiers of this city. These include: (i) the Eiffel Tower, which is perhaps the most famous symbol of the city; (ii) historical monuments, including landmarks such as Notre Dame Cathedral, the Louvre Museum, the Arc de Triomphe and the Sacré-Cœur Basilica; (iii) special places with a romantic ambience such as the Champs-Élysées or the Seine, the Moulin Rouge and the Montmartre district, restaurants and cafés; (iv) French cuisine, baguettes and red wine; (v) fashion and shopping; and (vi) a wide range of 'art and culture' as the city has been a centre for art, literature and culture for centuries.

Many scholars believe that social representations can not only shape but also change perceptions of many things over time (e.g. judgements about the beauty and authenticity of a place, its status, tourists' preferences), as they are social constructions shaped by social interaction and follow their own path, spread and repeated through different communication systems (Deaux and Philogene, 2001).

Natural things like a rock, a beach, a rainbow do not exist by themselves as objective entities that are attractive or unattractive. The concepts of 'beautiful' or 'ugly' arise at the moment we arrive at judgements. Thus we can say that a Greek beach undoubtedly has a timeless natural value, as does a Greek marsh, if we accept that there are no ugly or beautiful images in nature. However, the collective opinion that recognizes a geographical area such as a Greek beach as an attractive destination (as opposed to a marsh) can change depending on what is presented as natural beauty in art, in the media and, of course, in the texts and images used by the tourism industry. Often this process is reinforced by an intense combination of different media controlled by the tourism industry.

However, since the development of tourist activities requires the constant presentation of new products that are symbolically different from everyday life, the swamp could one day be presented in the Greek media as an attractive destination for special forms of tourism.

Indeed, swamp tourism is now considered a form of ecotourism in many countries, visiting and exploring swamps, marshes, wetlands and other similar environments, such as the New Orleans Swamp Tour and Everglades Airboat Tours in the USA, which offer visitors the opportunity to experience unique ecosystems, observe diverse plant and animal life, and learn about the importance of wetlands to the environment. So the creativity of tourism marketers and the combination of each era's media mix construct the trends of what is beautiful and attractive for tourism imagery, and if we want to go a little deeper, each era's tourism marketing and media mix construct what the term 'tourism' ultimately means.

Be that as it may, the commonalities between the various groups that make up the final 'image of a destination' remain relatively stable, or rather, some of the core elements that are part of the collective memory may be very fixed, but in reality, they are not completely immutable in the long run. Interaction between individuals, groups and social institutions reinforces a process that slightly alters the meaning of social representations (Hogg and Abrams, 1988; Pearce *et al.*, 1996).

Normally, social representations are decisively altered when their relations to the meanings of collective action change, i.e. when they come into complete conflict with social practices (Flament, 1989).

However, as we argue in this book, we must always bear in mind that certain ideas about certain countries have developed and are cultivated by the media of each era over a very long period of time and can become deeply imprinted in the collective memory.

In any case, the construction of the general 'image of a country' in contemporary society can be summarized by various factors, including collective memory, the media, the tourism industry, the state and political power, and civil society, which influence social norms and social practices through institutionalized or informal processes.

Finally, we can also argue at this point that the group that has the greatest power to manipulate the media has the greatest potential to change and define the 'image of a destination'.

## 6.3 Impact of Modern Mass Media on Image of a Destination

At this point, the question arises whether channelling through today's mass media in the desired direction, the construction of the desired image, is as easy as image managers and public relations people believe. As has already been mentioned, we must always remember that the core images of social representations of certain countries have existed since ancient times and are based on our schooling, mythology, history, tradition, art and so on.

In simpler terms, if we believe that tourism advertising can completely shape a country's modern image and drastically change an established perception, then we are essentially embracing the age-old idea that modern mass media can have a direct and significant impact on individuals and reshape their existing views. This points to a world where the media exerts immense influence, meaning that our understanding of reality is constructed by the media (Thibault, 2016).

We must not forget that imaginative travel, virtual travel and communicative travel, (i.e. three of the five forms of mobility defined by Urry (2016))[5] are highly dependent on the media.

Narratives in murals, scrolls, paintings, scientific and religious texts, literary works, popular novels, newspapers, advertisements, song lyrics, films, TV and feature films stimulate individuals' inclination to form mental images with imaginary journeys. In addition, the use of wireless digital technologies and mobile telephony, social media and travel agency platforms on the Internet enable intensive virtual and communicative journeys.

Finally, it cannot be denied that the various narratives disseminated by the numerous and ever-growing media using new and advanced digital and network technologies and their applications have a much greater potential to influence tourism demand than the media of the past.

Both the frequency and the intensity with which people receive information from the new media today would have been unimaginable to someone looking for information about a place at the beginning of the 20th century.

We can say that new media platforms offer a diverse, dynamic, polyphonic and often impartial approach to influencing perceptions of destinations by bringing a variety of voices and perspectives to the narrative.

The tourism industry has been quick to recognize the power of new technologies and is benefitting from the functionality and popularity of social media platforms. Many businesses and tourist boards use them as a basic tool to promote their products and services, and in this way these new devices play a crucial role in shaping the image of tourism today.

Social media such as Facebook, Instagram, Twitter and TikTok contain not only tourism industry advertisements, but also user-generated content, photos and videos of places and personal stories of people who have visited certain destinations.

They offer 'authenticity' to potential travellers, through real-time postings and visual and personal information about what a place has to offer by the visitors themselves, making tourist images accessible worldwide.

In general, it can be said that the representation of a destination on social media also serves as an educational tool that enables tourists to improve their understanding of a place, which in turn significantly influences their decision-making behaviour. Research shows that when social media platforms act as intermediaries, a composite image of a destination emerges through the blending of cognitive and emotional insights from suppliers, consumers and third parties. This effect is particularly pronounced for tourists who have no preconceived notions about the destination they wish to explore (Ghazali and Cai, 2013; Pan *et al.*, 2021).

Through the new media, the creation of innovative features in the representations of tourism becomes much easier and has assumed a great scale, as the tourism industry is constantly producing experimental signs at very low cost and creating new products that give symbolic meaning to different places in order to attract the interest of the public.

Moreover, it has been observed that young consumers of tourism products, who constitute the highest percentage of users of new technologies, also actively participate in various processes to promote the image of tourism and sometimes in the creation of new tourism products (Bizirgianni and Dionysopoulou, 2013).

Be that as it may, in the 'age of social media', people are now exposed to a wide range of information and images related to travel decisions and also actively participate in the creation of tourism image.

The continuous flow and mass reproduction of symbolic tourist associations, disseminated daily over a relatively long period of time through the new media, cause them to become imprinted in people's minds, and eventually tourist significations become increasingly important mechanisms for the process of forming social representations about places.

In this case, tourist images can effectively change not only tourist demand, but also the attitude of the inhabitants of a place towards it. Therefore, we can claim that the promotion of tourism images in the new media through the continuous construction of new forms of 'authenticity' by users and the tourism industry, adding new tourist features to the identity of a place, can not only influence the construction of the 'image of a destination', but also activate and decisively change the 'sense of place' as perceived by its inhabitants themselves.[6]

## 6.4 Summary

To summarize, the complex dynamics of collective memory, individual experiences and ubiquitous media narratives shape our perception of destinations in intricate ways, resulting in a complex structure that goes beyond the purely visual representation. The media in its many forms – from news, advertising, artistic expression and entertainment to the dissemination of information and historiographical interpretations – exerts a profound influence on our perception of different places and constructs 'tourism images'. This symbiotic relationship between organic and induced tourism images, coupled with the influence of cultural diplomacy and strategic marketing, emphasizes the multi-layered nature of destination image creation and dissemination.

## Notes

1 The concept of memory refers to a complex process that involves the encoding, storage and retrieval of events and experiences. Halbwachs (1950) argues that memories are inextricably linked to place and time and emphasizes the role of social groups in shaping individual and collective memory. Human memory is inherently socialized and develops through interactions within groups such as the family and ethnic communities, but also through contact with mass media. Even personal memories are influenced by dialogue and communication and are therefore constantly changing. Memories of distant lands are also preserved and reconstructed through conversations, emphasizing certain elements and neglecting others. This dynamic process helps to shape the image of a destination. Collective memory, which summarizes shared experiences, interpretations and representations, is a key component of a group's understanding of historical, geographical and cultural elements of a place. Finally, when we talk about collective memory and the 'image' of a country, a nation, a place or a city, we must always keep in mind the inherent ambiguity and dichotomy between this 'image' and the actual reality. Simply put, even if we philosophically assume that there is a concrete, absolute reality, representations are selective and do not fully depict reality, which leads to a fundamental distinction between 'image' and 'reality'.

2 Although we will not go into the distinction between these terms in detail in this book, we can provide some brief explanations of their interpretations here. First, the term 'image of a country' can be understood as a construct that a communicator wishes to promote. However, it can also refer to people's personal perceptions of a particular country over the course of their lives or to the sum of their beliefs, ideas and impressions about that country (Kotler *et al.*, 1993). The term 'image of a country' has long been used in business administration and marketing, but has always been in common use and is therefore easier for us

to use in this book. In these areas of research, the term 'image of a destination' can also be used as an alternative, which can also refer to cities, regions, villages, etc. Researchers generally agree that the image of a destination can be divided into two main aspects: (i) the cognitive dimension; and (ii) the emotional dimension. The first involves beliefs and understanding of the physical, historical and geographical features of a destination, while the second relates to the emotions that arise when people make a connection with the place and subsequently evaluate it. However, a group of scholars recognize a third dimension that relates to how individuals manage their knowledge and feelings about a particular destination and reflects their behavioural intentions (Gartner, 1994; Echtner and Ritchie, 2003; Hosany *et al.*, 2006; Michaelidou *et al.*, 2013). In retrospect, it can be said that image research in business studies has developed steadily since the 1930s, following a series of relatively well-organized studies that measured people's attitudes. With the major development of consumer society after the Second World War, marketing research began to focus more on measuring product attributes and their relationship to image. The topic underwent an interesting development with the publication of Boulding's (1956) book *The Image: Knowledge in Life and Society*, which examined image as a key to understanding society and human behaviour. In the early 1970s, organized tourism research on the image of destinations was dynamically initiated and has been very productive ever since. Secondly, the concept of 'image of a nation' is widely used in the social sciences, particularly in fields such as political sociology, political science, international relations and cultural studies. Its importance also extends to the field of international relations, where it encompasses the carefully crafted impressions of nations (states) as political entities perceived on a global scale and projected by actors such as policy makers, cultural diplomats and the media. Thirdly, the term 'image of place' is widely used in geography, especially in the field of human geography. It encompasses a variety of elements, including not only the physical attributes such as landscapes, landmarks and architectural features, but also the cultural and emotional connections associated with a particular geographical place. In some cases it is associated with 'geosemiotics', an interdisciplinary field concerned with the complex interplay of signs, symbols and meaning in the context of geography and the constructed environment. Finally, the term 'image of the city' is primarily used in the field of urban planning and urban design. It refers to a collection of mental impressions that become fixed in the minds of people who encounter the city. Essentially, it serves as a tool for analysing the spatial relationships, prominent landmarks, pathways and mental images that people have of urban environments. In 1960, Kevin Lynch analysed cities such as Boston, Jersey City and Los Angeles in his work *The Image of the City*. He suggested that one can gain valuable insights into a city by creating a mental map that includes elements such as routes, boundaries, neighbourhoods and nodes. Barthes (1967) also chose semiotic analysis as a systematic approach to deciphering the city. He argued that the city functions like a language, encoded by its built environment, and that an astute observer can decode it through a careful process of cognitive mapping. We should mention that the legacy of the Situationists' approach to deciphering and criticizing the urban landscape and the 'image of the city' has also left a lasting impact on discussions about urbanism, architecture and the relationship between people and their built environment (Debord, 1955).

[3] The relationship between the tourist 'image of a destination' and the 'tourist imaginary' is often unclear, and it is important to clarify these terms. The 'image of a destination' is about how people perceive and emotionally connect with a place, while the 'tourist imaginary' is about imaginative conceptions of places, people and services that resemble a realm of fantasy. The tourist 'image of a destination', which has to do with people's reception and understanding of a place, is more related to the terms 'impression', 'opinion', 'belief', 'view', 'emotional response', etc. without denying that it can also be related to imaginary elements, while the 'tourist imaginary' is closer to terms such as 'imagination', 'fantasy', 'feeling', 'sensation', and 'creativity', but without denying that it can also contain some logical elements or lead to logical assumptions. The 'imaginary' refers to non-physical, abstract concepts. The 'tourist imaginary' often feeds on historical stereotypes, myths and fantasies that influence travellers' perceptions. These 'mythtopias' can overshadow the true culture, geography and historical identity of a place. 'Tourist imaginaries' emerge from the interaction between information and people's imagination and shape new meanings related to tourism. Artistic expressions such as literature, films and art play a crucial role in stimulating these imaginations. The 'tourist imaginaries', while influencing collective behaviour, may not coincide with the objectives of the tourism industry and exist as implicit cultural constructs. They draw from a variety of sources, including mythology, storytelling and modern media. These imaginaries can add new dimensions to the attractiveness of a destination, influenced by both the tourism industry and the tourists' imaginations (Carnegie, 2016).

[4] According to Halbwachs (2020), memory is largely a continuous reconstruction of the past with the help of data borrowed from the present, a reconstruction with the help of which other reconstructions have emerged that have their origin in earlier times and from which the image of the past has already emerged mutated.

[5] Urry (2016) identified five interdependent forms of mobility that reproduce and organize social life from different distances and shape and reshape the modern world. These five basic forms of mobility are: (i) 'corporal travel of people' for reasons of work, tourism, family relations, immigration, escape from bad situations, etc; (ii) 'physical movement of objects', products and any other form of organic or inorganic matter; (iii) 'imaginative travel' that transcends the boundaries of space and time through images from traditional media, by constructing and reconstructing representations; (iv) 'virtual travel' that takes place through the Internet or multimedia applications; and (v) 'communicative travel' through person-to-person messages with the exchange of personal messages in any form (postcards, letters, telegraph, telephone, e-mails, mobile phones, etc.).

[6] The term 'sense of place' is used to describe the relationship between people and their environment. It refers to a cognitive framework of values, ideas and concepts related to a particular place, which results from the combination of views expressed by different people, social groups and the media. We can generally say that it is the relevant social representations formed for a place and the emotions they evoke in individuals and social groups. It is a dynamic concept that changes over time and may contain additional elements to the usual ones for each person or group (Frank, 2004; Stahl, 2010).

## References


Antoniadis, K. (1995) *Latent Image – Photography Essay*. Moresopoulos Photographia, Athens, Greece. (in Greek).

Avraham, E. and Ketter, E. (2008) Media strategies for marketing places in crisis. *Journal of Vacation Marketing* 14(4), 373–374. DOI: 10.1177/1356766708094758.

Barthes, R. (1967) Semiologia e urbanistica. *Selezione della Critica d'Arte Contemporanea* 10, 10–17.

Benjamin, W. (1969) Paris: capital of the nineteenth century. *Perspecta* 12, 165–172. DOI: 10.2307/1566965.

Berger, J. (1972) *Ways of Seeing*. BBC and Penguin Books, London.

Bizirgianni, I. and Dionysopoulou, P. (2013) The influence of tourist trends of youth tourism through social media (SM) & information and communication technologies (ICTs). *Procedia - Social and Behavioral Sciences* 73, 652–660. DOI: 10.1016/j.sbspro.2013.02.102.

Bolan, P. and Williams, L. (2008) The role of image in service promotion: focusing on the influence of film on consumer choice within tourism. *International Journal of Consumer Studies* 32(4), 382–390. DOI: 10.1111/j.1470-6431.2008.00672.x.

Boulding, K.E. (1956) *The Image: Knowledge in Life and Society*. University of Michigan Press, Ann Arbor, Michigan. DOI: 10.3998/mpub.6607.

Boulding, K.E. (1959) National images and international systems. *Journal of Conflict Resolution* 3(2), 120–131. DOI: 10.1177/002200275900300204.

Carnegie, E. (2016) Myth. In: Jafari, J. and Xiao, H. (eds) *Encyclopedia of Tourism*. Springer, Cham, Switzerland, pp. 645–646. DOI: 10.1007/978-3-319-01384-8.

Cohen, G. (1972) The psychology of reading. *New Literary History* 4(1), 75–90. DOI: 10.2307/468493.

Dann, G. (1996) *The Language of Tourism: a Sociolinguistic Perspective*. CAB International, Wallingford, UK.

Deaux, K.E. and Philogene, G. (eds) (2001) *Representations*. Blackwell, Oxford, UK.

Debord, G. (1955) Introduction to a critique of urban geography [translated by Ken Knabb]. *Les Lèvres Nues* 6(2).

Echtner, C.M. and Ritchie, J.R.B. (1993) The measurement of destination image: an empirical assessment. *Journal of Travel Research* 31(4), 3–13. DOI: 10.1177/004728759303100402.

Echtner, C.M. and Ritchie, J.B. (2003) The meaning and measurement of destination image. *Journal of Tourism Studies* 14(1), 37–48.

Flament, C. (1989) Structures et dynamique des représentations sociales. In: Jodelet, D. (ed.) *Les Représentations Sociales* (in French). Presses Universitaires de France, Paris, pp. 204–219.

Fox, R. (2008) English in tourism: a sociolinguistic perspective. *Tourism and Hospitality Management* 14(1), 13–22. DOI: 10.20867/thm.14.1.2.

Frank, A.I. (2004) Sense of place. In: Caves, R.W. (ed.) *Encyclopedia of the City*. Routledge, London, pp. 590–591.

Gartner, W.C. (1994) Image formation process. *Journal of Travel & Tourism Marketing* 2(2–3), 191–216. DOI: 10.1300/J073v02n02_12.

Ghazali, R.M. and Cai, L. (2013) Social media sites in destination image formation. In: Munar, A.M., Gyimóthy, S. and Cai, L. (eds) *Tourism Social Media: Transformations in Identity, Community and Culture*, Vol. 18. Emerald Publishing, Leeds, UK, pp. 73–86. DOI: 10.1108/S1571-5043(2013)18.

Gunn, C. (1972) *Vacationscape*. University of Texas Press, Austin, Texas.

Halbwachs, M. (1950) *Mémoire Collective*. Presses Universitaires de France, Paris.

Hall, S. (1980) Encoding/decoding. In: Hall, S., Hobson, D., Lowe, A. and Willis, P. (eds) *Culture, Media, Language*. Routledge, London, pp. 117–127.

Hogg, M.A. and Abrams, D. (1988) *Social Identifications: a Social Psychology of Intergroup Relations and Group Processes*. Routledge, London.

Hosany, S., Ekinci, Y. and Uysal, M. (2006) Destination image and destination personality: an application of branding theories to tourism places. *Journal of Business Research* 59(5), 638–642. DOI: 10.1016/j.jbusres.2006.01.001.

Irimiea, S.B. (2018) The relationship between the language of tourism, tourism and sociology. *European Journal of Social Science Education and Research* 5(2), 128–141. DOI: 10.2478/ejser-2018-0041.

Ispas, A., Constantin, C.P. and Candrea, A.N. (2016) Evaluating the image of tourist destinations: a theoretical and empirical approach. In: Nedelea, A.M. (ed.) *Strategic Tools and Methods for Promoting Hospitality and Tourism Services*. IGI Global, Hershey, Pennsylvania, pp. 26–48. DOI: 10.4018/978-1-4666-9761-4.

Kotler, P., Haider, D. and Rein, I. (1993) *Marketing Places. Attracting Investment, Industry and Tourism to Cities, States, and Nations*. Maxwell Macmillan, New York.

Lopes, D. (2004) *Understanding Pictures*. Oxford University Press, Oxford, UK. DOI: 10.1093/acprof:oso/9780199272037.001.0001.

Lynch, K. (1960) *The Image of the City*. MIT Press, Cambridge, Massachusetts.

Mak, G. (2008) *In Europe: Travels Through the Twentieth Century*. Vintage, London.

Michaelidou, N., Siamagka, N.T., Moraes, C. and Micevski, M. (2013) Do marketers use visual representations of destinations that tourists value? Comparing visitors' image of a destination with marketer-controlled images online. *Journal of Travel Research* 52(6), 789–804. DOI: 10.1177/0047287513481272.

Molina, A., Gómez, M. and Martín-Consuegra, D. (2010) Tourism marketing information and destination image management. *African Journal of Business Management* 4(5), 722.

Pan, X., Rasouli, S. and Timmermans, H. (2021) Investigating tourist destination choice: effect of destination image from social network members. *Tourism Management* 83, 104217. DOI: 10.1016/j.tourman.2020.104217.

Pearce, P., Moscardo, G. and Ross, G. (1996) *Tourism Community Relationships*. Pergamon, Oxford, UK.

Samaras, A. (2014) *Images of Nations*. Kastaniotis Publications, Athens. (in Greek).

Stahl, G. (2010) Urban semiotics. In: Hutchison, R. (ed.) *Encyclopedia of Urban Studies*. Sage, Thousand Oaks, California, pp. 918–922.

Thibault, G. (2016) Needles and bullets: media theory, medicine, and propaganda, 1910–1940. In: Nixon, K. and Servitje, L. (eds) *Endemic: Essays in Contagion Theory*. Palgrave Macmillan, Basingstoke, UK, pp. 67–92. DOI: 10.1057/978-1-137-52141-5.

Thurlow, C. and Jaworski, A. (2010) *Tourism Discourse: Language and Global Mobility*. MacMillan, New York.

Tzumaka, E. (2005) *Cultural Diplomacy: International Facts and Greek Perspectives*. I. Sideris, Athens. (in Greek).

Urry, J. (2005) Gazing on history. In: Kleniewski, N. (ed.) *Cities and Society*. Blackwell Publishing, Oxford, UK, pp. 195–214. DOI: 10.1002/9780470752814.

Urry, J. (2016) Does mobility have a future? In: Griego, M. and Urry, J. (eds) *Mobilities: New Perspectives on Transport and Society*. Routledge, Abingdon, UK, pp. 3–19.

Winkler, K.P. (2005) *The Cambridge Companion to Berkeley*. Cambridge University Press, Cambridge, UK.

# Part Two – Media and Tourism

When we examine the media and their function in relation to another topic, such as tourism in this case, it is very practical to categorize the types of media so that we can better understand how each of their different forms is used to communicate in relation to that topic.

In order to give the reader as clear a picture as possible of the media in relation to the phenomenon of tourism, chapters in Part Two will attempt to present the media from a point of view that takes into account the format of their messages, their historical and technological development and the communication relationships they facilitate between the parties involved.

We will therefore take a general look at our topic, first analysing the media with static and then with dynamic forms of messages, taking into account the parameters mentioned above.

*Static messages* are those in which the signs and representations do not change, but remain fixed in the form of a text, image, etc. (i.e. their content remains unchanged). For example, an article in a newspaper or magazine is static because it always remains as it was printed. When we talk about communication with static messages in this book (e.g. in Chapters 7, 8 and 9), we are referring to a wide range of media, from ancient inscriptions and etchings on various materials to murals, graffiti, books, newspapers and any written or designed message conveyed through various surfaces and media.

Certainly, since prehistoric times there have always been a variety of representations, descriptions and stories about distant places and journeys, disseminated through various means with static messages. These static messages include carvings and inscriptions, visual representations of exotic places and means of transport in manuscripts, paintings and murals, a rather interesting production of maps and cartography as well as handwritten books, especially travelogues and chronicles.

In ancient Greece, for example, representations of the Olympic Games or the Panathenaic Games on pottery influenced people's thinking by triggering imaginary journeys that made it possible to construct the 'image of Olympia' or the 'image of Athens' and to awaken in them the desire to visit these places and take part in the events.

The invention of printing, in particular Johannes Gutenberg's movable type press in 1440, had a profound impact on the spread of travel fantasies. This typographic breakthrough revolutionized the exchange of information by enabling mass production and changing the landscape of content creation. Travel guides, maps and pamphlets were widely distributed and encouraged the spread of travel and geographical knowledge. The advent of printing changed the perception of travelling in society as production costs decreased and printed material was easier to transport compared with manuscripts. This efficiency made it possible to convey static messages and provide a relatively universal idea of places to a wide audience across vast geographical regions. Print therefore played a crucial role in shaping the cultural understanding of travel.

From then on, not only posters and simple pamphlets, but also books whose texts described specific places in one way or another could be printed quickly and in very large quantities. With the establishment of major publishing houses, such as Cambridge University Press, Oxford University Press, Leiden University Press, Schwabe Verlag, the Königliche Hof- und Staatsdruckerei, and the House of Elzevir, as well as the optimization of the mechanization of the printing process, the significant role of print media in the construction of social representations of travel gradually began to be established


DOI: 10.1079/9781800626034.PART2

(Pon, 2008; Eliot and Rose, 2017; Steinberg, 2017).

Advances in printing, combined with lower production costs and increasing literacy, accelerated the dissemination of geographical information and shaped a shared history of travel. In the digital age, dominated by screens, it can be difficult, especially for the younger generation, to understand the significant historical role of publications and literature in the development of tourism.

*Dynamic messages*, on the other hand, are those in which the signs and representations are not static during the time they are perceived by the public. This means that their content is always in motion. The broadcast of a radio advert, for example, follows a narrative that seems to be very much alive, changing and evolving within minutes, and when the message is over, it is gone; it does not exist in a static form that we can listen to again.

In this sense, the flow of sounds and any kind of visual movement or changing images can be seen as a form of communication that is transmitted dynamically in the sense of continuity. The liveliness of messages can thus range from an audio or visual flow such as a speech or a theatre play to advanced radio streams, films, television, interactive video games, computer graphics animations and multimedia applications. Dynamic messages such as these are discussed in relation to tourism in Chapters 10 and 11, this volume.

The language we speak is also a dynamic means of communication. Face-to-face oral communication is fleeting and instantaneous, it can be combined with gestures, facial expressions and body language and, above all, it can be spontaneous, as it can take place in a conversation between two or more interlocutors without much thought. Furthermore, the spoken word is not delivered neutrally, but is accompanied by other paralinguistic elements such as emphasis, pace of speech, intensity, hardness or softness of voice. Most of the time, spoken language is conveyed in a context of social interaction and is subject to constant feedback from all interlocutors.

Be that as it may, although dynamic, the direct spoken word was mostly ephemeral before the invention of writing and the only way to preserve it was through memory with its inherent limitations.

Although today the term dynamic media is mostly associated with new media, participation and interaction, we should not forget that when cinema was discovered, one of the most important characteristics that commentators attributed to the medium was its dynamism.[1]

Legend has it that during the screening of Auguste and Louis Lumière's short film *L' Arrivée d'un train en gare de La Ciotat* in Paris in 1896, when the audience saw an approaching locomotive on the screen, the dynamic movement of the images made people think it was going to run straight into them, whereupon they panicked and fled to the back of the hall. Russian film director Sergei Eisenstein celebrated cinema for the dynamism with which it moved people and used the dynamic transition of images in his montage to convey abstract ideas. Not to mention that many centuries before, the flow of the rhetorical speeches of some ancient Greek politicians such as Pericles was said to be so dynamic that it could captivate the crowd (Sirois-Trahan, 2004; Tsakmakis, 2006; Eisenstein, 2014).

New media technologies (the focus of Chapter 11) have made the distinction between static and dynamic messages more difficult, offering not only interactivity but also a combination of static and dynamic message formats (text, audio and video), with all forms presented simultaneously in the same context. That's not all: recipients can access the dynamic message as often as they like, as it is not broadcast as an ephemeral radio or television programme, but is stored on the website.

In this context, the focus should be on evaluating the interactivity of the communication process and the facilitation of interaction between users. It is important to assess whether the communication structure allows interventions and active participation of all participants. In addition, the historical dimension should be consistently considered in any categorization.[2]

The following discussion in the chapters of Part Two will focus exclusively on the critical components within a broad spectrum of media with static and dynamic messages on tourism. This selective approach aims to enhance understanding and emphasize certain themes due to their notable importance. A comprehensive analysis of all conceivable media used to convey tourism messages would be impractical, if not unachievable.

## Notes

[1] When we speak of media with dynamic messages in this book, we are not referring to the now common notion of dynamic media, which only includes those whose content is constantly updated and interactive, such as websites, online forums and social media, but to any medium in which the signifiers of its messages are transmitted to the audience in a constant flow, or rather follow a temporal flow and finally come to an end when the message is completed.

[2] From an anthropological or philosophical point of view, the historical course shows that people's communication relationships have not changed through the means they use, with the logic that, as far as communication is concerned, all types of interaction already existed in antiquity. For example, there were means that conveyed messages in static form, such as the papyrus or the pediments of the temples, but there were also means that conveyed messages in dynamic form, such as the singers (*aoidoi*), the theatre and music. Finally there was the agora, the central public space, which can be described as a model form of means that allowed direct interaction and participation of all in the communication process. In short, the historical perspective we adopt here in relation to media technology emphasizes for the reader the role of technology in enlarging the audience, or rather in enlarging the number of participants in the communication process, and not so much in its contribution to enriching the types of communicative relationships between people through the media.

## References

Eisenstein, S. (2014) *Film Form: Essays in Film Theory*. Harcourt, Brace and World, Inc., New York.

Eliot, S. and Rose, J. (eds) (2017) *A Companion to the History of the Book*. Blackwell Publishing, Malden, Massachusetts.

Pon, L. (2008) Prints. In: Donsbach, W. (ed.) *The International Encyclopedia of Communication*. Blackwell Publishing, Malden, Massachusetts, pp. 3893–3898.

Sirois-Trahan, J.P. (2004) Mythes et limites du train-qui-fonce-sur-les-spectateurs. In: Innocenti, V. and Re, V. (eds) *Limina: le Soglie del Film/Film's Thresholds*. Forum, Udine, Italy, pp. 203–216.

Steinberg, S.H. (2017) *Five Hundred Years of Printing*. Courier Dover Publications, New York.

Tsakmakis, A. (2006) Leaders, crowds, and the power of the image: political communication in Thucydides. In: Tsakmakis, A. and Rengakos, A. (eds) *Brill's Companion to Thucydides*. Brill, Leiden, the Netherlands, pp. 161–187.

# 7 Painting and Tourism


**Abstract**

Over the centuries, painting has strongly influenced our understanding of travelling. From the rudimentary hunting motifs of early cave art to the intricate illustrations of the Renaissance, painting has depicted images of human mobility. In the 19th century, artists such as Monet and Renoir created a characteristic tourist image for Paris, and in the 20th century, the legacy of abstract painting and the rise of digital art continue to disseminate images of place. Painters who skilfully combine art and space fuel our enduring wanderlust. This chapter looks at another aspect of cave painting, ancient murals and decorative designs on clay vessels, and explores the symbiotic relationship between tourism and art from the Middle Ages to the Renaissance. The narrative extends to the present day and shows how painting continues to shape our perception of distant worlds and different modes of transport.


## 7.1 History of Representing Travel Through Painting

The visual arts, especially painting, played an important role in the development of travel by offering people pictorial representations and helping them to imagine the places they would like to visit. On the other hand, the adventure of travel and the places visited inspired many painters.

For many people, painting is almost synonymous with the concept of art. It can be said that painting has a long history in terms of the development of its expressive and decorative applications. The history of artistic expression through painting is a tapestry from prehistoric times in the caves of Lascaux and Altamira, the intricate beauty of Egyptian tombs, the grace of Greek vases, the vivid narratives on Roman frescoes, the solemn sanctity of medieval church walls, the transcendent masterpieces of the Renaissance in the 15th and 16th centuries, the paintings of modern times and so on.

One could argue that the origins of graphic representations and human creativity are directly related to symbolic thought. Barnard (2012) locates the possible origin of symbolic thinking as early as 130,000 years ago and argues that it is related to the development of an aesthetic consciousness, but also to the desire of humans to communicate with others, perhaps with the aim of influencing their perceptions.[1]

Symbolic thinking encompasses a set of abilities that many believe fundamentally differentiate humans from other animals, as they are able to dance to rhythmic sounds by following a meaningful choreography, produce art by expressing emotions, and generally use signs by assigning more meanings to them. With the development of symbolic thinking, language, rituals, music, art, theatre and faith emerged. This eventually led to the linguistic complexity necessary for mythological and later philosophical thought (Tse, 2008).

Cave paintings, considered to be some of the earliest examples of symbolic thought and artistic expression excavated by archaeologists, were created independently in different places around the world nearly 39,900 years ago (Roebroeks, 2014).

For example, the famous cave paintings in the Lascaux Cave in southern France, which depict various animals such as aurochs, horses and deer, were created somewhat later in the late Palaeolithic period (about 21,000–10,000 BC). The purpose of these paintings remains unknown, but it is generally assumed that they served ritual purposes related to hunting. The question arises, however, whether these paintings were placed before the hunt to encourage hunters to move into the areas and hunt the animals, or whether they were placed after the hunt to replenish the herds by immortalizing them (Jones, 1992). In the first case we are



DOI: 10.1079/9781800626034.0007

dealing with the first graphic representations of people suggesting mobility and action, while in the second we have the first narratives about the movement of individuals to hunt.[2]

The Egyptians, in addition to their gods and rulers, also liked to draw ships and boats, and in several cases they had some spiritual significance for them. For example, the most important ship for the ancient Egyptians was the 'barque of Ra', which crossed the sky every day, carrying the sun and giving light to the world.

Be that as it may, most forms of Egyptian art were created primarily for metaphysical reasons in the tombs of their great leaders. One could say that a large part of the pictorial representations were created for the pleasure of the ruling class to be immortalized, that is to live on in the collective memory after their death.

However, some frescoes found in tombs, such as that of the tomb of In-Snefru-Ishtef in Dahshur depicting two ships with bipedal masts, can sufficiently attest to the extent to which water travel dominated the lives of the ancient Egyptians. It can therefore be assumed that the numerous depictions of ships in ancient Egyptian wall paintings contribute to the assumption that those people regarded water travel as a very important part of their lives.

In Greek and Roman art we also find representations of journeys with various means of transportation such as horses, ships and carts, as well as scenes related to hospitality, especially on vases, but also on many wall paintings. We can say that these works prove the importance of these two practices (i.e. travelling and hospitality) in Greek and Roman society, and that all these representations acted as mechanisms that established them as fundamental parts of people's lives.

In the Middle Ages, images with religious themes dominated, especially frescoes, panel paintings, but also other forms of illustrative representations such as mosaics or stained glass. All these images depicting God, Jesus Christ, the Holy Spirit, the Virgin Mary, the apostles and the saints mainly decorated churches, cathedrals and chapels. In the great castles and palaces, however, among the religious paintings were also those depicting various means of transport such as horses, chariots or ships, but mostly in connection with campaigns, battles or the transport of clergy and nobles.

Be that as it may, it could be argued that most religious paintings depicting the lives of Christ and the apostles in the Holy Land constructed for the European population the core images of social representations for places in the Middle East and in a sense conveyed the idea of a pilgrimage there. Several medieval paintings even depict pilgrims on their journey to the Holy Land.

Some of these works of art can be found in Italian churches, such as the Cathedral of Monreale in Sicily or the Abbey of St Scholastica in the province of Rome, but also in Spanish churches, such as the Hermitage of San Baudelio de Berlanga in the province of Soria or Santa Maria de Mur in Catalonia.

A famous example that inspires the faithful to travel to the East is the painting *The Pilgrims of Emmaus* by Duccio di Buoninsegna, a 14th-century Sienese painter. Although the painting refers to the 'aftermath of Christ's death' and does not directly depict the journey of medieval pilgrims to the Holy Land, it illustrates this idea of a sacred journey. So, we see that much of the virtual representation, even in this period, was still a mechanism that provoked human mobility.

It is also worth mentioning the medieval graffiti, made spontaneously and without authorization by the common people, depicting ships, which in a way is related to our theme. Many of these ship images can be found on the exterior walls, but also inside European coastal churches, especially near medieval ports, and inland.

Ship graffiti was a tradition practised everywhere since the early days of seafaring, but most commonly in the regions that now comprise the nation states of Cyprus, Israel, Egypt, Greece, Bulgaria, Britain, France, Norway, Denmark, Spain and Italy. It is likely that these images reflected the existential problems of seamen struggling with the dangers of seafaring or the fear for their loved ones left behind (Muscat, 2000).

Although this custom of ship graffiti is older, it continued into the fifth and sixth centuries AD and was usually carried out in sacred places such as churches, tombs and catacombs, which had an inherent mystical power. There is no clear indication of whether the authorities categorically rejected this, but most likely it was performed without the consent of the Church (Tsangaris, 2022).

Renaissance art is characterized by a gradual transition from the theological themes of the Middle Ages to the new representational forms of the 15th century, such as linear perspective, chiaroscuro, *sfumato*, *trompe-l'œil*, realistic depictions of the human body, landscapes with depth and portraits.

Many paintings continued to depict biblical scenes and events from the Christian religion, but artists and their patrons began to develop a new sensitivity to themes from classical antiquity. Moreover, painters gradually began to paint portraits of the nobility or the wealthy bourgeoisie and events from contemporary life in the city and countryside.

Thus, despite the numerous paintings with religious or ancient references, the depiction of real existing places seems to develop during this period, which will later strongly support the mechanisms for the construction of the 'image of place'.

## 7.2 Depictions of Real Places in Paintings

In the depictions of Rome as made by Taddeo di Bartolo (1363–1422), Leonardo Bufalini (*c.*1486–1552) and Mario Cartaro (1540–1620), artists sought to convey the appearance of the city in their time alongside that of its glorious past, reconciling what they could see with an imaginary, timeless ideal (Maier, 2012).

Today we can claim that the contemporary tourist image of Rome contains some core features that were also present in the representations of the time, such as themes from ancient Roman history, monuments, myths, the Vatican City and Renaissance architecture.

Nevertheless, the image of Rome as a prominent tourist destination has changed over the centuries. In addition to the aspects already mentioned, it now includes broader themes interwoven with art and culture. These include the delicacies of Italian cuisine, charming Italian restaurants and cafés, papal gatherings, vibrant festivals, the appeal of Italian fashion and shopping, the cinematic world of Cinecittà, the rhythms of Italian music and a host of other cultural dimensions.

Vittore Carpaccio's (*c.*1460–1525) painting *The Miracle of the Relic of the Cross on the Rialto Bridge* (1495–1496) depicting Venice, the Flemish landscapes of Pieter Bruegel (1525–1569) or El Greco's (1541–1614) *View of Toledo* (1599–1600) are some other of the many examples that can convey to us the specific way of representing spatial thinking as it developed during this period. In general, the treatment of space and place in painting of this period changed considerably. Artists increasingly emphasized the use of space and place to create a more realistic and intense experience for the viewer. This development in the understanding and use of space was a central aspect of the artistic and cultural progress of the Renaissance.

When we use the term 'landscape' today, we primarily think of its aesthetic qualities. We long for leisurely walks in nature, discovering beautiful vistas characterized by the aesthetic standards of artists and critics. It can be said that early paintings depicting real natural landscapes, historic cities and seascapes laid the foundations for the social imaginaries associated with wanderlust among those fortunate enough to have access to these vivid images (Terkenli, 2001; Bonarou, 2012).[3]

According to Moraitis (2015), the Greek landscape was a popular model for Western art and architecture. It represented a kind of natural paradise known as 'Arcadia' and played a role in shaping Western culture and politics, especially during the Renaissance, Baroque and Enlightenment. With the advent of Romanticism, however, Western societies became more interested in their own cultures and local landscapes, but the Greek landscape still had an influence. As we shall see, it became a central theme in the pro-Greek movement before and after the Greek Revolution. It influenced the demand for Greek culture and the way Greeks interacted with Europe and the West.

In this context, however, it should be noted that unless the representations in question were displayed in a public place such as a church, they were not accessible to large numbers of people. Their value as a means of forming social representations may therefore not have been so great at the time, but with the development of printing technology (and later photography), the mass reproduction of works of art became an important source of 'ways of seeing' and of the formation of core images of the places depicted a few centuries later.

Be that as it may, ultimately these works of art are means that convey to us the 'ways of seeing' of the people of the time and presumably also the corresponding perception of contexts such as 'the image of the place', or the journey.

The Dutch cityscapes of Johannes Vermeer (1632–1675), the iconic depictions of Venice by Francesco Guardi (1712–1793) and the atmospheric landscapes of various European artists such as Claude Lorrain (1604–1682), Jan van Goyen (1596–1656), Aelbert Cuyp (1620–1691), Jacob van Ruisdael (*c.*1628–1682) and Salvator Rosa (1615–1673) offer us unique insights into the visual habits of their respective eras, whether in the context of urban life or the natural world.

It is truly remarkable that these ordinary urban and natural landscapes had already acquired great importance in the world of art. It is worth pointing out once again that any artistic representation of a place can inspire the viewer to visit it for themselves. Indeed, there are many earlier modern artists who depicted European cityscapes and landscapes, but to discuss them all here would require several pages.

In the mid-18th century, interest in classical antiquity reawakened in Rome's global art scene, establishing the era of neoclassicism. It is believed that this cultural shift was primarily related to the excavation and exploration of the buried cities of Herculaneum and Pompeii, and was also spurred by the scholarly efforts of historians and archaeologists such as the German scholar Johann Joachim Winckelmann (1717–1768).

Neoclassicism left an indelible mark on various creative fields, especially architecture. In painting, it led to less emphasis on colour in favour of precise delineation, with compositions often following the classical principles of balance, symmetry and uniformity. Sometimes the chiaroscuro was completely abandoned in favour of an exclusive concentration on the lines (*Phaidon Encyclopedia of Art and Artists*, 1979).

The paintings of Pompeo Batoni (1708–1787), Gavin Hamilton (1723–1798), John Flaxman (1755–1826), Anton Raphael Mengs (1728–1779), Angelica Kauffmann (1741–1807), Jacques-Louis David (1748–1825) and many other artists who were part of this movement revived the ancient Greco-Roman aesthetic in the collective memory of the Western world.

Since neoclassicism was associated with these two ancient cultures, it motivated many Europeans to travel to these Mediterranean regions. Moreover, we can claim that the perspective and 'staging' of these paintings formed the basis for the aesthetics of the representation of the ancient world in tourism advertising in the following centuries.

As we have seen, there was already a relative attraction to the depiction of nature in the 18th century, but the enormous popularity of wild nature was mainly associated with the Romantic movement that began in the late 18th and early 19th centuries. The main characteristics associated with Romanticism were sensitivity, imagination against logic and order, spontaneity, freedom from rules, admiration for the Middle Ages, for national but above all for primitive values associated with a return to what was considered 'real nature'.

Although Romanticism differed from neoclassicism in its characteristics and ideology, there were sometimes also many similarities in the visual arts. These commonalities often arose from the broader historical and cultural context in which both movements developed, and included elements such as historical references and the use of mythology, human emotion, national identity, landscape depictions, portraiture and so on (*Phaidon Encyclopedia of Art and Artists*, 1979).

Romanticism placed great emphasis on the emotions evoked by unspoilt landscapes, with a focus on mountainous regions. However, the influence of Romanticism went beyond the mere appreciation of mountains; it also contributed to a significant appreciation of coastal areas and seascapes. As history shows, many elements of such representations of mountain and coastal regions were later brought into focus by the burgeoning tourism industry.

We can say that in the paintings of the Romantics Caspar David Friedrich (1774–1840), Francisco Goya (1746–1828), Eugène Delacroix (1798–1863), William Blake (1757–1827), John Constable (1776–1837) and many others, a specific way of seeing was cultivated that was later adopted by the 'tourist gaze'.

In essence, Romanticism laid the foundation for the development of the modern methodology for constructing the tourist image of places. It can be argued that some of the principles of depicting untouched and untamed landscapes as alluring destinations, which are a hallmark of modern tourism representations, originated in this artistic movement.

At this point, mention should also be made of Thomas Cole (1801–1848), an English-born American artist whose aesthetic vision was influenced by Romanticism. Cole vacillated between the need to base his art on a rigorous observation of nature and the desire to exaggerate this naturalism through 'fantasy'. He achieved this through dramatic lighting effects and swirling vapours, but his works often included ancient monuments and small figures. The works *Aqueduct at Rome* (1832), *Sunset, View of the Catskill* (1833), *The Course of Empire: Desolation* (1836), *Shroon Mountain, Adirondacks* (1838) and *The Architect's Dream* (1840) could be considered models that were followed years later by artistic directors of tourism promotion. Some of Cole's paintings such as *A View of the Two Lakes and Mountain House, Catskill Mountains, Morning* (1844) and *A Pic-Nic Party* (1846) depict individuals, travellers and excursionists in the landscape. We believe that Cole's way of depicting the world had a decisive influence on tourism advertising.

As the travel activities of the bourgeoisie increased in the 19th century, there were also a considerable number of artists in their ranks. Above all, many poets, painters, illustrators, sculptors and others set off on journeys to the Orient and North Africa. Their aim was to be inspired by the visual representations of the biblical narratives and other famous literary works they encountered and thus gain a deeper understanding. This cultural exchange contributed significantly to the development of Orientalism, as described by Edward Said in 1978, and became a prominent feature in the compositions of the era (Assunção and Paschoal, 2021).

The concept of Orientalism is primarily associated with the work of the Palestinian-American scholar Edward Said (1935–2003), who argued that cultural or geographical entities such as the 'Orient' are not fixed and natural, but rather should be understood as specific constructions based on a particular history and tradition. It is a paradigm in which Eastern cultures are understood by the West through specific representations, expressions and vocabulary that shape a certain kind of reality and image in Western culture.

In this way, Orientalism leads to a distinction between West and East, emphasizing the supposed rationality, reflexivity and dynamism of the West and the inferiority of the Orient. It can also be said that the idea of Orientalism suggests that racism is not just a matter of individual psychology, but is constituted by cultural patterns of representation that are deeply rooted in the practices, discourses, art and more generally in the media of Western societies (Barker, 2004; Turner, 2006).

To see how Orientalism influences tourism, it is important to understand how representations in paintings, but also in all other images and writings, or rather all references in the media in general about the East, influence the actions of the host country and the tourists.[4]

## 7.3 Recreation Depicted in Paintings Contributed to Creating the Tourist Image

To return to our discussion, it must be emphasized once again that epics and historical scenes in famous places of ancient Greece and the Roman Empire were among the popular motifs of Western painting of those years, which had become ingrained in the collective memory and created a 'habitus' for travelling to these regions.

However, it is interesting to note that there were also many paintings with holiday, travel and leisure themes, often painted by famous artists. These included: (i) Honoré Daumier (1808–1879) with *The Third Class Carriage* (1862–864); (ii) Édouard Manet (1832–1883) with *Lunch on the Grass* (1863), *The Railway* (1873), *On the Beach* (1873) and *The Navigation* (1874); and (iii) Eugène Louis Boudin (1824–1898) with *The Beach at Trouville* (1863), *The Empress Eugénie on the Beach at Trouville* (1863) and *Bathing Time at Deauville* (1865). These artists and many others shaped the modern concept of tourism by aesthetically imprinting these practices on the public consciousness as typical forms of recreation.

Boudin's depictions did indeed go beyond the purely artistic; they became powerful instruments for shaping public consciousness. By depicting scenes of people engaging in leisure activities such as going to the beach and bathing, Boudin created a visual narrative that turned these practices into iconic symbols of relaxation and escape. The charm and vitality exuded by his paintings came to symbolize a new cultural phenomenon: the pursuit of leisure and pleasure through travel (Fig. 7.1).

It can be argued that the Impressionist and post-Impressionist painters had a special relationship with tourism, as many of their motifs are clearly related to tourism and leisure, while many others contribute to the formation of the tourist image for certain places such as Paris.[5]

The Impressionists' preoccupation with outdoor activities, cafés, theatres, concerts, ballet and dance, festivals and celebrations, landscapes, bathing and beach scenes and the indescribable splendour of their works make us think of many commercial images from tourism advertising (Costa, 2010).

Perhaps this is because they were working in an environment where the idea of commercial mass tourism was just beginning to develop strongly. A few artworks from this period that are closely associated with the representation of tourism and recreation are: (i) Claude Monet's *Quai du Louvre* (1867); (ii) Frédéric Bazille's (1841–1870) *Summer Scene* (1869); (iii) Pierre-Auguste Renoir's (1841–1919) *Pont Neuf* (1872), *Dance at Le moulin de la Galette* (1876) and *Lunch of the Boating Party* (1881); and (iv) Georges Seurat's (1859–1891) *Bathers at Asnières* (1884) and *A Sunday Afternoon on the Island of La Grande Jatte* (1884–1886).

The early paintings of the French Impressionist Claude Monet (1840–1926) in particular reflect the changes in everyday life that accompanied the rise of tourism. They construct tourist identities by showing the different social groups, locals and visitors, in the same space, and in his later works the locals were even removed from the scene (O'Dowd, 2013).

**Fig. 7.1.** Eugène Boudin, *Bathing Time at Deauville* (1865) (image used under licence CC0). Boudin's works like this laid the foundation for the visual establishment of stereotypical leisure activities associated with the bourgeoisie. This association became even more pronounced in the 20th century, especially with the advent of mass tourism.

Several of Monet's works contributed significantly to shaping the image of the bourgeois holiday, as they relate directly to tourism and leisure, such as *Lunch on the Grass* (1865–1886), *Garden at Sainte-Adresse* (1867), *La Grenouillère* (1869), *River Scene at Bennecourt* (1868), *The Beach at Trouville* (1870) and *The Hôtel des Roches Noires, Trouville* (1870).

In connection with our theme, we should also mention the work of the French painter Toulouse-Lautrec (1864–1901), who created a series of posters for the Moulin Rouge, a famous Parisian cabaret, depicting the garish Parisian nightlife in a unique and outstanding style that was to shape the way many tourists would later perceive Paris. Today, the Red Windmill still towers over the Pigalle district and, together with Toulouse-Lautrec's paintings, forms a core element of the tourist image of the 'City of Light'.

In the 20th century, as art evolved towards more abstract forms, artists nevertheless created some paintings that influenced the tourist gaze with illusionist elements of visible reality, adapting to the spirit of the times. Such paintings include: (i) André Derain's (1880–1954) *Banks of the Seine* (1905), *The Drying Sails* (1905) and *Big Ben* (1906–1907); (ii) Pablo Picasso's (1881–1973) *On the Beach* (1937); (iii) Fernand Léger's (1881–1955) *Leisure* (1948–1949); (iv) Marc Chagall's (1887–1985) *Paris Through the Window* (1913); and (v) Giorgio de Chirico's (1888–1978) *Piazza d'Italia* (1913).

It is also worth mentioning some important American artists whose paintings often deal with aspects of leisure and tourism, such as: (i) Maurice Prendergast's (1858–1924) *Sunny Day at the Beach* (nd) and *Summer Visitors* (1897); (ii) Robert Henri's (1865–1929) *At Far Rockaway* (1902); (iii) Edward Hopper's (1882–1967) *Sea Watchers* (1952) and *Gloucester Beach* (1924); and (iv) Marsden Hartley's (1877–1943) *On the Beach* (1940).

Edward Hopper's penchant for travel, especially seaside holidays, adds a fascinating dimension to his artistic output. His paintings can be compared to visual postcards that capture the essence of a place and invite others to share in his experiences. A closer look at Hopper's work reveals a recurring theme rooted in his adventures as a traveller. Gas (petrol) stations, automobiles (cars), trains and boats frequently grace the canvas in works such as *Compartment Car* (1938), *Gas* (1940) and *Jo in Wyoming* (1946), to name a few. It is therefore not surprising that the Whitney Museum in New York aptly titled one of its Hopper exhibitions 'Holiday in Reality: Edward Hopper' (June–December 2006), recognizing the deep connection between Hopper's art and his wanderlust (Costa, 2010).

## 7.4 Present-Day Artworks and Tourism

### 7.4.1 Graffiti and street art

For a more comprehensive exploration of the theme of 'painting and tourism', we will look at related motifs, including graffiti and street art, as well as the fascinating field of tourist art.

In addition to the official visual communication systems used by the state, municipalities and businesses to regulate urban life and convey messages, such as street signs, traffic signs, billboards, advertisements and shop signs, there are also some forms of state-banned alternative media, usually surreptitiously placed on walls. These include spray-painted slogans, various messages written in felt-tip pen with existential statements, political positions or obscenities, and of course the classic American graffiti and street art.[6]

At this point it is interesting to note that in many cases subcultural individuals and social groups also used their alternative drawings in connection with their travels. For example, hobos, a marginalized group of migrant workers who lived on the fringes of American society since the late 19th century, spread travel information through their graffiti. In search of temporary jobs, they travelled by train across the continent, leaving their enigmatic signs on remote train carriages and stations. They included pictures, sketches, logos, nicknames, dates and sometimes specific symbols that assessed the character and atmosphere of a place, indicating whether it was marked as hostile or suitable for overnight stays, with friendly or unfriendly inhabitants, easy-to-find food, etc. (Austin, 2001; Lennon, 2016).

In this sense, the hobos, though poor and marginalized, never suppressed their inner human need to travel to gain experience and expand their consciousness. Instead, they travelled throughout America, and even with the numerous incomprehensible signs, they affirmed their presence in a special way for all to see.

Be that as it may, the beginnings of graffiti as we know and understand it today were invented and unfolded by subcultural groups in the late 1960s and 1970s, initially in major cities in the USA such as Philadelphia and New York City. After this graffiti spread from the streets of American cities around the world in the late 1970s, and later with the concept of street art, things took a strange turn.

These two forms of expression were initially seen as detrimental to tourism in a region, as they suggested civil disobedience and were associated with crime – situations that tourists avoid. More or less in the last decades of the millennium, however, both classic American graffiti and street art were increasingly promoted by parts of the art world, the tourism industry and city councils as new art forms worthy of being seen by visitors and therefore considered attractions, and at the same time numerous street galleries emerged.

Tourism agencies of prominent cities such as New York City, Melbourne, Berlin and even Athens have introduced tailor-made tours that offer tourists the opportunity to engage directly with these unique forms of communication.

Consequently, it is clear that a kind of 'recuperation' has taken place, whereby the tourism industry of capitalist society has absorbed these once rebellious forms of expression, removing their more subversive aspects, such as their revolutionary, guerrilla-like character, and integrating them into the mainstream of commercial urban culture and the realm of civic tourism (Debord and Wolman, 1956; Knabb, 1981).

Nevertheless, graffiti and street art, once regarded by the tourism industry as barbarism or trivial forms of communication by subcultures, are now perceived as elements that lend 'authenticity' to the atmosphere of a city and should be understood as art products that promote urban tourism. Here we can refer to Cohen (1988) who recognized the existence of an emerging authenticity as an active and changing force that adapts to the data of postmodernity.[7]

### 7.4.2 Tourist art

The terms ethnic art, commercial visual art, tourist art and souvenirs are regularly confused by most people, and their meanings are interchanged. Graburn (1976), in his work *Ethnic and Tourist Arts*, has attempted to delineate these and other categories through focused essays on indigenous art from a cross-cultural and cross-disciplinary perspective. We will not elaborate on the different terms here, but only mention that Graburn (1976) defines commercial visual art as those objects that adhere to culturally embedded aesthetic and formal standards, even if they are produced with the aim of selling them later.

In this book, we consider tourist art as a lower form of contemporary art, usually produced by local artists at tourist sites, often in too many copies for consumption by visitors to these sites (Jules-Rosette, 1984).

It should be stressed that tourist art is very close to the concept of souvenir (if we do not include it as a sub-category) and it is often difficult to find its boundaries. Thus, tourist art includes various products such as: (i) ink and charcoal drawings of the Eiffel Tower or Notre-Dame de Paris made by local French artists for mass sale to tourists; (ii) woodcarvings in the Alps; (iii) handmade ceramics from Italy, Portugal or Spain; (iv) miniature marble sculptures from Greece; (v) painters portraying tourists at tourist sites; or (vi) statuettes made of wood in Yucatán.

In general, the authenticity of tourist art is considered important by tourists, as they want to be sure that the product they are buying is unique and made by a local artisan according to traditional methods and patterns (however, this is a fallacy, as we all know that these pieces are made according to a pattern in very large quantities). Tourist art can be seen as a typical tourist medium that serves both to build the tourist image of a place and to prove that someone has been to that particular place.

### 7.4.3 Digital art

Many contemporary artists are now working in the digital realm, creating new forms of digital art that are both influenced by and related to the legacy of traditional painting. This fusion of old and new forms of artistic expression offers travellers new perspectives on familiar subjects and encourages exploration and curiosity.

In the digital age, tourists continue to be fascinated by the power of images from paintings and play an active role in documenting their experiences and sharing them with a global audience. Famous paintings and the artistic heritage they represent have seamlessly integrated into this modern landscape, making art an integral part of the way contemporary travellers engage with the world and express their constant desire to explore and discover new horizons (Kennedy, 2016; Fang, 2021).

### 7.4.4 Art tourism

Finally, if we look at the relationship between painting and tourism from a different angle, we should underline the fact that paintings and artworks in general are the main reason for many people to visit a place, and this is exploited by the tourism industry.

The cave paintings at Lascaux and Altamira, even though they are very vulnerable to destruction due to their display to tourists, were a big attraction for tourists who wanted to see the first artistic expressions up close. Many people travel to Florence to see Renaissance works of art in the Galleria degli Uffizi, the Galleria dell'Accademia, Palazzo Pitti or the Museo del Bargello; or to Paris to spend endless hours in the Louvre Museum; or to London to see works of art from antiquity in the British Museum. How many of us do not dream of one day visiting the Museum of Modern Art in New York?

In this context, the tourism industry invested very early in the promotion of painting in relation to tourism, through various strategies and structures such as the promotion of museums, galleries, street art galleries, and private and state cultural promotion agencies. Unfortunately, this fact has of course contributed to the increasing commercialization of the idea of art.

More generally, 'art tourism' has become a special category of cultural tourism, which is now an essential part of the tourism industry. 'Art tourism' has been broadly defined as any tourism activity that results from a person's desire to see works of art, and includes both those who travel to another place specifically to see art and those who plan frequent or occasional visits to see works of art in addition to their actual tourism purposes (Franklin, 2018).

## 7.5 Summary

To summarize, the art of painting has always played a central role in shaping our understanding of travel throughout human history. From the first cave paintings, which served as basic travel narratives for the ancients, to the elaborate illuminated manuscripts and religious art in advanced civilizations, art has been both a storyteller and a conveyor of knowledge, decoding tales of distant lands and exotic encounters. As societies developed, so did the art of painting, which reached a high level during the Renaissance. Artistic skills merged seamlessly with scientific curiosity, resulting in intricate travel illustrations and maps that not only reflected the spirit of discovery but also encouraged further travelling.

The Impressionist movement of the 19th century played a decisive role in defining the tourist appeal of Paris. Through innovative techniques that captured everyday life, light and atmosphere, these artists presented Paris as a captivating and culturally opulent destination. In the 20th century, characterized by mass tourism, the concept of the 'tourist gaze' was introduced, formulated by John Urry, describing how travellers perceive and experience destinations from the perspective of the tourist visitor. This phenomenon is influenced by various factors, including photography, art, general media and popular culture, all of which help to shape and mould the impressions of people exploring new places. The tourist gaze emphasizes the role of external influences in shaping the way tourists see and interpret the destinations they visit. Painting in particular contributed to the creation of idealized images of destinations, shaping the way tourists engaged with and interpreted the places they explored.

In the digital age, the influence of painting has transcended its traditional boundaries and entered the realm of digital art, reinforcing the profound impact of images on the design of the tourist experience. Today's travellers are not just passive consumers of travel imagery, but have become active participants, sharing their encounters with famous works of art while producing their own tourism imagery. This dynamic interplay between consumption and production drives the ever-evolving narratives of travel and discovery that are disseminated in real time across the digital landscape. The enduring power of painting in all its many forms continues to shape our perception of the world, fuelling our insatiable desire to explore and discover.

## Notes

[1] At this point, however, it must be emphasized that scholars' opinions differ considerably about the exact period in which symbolic thought developed.

[2] However, many anthropologists also argue that the Palaeolithic 'artists' believed they were performing a magical act with this particular practice. According to this logic, painting was an act of taking possession of representations and controlling reality. That is, by drawing the animal, the hunters believed they already possessed it. In this case, therefore, there is no particular link between these works and human mobility (Uscatescu, 1986).

[3] The term 'landscape' has gradually evolved over the centuries, initially referring to the physical characteristics of the land and eventually to aesthetic, artistic and philosophical aspects. The origin of the term can be traced back to the Middle Dutch word *landschap*, which referred to a specific geographical region or place. The use of the term developed in response to changes in culture, art, philosophy and the relationship between the individuals and their environment. As for the tourist 'landscape of a country', we can argue that the territory of a country comprises a space that can be visually represented with millions of images from countless angles, selected in frames that depend on the mood, cultural capital, purpose, etc. of the people who frame the space, and of course on the time and culture of the society from which these people come. In this sense, then, certain images of a country's territory (i.e. certain framings of its space that are selected and determined by various factors) are ultimately regarded as typical 'landscapes of the country'. Philosophers, writers and poets of the Romantic period, such as the British Romantic poet William Wordsworth (1770–1850) and the British painter John Constable (1776–1837), emphasized the emotional and spiritual connection between people and the land, so that the term 'landscape' began to encompass not only the physical landscape but also the emotional and psychological responses to nature. Artists who represent a range of styles and periods in the history of landscape painting and whose contributions have had a lasting impact on how we perceive the image of a landscape include: (i) the French painters Claude Lorrain (1600–1682), Charles-François Daubigny (1817–1878), Gustave Courbet (1819–1877), Eugène Boudin (1824–1898) and Claude Monet (1840–1926); (ii) the Danish-French painter Camille Pissarro (1830–1903); (iii) the German Romantic painter Caspar David Friedrich (1774–1840); (iv) the Dutch post-Impressionist painter Vincent van Gogh (1853–1890); (v) British painters such as J.M.W. Turner (1775–1851) and Thomas Chambers (1808–1869); and (vi) American painters such as Asher Brown Durand (1796–1886), Sanford Robinson Gifford (1823–1880), Thomas Cole (1801–1848), John Frederick Kensett (1816–1872), Frederic Edwin Church (1826–1900) and, more generally, the Hudson River School, Albert Bierstadt (1830–1902) and Georgia O'Keeffe (1887–1986).

[4] Numerous studies have looked at paintings and works of art in general and their power to construct and reinforce images of places and local cultures and to reproduce Orientalism. In essence, many of these studies show that paintings were a successful institutional support for tourism because they made distant places visible in a special way to those who would otherwise never have seen them (Nochlin, 1989; Mason, 1998; Bohrer, 2003; O'Dowd, 2013; Ismailinejad, 2015).

[5] Impressionism, a lively artistic movement that flourished in France in the 19th century, was characterized by its distinctive features. The artists used small, disjointed strokes of colour, created a sense of absolute forms and placed great emphasis on capturing the nuances of natural light. The Impressionist painters who emerged around 1860 were a group of young and innovative people from different backgrounds who came together in Paris. They freed themselves from the constraints of traditional art schools and explored the forests of Fontainebleau before setting off for the estuaries of the Seine and the banks of the English Channel. These places, flooded with light and bathed by water, became a source of inspiration for the Impressionists and characterized their revolutionary approach to art (Phaidon Encyclopedia of Art and Artists, 1979).

[6] The term 'alternative media' refers to various forms of media that have historically followed an alternative path that differs in one or more areas (organization, content, etc.) from the practice of traditional mass media (mainstream media). Alternative media are therefore usually small-scale communication systems such as community radio stations, independent and small-scale film productions, fanzines or graffiti, often run on a voluntary basis by individuals or groups of local communities or other minority social groups who share their thoughts and opinions without aiming for commercial gain (Atton, 2001; Downing, 2008).

[7] To revive a mature destination that is in danger of entering a declining phase because the mass of tourists begin to go elsewhere after repeated visits, the local tourism industry must develop various alternative offerings. In this sense, the tourism industry soon discovered the fertile field of unauthorized urban graphics, which is a controversial part of authentic urban daily life, and organized new tourism products to attract tourists again by elaborating the commercial perspective of American graffiti and street art (Gartner, 1996; Tsangaris, 2022).

## References


Assunção, N.M.R.G.D. and Paschoal, N.I.C. (2021) "Orientalism in motion: representations of "belly dance" in paintings and travel literature (19th century)". *Revista Brasileira de Estudos da Presença* 12(1), e113430.

Atton, C. (2001) *Alternative Media*. Sage, London.

Austin, J. (2001) *Taking the Train: How Graffiti Art Became an Urban Crisis in New York City*. Columbia University Press, New York.

Barker, C. (2004) *The Sage Dictionary of Cultural Studies*. Sage, London.

Barnard, A. (2012) *Genesis of Symbolic Thought*. Cambridge University Press, Cambridge, UK.

Bohrer, F.N. (2003) *Orientalism and Visual Culture: Imagining Mesopotamia in Nineteenth-Century Europe*. Cambridge University Press, Cambridge, UK.

Bonarou, C. (2012) *Visual Culture and Tourism. Representations of Greece on Tourist Postcards*. Papazisi Publications, Athens. (in Greek).

Cohen, E. (1988) Authenticity and commoditization in tourism. *Annals of Tourism Research* 15(3), 371–386. DOI: 10.1016/0160-7383(88)90028-X.

Costa, T. (2010) Edward Hopper: glancing at gaze with a wink at tourism. In: Burns, P., Palmer, C. and Lester, J.A. (eds) *Tourism and Visual Culture*, Vol. 1. *Theories and Concepts*, CAB International, Wallingford, UK, pp. 43–51. DOI: 10.1079/9781845936099.0000.

Debord, G. and Wolman, G.J. (1956) A user's guide to detournement. *Les Lèvres Nues 8. Situationist International online*. Available at: https://www.cddc.vt.edu/sionline/presitu/usersguide.html (accessed 10 March 2021).

Downing, J. (2008) Social movement theories and alternative media. *Communication, Culture and Critique* 1(1), 40–50. DOI: 10.1111/j.1753-9137.2007.00005.x.

Fang, L. (2021) The application of digital media art in art museum public education based on human-computer interaction technology. In: *2nd International Conference on Information Science and Education (ICISE-IE), 26–28 November 2021*, Chongqing, China. DOI: 10.1109/ICISE-IE53922.2021.00317.

Franklin, A. (2018) Art tourism: a new field for tourist studies. *Tourist Studies* 18(4), 399–416. DOI: 10.1177/1468797618815025.

Gartner, W.C. (1996) *Tourism Development: Principles, Processes, and Policies*. Wiley, New York.

Graburn, N.H.H. (ed.) (1976) *Ethnic and Tourist Arts: Cultural Expressions from the Fourth World*. University of California Press, Berkeley, California.

Ismailinejad, Z.S. (2015) Orientalist paintings and said orientalism. *International Letters of Social and Humanistic Sciences* 50, 68–81. DOI: 10.18052/www.scipress.com/ILSHS.50.68.

Jones, A.F. (1992) *Introduction to Art*. HarperCollins, New York.

Jules-Rosette, B. (1984) *The Messages of Tourist Art: An African Semiotic System in Comparative Perspective*. Plenum Press, New York.

Kennedy, B. (2016) Zones of seeing: artistic, touristic and digital images of the DMZ. *East Asian Journal of Popular Culture* 2(2), 287–303. DOI: 10.1386/eapc.2.2.287_1.

Knabb, K. (1981) *Situationist International Anthology*. Bureau of Public Secrets, Berkeley, California.
Lennon, J.F. (2016) Trains, railroad workers and illegal riders. In: Ross, J.I. (ed.) *Routledge Handbook of Graffiti and Street Art*. Routledge, New York, pp. 27–35.
Maier, J. (2012) A true likeness: the Renaissance city portrait. *Renaissance Quarterly* 65(3), 711–752.
Mason, P. (1998) *Infelicities: Representations of the Exotic*. John Hopkins University Press, Baltimore, Maryland.
Moraitis, K. (2015) *The Art of Landscape: a Cultural Survey of Modern Landscape Considerations and Formations*. Kallipos Open Academic Editions, Athens. (in Greek).
Muscat, J. (2000) The Tarxien ship graffiti revisited. *Melita Historica* 13(1), 49–57.
Nochlin, L. (1989) *The Politics of Vision: Essays on Nineteenth Century Art and Society*. Harper and Row, New York.
O'Dowd, S. (2013) *Imagining Destinations: Art Posters and the Promotion of Tourism*. Arizona State University, Tempe, Arizona.
*Phaidon Encyclopedia of Art and Artists* (1979). Phaidon Press Limited, Oxford, UK.
Roebroeks, W. (2014) Archaeology: art on the move. *Nature* 514(7521), 170–171. DOI: 10.1038/514170a.
Terkenli, T.S. (2001) Towards a theory of the landscape: the Aegean landscape as a cultural image. *Landscape and Urban Planning* 57(3–4), 197–208. DOI: 10.1016/S0169-2046(01)00204-3.
Tsangaris, M. (2022) *Radical Communications: Rebellious Expressions on Urban Walls*. Lexington Books, Lanham, Maryland.
Tse, P.U. (2008) Symbolic thought and the evolution of human morality. *Moral Psychology* 1, 269–297.
Turner, B.T. (2006) Orientalism. In: Turner, B. (ed.) *The Cambridge Dictionary of Sociology*. Cambridge University Press, Cambridge, UK, p. 427. DOI: 10.1017/9781316135334.
Uscatescu, G. (1986) *Brief Theory and History of Culture*. Ekdosis Grigori, Athens. (in Greek).

# 8 Literature and Tourism

**Abstract**

Throughout history, the written word and literature have had a lasting influence on our perception of distant places and the development of travelling. This chapter examines the contributions of writing from ancient civilizations to the present day and explores how literary figures not only described distant lands and exotic cultures, but also ignited readers' innate desire to travel. Throughout history, journeys became fascinating narratives documented by those who wished to share their experiences. The 20th century was a pivotal moment when educational reforms and improved access to schooling enabled a large proportion of the European population to read and write. At the same time, advances in printing technology increased the reach and influence of written works. This change, combined with advances in transport and communication technology and social changes brought about by industrialization and capitalism, accelerated the historical development of modern tourism.

While it is widely recognized that spoken language predates written language, the link between writing and the transmission of knowledge was not always as obvious as it is today. For a long time, the preservation of a group's identity relied primarily on human memory, followed by architectural remains and art. The relatively recent emergence of writing, when considered in the context of the entire existence of the human species on Earth, has led many scholars to suggest that the written word is a cultural triumph rather than a direct result of human biological evolution. Nonetheless, the spatial and temporal limitations that accompanied the spread of the spoken word necessitated the emergence of writing.

Writing was able to visually replace the spoken message with a material on which it was recorded, thus giving language the permanence of this material. Through writing, the word was freed from the limited spatial and temporal dimension of its utterance and the message became a visible and observable physical object. However, because written text is detached from the direct relationship between speaker and receiver during communication, it allows people to give different meanings to the content of messages, which is easier to avoid when two or more people are in direct face-to-face interaction. Consequently, texts can be interpreted differently by different readers and social groups, depending on their state of mind, their psychological and social conditions, the particular time in which they are read and, above all, their cultural capital.

In reality, it is very difficult to define a specific time and place as the cradle of pure travel literature. Undoubtedly, texts demonstrating the close relationship between travel narratives and the written word, as well as the virtual representation of space, can be traced back to very early times, with finds from antiquity in different parts of the world.

Besides texts engraved on marble or other stones, ancient people used many other materials to record their thoughts, such as clay, shells, wood, metal plates, skin, various leaves or barks of plants and so on. For longer texts, however, papyrus and parchment were used. These materials were crucial for the preservation of knowledge and remained an indispensable means of writing until the introduction and spread of paper, which appeared in China towards the beginning of the second century CE and gradually spread to other parts of the world (Kakridis, 2006).

With their various collections of books, archives, libraries and monumental inscriptions, people shaped the collective memory of societies, but especially the memory of rulers


DOI: 10.1079/9781800626034.0008

who could enforce their version in chronicles describing their exploits, which brings us to the limit where memory becomes 'history' (Le Goff, 1992).

## 8.1 Origins of Travel Literature: Homer's *Odyssey*

It is often claimed that the origins of travel literature can be traced back to the poem *Odyssey*, attributed to Homer, which describes a mythical journey to imaginary and (almost) real places in the Mediterranean after the Trojan War, estimated to have taken place around the 12th or 13th century. This epic poem by Homer, which describes the travelling adventures of Odysseus, may have somehow motivated the people of the time to follow his example and turn to the vast sea that can bring riches.[1]

Be that as it may, the epic not only describes the adventures and voyages of Odysseus, but also gives us the nautical knowledge of its time. At this point, however, it should be noted that the situation is somewhat complicated, since the systematic collection and recording of Homer's poems did not take place until later in the sixth century BC, when the city of Athens was ruled by the tyrant Peisistratos (600–527 BC). Homer's poems can thus give us mixed information about the social representation of the Greek world in relation to three periods: (i) during the mythological years; (ii) during Homer's time; and (iii) during the time in which the poems were recorded.

Although the *Odyssey* speaks of travel, it refers to the journeys of a king and his army returning from war, and is therefore characterized by the element of coercion and expediency, which is far removed from the idea of tourism in the sense we understand it in this book. Of course, military movements have nothing to do with tourism, but the poem refers to complex travel adventures and certainly resists the idea of the soldier returning home from war.

It is important to point out that the *Odyssey* seems to be based on a cultural background in which the idea that men must travel, and women must stay at home is taken for granted. However, it is absurd to assume that there was no female mobility at all at that time, as if women had roots that tied them to a point in the earth and therefore there were no female travel adventures to be told in poetry. Societal values and beliefs such as that the adventure of travel is exclusive to men while women must stay at home have been embedded in literature since ancient times, starting with this work. It was only natural that a man, Odysseus, lost on his way back to his kingdom from a war, and his son, Telemachus, who is looking for him, should both travel, while the woman, Penelope, who occupies a passive and immobile position, must wait for them at home.

Penelope's destiny does not seem to be to wander and explore the world and have all kinds of experiences, but she is presented as a destination and a refuge. In this respect, it can be said that this narrative became a kind of paradigm that determined gender behaviour in literature, even if it varied greatly depending on the historical period, context and individual circumstances.

Be that as it may, this scenario had a long life and shows how long the profound social prejudices against female travel lasted. Yet despite these limitations, which have been reflected in literature since antiquity, women have always travelled further than the 'male myth' just outlined. Throughout the centuries, women accompanied their husbands, fathers and brothers on journeys, but they also travelled alone, although their presence in these ventures is rarely mentioned by contemporary and later commentators (Casson, 1994; Bowers, 2000; Thompson, 2011).

With regard to the themes discussed in this chapter, we can simply say that Homer's two epic poems, the *Iliad* and the *Odyssey*, became constant points of reference in later times and influenced many writers worldwide. The ancient Roman poet of the Augustan period Virgil used these works very early on as the basis for his own epics, the British poet John Milton (1608–1674) was so influenced during the English Civil War that he wrote his own epic poem *Paradise Lost* (1667), the Irish novelist and poet James Joyce (1882–1941) wrote the novel *Ulysses* (Joyce, 1922) under the influence of Homer, as did many other writers. It is also interesting that Homer's work has been transformed into a female discourse by writers such as the Canadian poet and novelist Margaret Atwood with *The Penelopiad* (2005), the British writer Pat Barker with *Silence of the Girls* (2018) and the Italian writer Marilù Oliva with *The Odyssey: Told by Penelope, Circe, Calypso and the Others* (2020).

Finally, dozens of films and television productions have been made based on the themes of Homer's epics. Both readers and cinema and television viewers who follow the narratives of these works undoubtedly refresh their conception of Greece, which benefits Greece's cultural reputation, which in turn has a very positive effect on the country's tourist image, with the *Odyssey* in particular being able to arouse the desire to travel (Beaton, 2021).

## 8.2 Depiction of Distant Places by the Ancient Greeks

It is believed that as early as the sixth century BC, there were several geographical sketches by the ancient Greeks in which their creators depicted distant places so that anyone who had access to them could consult them if they needed information for their future travels.[2]

Scholars believe that the Milesian natural philosophers were among the first to attempt to map the Earth according to recognizable scientific principles. These innovations are said to date back to Anaximander (*c.*610–546 BC), who was a student of Thales of Miletus (*c.*626–545 BC), one of the Seven Sages of ancient Greece (Aujac *et al.*, 1987).

Certainly, people living in the larger area where Greek civilization developed learned much about the then-known world from Homer's poems, especially about part of the Mediterranean and its coasts, as well as some other inland areas, but the wealth of geographical and ethnographic information recorded a few centuries later by Herodotus (484–425 BC) enriched knowledge of faraway places even more.

Herodotus travelled for many years through Egypt, Scythia, mainland Greece, Asia Minor and the Levantine coast, writing down his experiences and observations to provide the people of his time and later generations with detailed accounts of important historical events and geographical details. Although Herodotus generally followed some basic ethnographic and anthropological principles, he described other peoples and their culture through a Greek prism, looking at others through the eyes of his countrymen and often applying analogies or inversions in relation to local customs. In other words, he defined diet, marriage, travel and all other customs in terms of a Greek norm. Be that as it may, he also tried to show that customs could be transferred from one culture to another, and to convince the Greek reader, who thought Delphi was the navel of the world, that the cultures and customs of strangers also had value (Coulet, 1996).

According to Casson (1994), Herodotus is not only the world's first historian, but also the first travel writer. Of course, after Herodotus, but also before, there were other scholars such as Xenophon (430–355 BC), Thucydides (460–399 BC) or Strabo (63 BC–24 AD) and many others whose works stimulated the imagination of people in the Greek world to form a picture of the different places they described.

A typical example of the influence of the media of the time on the tastes of curious and adventurous travellers was the recording of the Seven Wonders of the World. Following the popular tradition from Hellenistic times, various authors and poets such as Diodorus Siculus (first century BC), Callimachus of Cyrene (*c.*310–240 BC) and Antipater of Sidon (second century BC) recorded seven monuments that stand out in size and splendour among all human constructions on the edge of the Mediterranean and in the Middle East (Sacks, 2014).

In this case, it is clear that the literary world of the time set the agenda of preferred destinations. It is inconceivable that anyone living at the time would have learned of the Seven Wonders of the World and not have been filled with a great desire to visit them.

In the second century AD, Pausanias (*c.*110–180) wrote a very authentic travel guide entitled *Description of Greece*, in which he reproduced in ten books exactly what he had seen on his numerous journeys, as well as information he had gathered from secondary sources about the Peloponnese and some other regions of Greece (Themelis, 2016).

However, the idea of writing a travel guide did not come from Pausanias. Other writers before him, such as the Stoic Polemon of Ilium (second century BC) and Heliodorus of Athens (second century BC), had already written some kind of travel guide (Casson, 1994).

Pausanias travelled through Syria, Palestine, the Dead Sea, Egypt, Libya, Rome, all of Greece and the Aegean Islands. His aim was to trace and describe all the memorable sites and monuments of the Greek world, mentioning the various mythological, historical, religious or folk traditions and stories associated with them. According to Casson (1994), Pausanias is the milestone in the history of tourism.

## 8.3 The Romans' Travel Literature

For the sake of brevity, we shall refrain from going into the extensive literary work of the Romans, which also contributed greatly to conveying to the public images of various distant places and the practice of travelling. We will limit ourselves to mentioning the names of the great Roman historians such as Sallust (*c.*86–35 BC), Julius Caesar (100–44 BC), Livius (59 BC–17 AD), Tacitus (*c.*56–120 AD) or Suetonius (69–122 AD), whose descriptions formed the basis for the construction of social representations of many near and far places during the Roman era. Also worth mentioning is Ptolemy (*c.*100–170 AD), who belongs to the field of historical and regional geography and contributed significantly to the conception of topographical representations. But also, many other scholars, such as Pomponius Mela (first century CE) and Pliny the Elder (*c.*23–79 AD), contributed with their works in one way or another to the formation of images of many places in the minds of people in the Roman Empire.

## 8.4 Accounts of Travel From the Fourth Century Onwards

After a brief look at the history of travel literature, it becomes clear that until recently there was a clear imbalance between the number of travel books written by women and those written by men. Several factors contribute to this imbalance. One factor was that a higher percentage of men were literate, and their writings were more actively preserved, resulting in a lack of historical records of travel writing by women. In addition, societal stereotypes prevented women from expressing themselves freely, and prevailing social norms often discouraged women from publishing their letters and chronicles due to insecurity or external pressures. These multiple challenges help to shed light on the historical underrepresentation of women in the field of travel writing.

Be that as it may, one of the earliest texts describing the travels and experiences of pilgrims in the Holy Land is the *Itinerarium Egeriae*, written in the late fourth century by a Western European pilgrim, Egeria, who was a nun and probably from the Iberian Peninsula (McGowan and Bradshaw, 2018). Her work was essentially a letter to the women of her spiritual community, a wonderful narrative text characterized by a simple style of expression with the most elementary features.

A notable medieval work that has characteristics of a travel literature is the *Confessions* of Augustine of Hippo (354–430) (St Augustine, 2017). Augustine, a theologian and philosopher of Berber origin, played a significant role in early Christianity and excelled as a theologian and philosopher. In 386, he embarked on a journey from his native North Africa to Italy to visit the major Christian centres of his day and engage in theological discussions with other scholars and religious leaders. Although his writings capture the essence of travel, they differ from contemporary travelogues. Instead, they serve as accounts of St Augustine's spiritual journey and the reflections he made on his travels (Brown, 2013).

Another well-known early account of a journey to the Holy Land is that of the Irish monk Adomnán (*c.*624–704), *De locis sanctis*, written in the late seventh century and considered by many scholars to be one of the earliest and most important works of travel literature of the Middle Ages (Campbell , 1988).

According to Legassie (2017), travel writing developed in many forms in remarkable ways in the years between 1200 and 1500. During these centuries, in addition to all the activities associated with travel, such as trade, emigration, diplomacy, pilgrimage, festivals and fairs, the desire to travel with the aim of writing a literary work grew, as it did in antiquity with Herodotus and Pausanias.

Thus, one of the reasons why some people travelled became to describe distant places and the cultures they encountered, and more generally to write down their experiences. The claim that travel writing increased to a particular extent is based on the fact that numerous literary works, travel diaries and memoirs by travellers, explorers, seafarers, diplomats, preachers and traders who wrote about their journeys have survived from this period. Many of these works were written by individuals who took part in group pilgrimages to the Holy Land or travelled to Europe, Africa and Asia on their own for other reasons.

These writings played a decisive role in shaping the perception of distant lands and formed the core images for social representations of travelling in this era. For those who did not have the opportunity to travel to these faraway places, the narratives became a valuable source of insight into the cities, natural landscapes, geographical features and cultures of these distant regions. These accounts also provided insights into the means of transport, the excitement of travelling adventures and the thirst for knowledge associated with discovery. In particular, the memoirs of famous navigators who had travelled far and wide, such as Marco Polo (1254–1324), whose experiences were recorded by Rusticello da Pisa in the 13th century, significantly influenced the emergence of travel trends among European explorers.

It is worth noting that another late medieval Christian mystic from Britain, Margery Kempe (*c.*1373–1438), left writings describing her pilgrimage to various religious sites in Europe and the Holy Land.

As Bowers (2000) points out, Kempe was a unique traveller who practised travel like a performance art, as a means of expression to convey a particular social meaning. Kempe's travels were a violation of what was 'expected' of women. Her actions went against the social view of the time that a woman should remain in the safety of her home as a housewife and maid. For her, travelling was more than just getting from one place to another. It was a means of expression, like writing, painting and sculpting, through which she could explore but also challenge and change the social norms, values and beliefs of her time.

In addition to the pilgrims' travel diaries and travelogues, the growing number of Bible manuscripts produced also influenced the flow of believers travelling to the Holy Land for prayer. The Gutenburg Bible was the first book to be mass-produced with movable type in the West in the mid-15th century, and from then on, letterpress printing became much easier, leading to a much wider distribution of this medium.[3]

The 'Grand Tour', which more or less emerged in the late Renaissance, was directly related to literature. As already mentioned (Chapter 2, this volume), it was a traditional journey of young aristocrats to continental Europe and the Mediterranean coasts to complete and refine their education by acquiring cultural knowledge. The ancient and medieval works of Greco-Roman literature, which were part of the education of British aristocrats at the time, but also generally studied throughout Europe, played an important role in motivating the travellers who took part in the 'Grand Tour' to reach Rome, Florence, Venice, Naples, Milan, Verona, Turin, Genoa, Pisa, etc., but also Athens and other places in Greece.

In addition, several books by people who already had experience of travelling to Europe, such as the travel book *Coryat's Crudities* (1611) by Thomas Coryat (*c.*1577–1617) and *The Voyage of Italy* (1670) by Richard Lassels (*c.*1603–1668), were important sources for the emergence and spread of the idea of the 'Grand Tour' (Chaney, 1985; Parr, 1992).

Many other scholars also wrote about the 'Grand Tour', either because they participated in the event themselves or because characters in their novels experienced the adventure. These, among many others, included: (i) Sir Francis Bacon (1561–1626) with *Essays: Of Travel* (1625); (ii) Tobias Smollett (1721–1771) with *The Adventures of Peregrine Pickle* (1751) and *Travels through France and Italy* (1766); (iii) Laurence Sterne (1713–1768) with *A Sentimental Journey Through France and Italy* (1768); (iv) Olaudah Equiano (*c.*1745–1797) with *The Interesting Narrative of the Life of Olaudah Equiano, or Gustavus Vassa, the African* (1789); (v) Johann Wolfgang von Goethe (1749–1832) with *Italian Journey: 1786–1788* (1816); and (vi) Mary Wollstonecraft Shelley (1797–1851) with *History of a Six Weeks' Tour* (1817) (Blackmore, 2020).

Interestingly, many contemporary scholars have examined the role of medieval and Renaissance travel writers in shaping the intellectual foundations of modern European imperialism, drawing on Said's (1978) concept of *Orientalism*. In doing so, they have examined how these writers created exotic representations of faraway places, described local communities and provided ethnographic accounts, all of which contributed to the wider discourse (Akbari, 2009; Phillips, 2014; Khanmohamadi, 2014).

In general, it can be said that the spirit of travel literature can be found in many famous writers from a long period, including the Renaissance and the Enlightenment. This is the case especially in the works of the following: (i) the Spanish writer Miguel de Cervantes (1547–1616); (ii) the French countess and writer Madame de La Fayette (1634–1693); (iii) the French writers and playwrights Alain-René Lesage (1668–1747) and Pierre de Marivaux (1688–1763); and (iv) the British writers Samuel Richardson (1689–1761) and Henry Fielding (1707–1754).

### 8.4.1 The 'silver age of travel literature' and the 'golden age'

As travellers' inclination to write down and publish their impressions increased over the years with the development of printing technology, many scholars refer to the period between the 17th and 18th centuries as the 'silver age of travel literature'. During this period, more and more merchants, explorers, diplomats, soldiers, scientists, missionaries or simply wanderers travelled the world and wrote accounts, letters, memoirs, poems and articles about faraway places, sometimes due to the demands of their work, but sometimes for personal satisfaction.

In addition, many early travel accounts were written by individuals who became known primarily for their travels, such as the British explorer William Dampier (1651–1715), the French Baron de Lahontan (1666–1716), the British writer Daniel Defoe (1660–1731), the French archbishop and writer François Fénelon (1651–1715), the Irish poet and essayist Jonathan Swift (1667–1745) and hundreds of others (Adams, 1978).

However, it could be argued that the 18th century marked the peak of an era in which intrepid and adventurous travellers flourished, often referred to as the 'golden age'. This period ended just before the advent of steam engine technology, which enabled steamships to outperform sailing vessels, and before the emergence and growth of the modern tourism industry.

Travel books fascinated the European public, and the publication of each book was most likely a commercial success. Many of these books were not only loved by the general public as entertainment that made them dream of adventures in faraway places, but they were also valuable sources for writers, historians, sociologists and poets. They provided material for literary works, plays, essays, scientific studies, interpretations of historical events, philosophical ideas, political reflections and cultural references. Travel books were also of particular interest to governments, politicians and diplomats, as they sometimes dealt with political and social issues of the countries visited.

Most of the travellers who wrote in this period belonged to the upper classes and were highly educated. The heyday of this type of travel literature can also be seen in the expensive and monumental illustrated editions that appeared. Often, the publication of a work was the result of long journeys and the collaboration of the author with painters, draughtsmen, archaeologists and printing technicians. These were huge books, unimaginable in today's graphic arts industry, for they impress with their magnificence, their dimensions, the quality of the images, the typographical care and the richness of the material (Simopoulos, 2003).

Against this background, it is not surprising that the later development of the travel novel in Europe contributed to the association of travel as a leisure activity with an educated elite, especially from the bourgeoisie. This privileged social group became the role model for the general population, stimulating the world's imagination about travel and encouraging the desire for tourism (Newby, 1981).

Although travelling was more accessible to the privileged classes, it would of course be wrong to assume that people from the lower classes did not travel. As we noted at the beginning of the book (Introduction, this volume), people have a need to move around in space. So in the less affluent and the lower classes there was certainly some mobility, perhaps over shorter distances, even if there is not so much historical data, in the sense that people did not leave behind an extensive body of work in which they recorded their experiences, how they perceived mobility, how they travelled, how they lived while travelling, what impressions they gained of foreign places and so on.

## 8.5 Nineteenth-Century Literature Fuels People's Desire to Travel

Subsequently, many famous 19th-century literary figures, such as the French novelist Jules Verne (1828–1905) and the Scotsman Robert Louis Stevenson (1850–1894), used travel as a key element in parts of their works. In those years, the growing interest in tourism also led some publishers to specialize in travel literature and guidebooks, such as the publishing houses of Karl Baedeker (1801–1859) and John Murray III (1808–1892) or the American publisher and author George Palmer Putnam (1814–1872), who published *The Tourist in Europe* (1838), among other books.

Particularly noteworthy, however, are the European writers and poets of the Romantic period, who, like their painter colleagues, passionately praised the beauty of nature, idyllic landscapes and Greco-Roman antiquities in their works. Journals such as *Das Athenaeum* by the Schlegel brothers or *La Muse Francaise* by Émile Deschamps (1791–1871) helped to spread these ideals (Bled, 2003).

In this way, the creators of Romanticism directed the thoughts and ideas of a large part of the European population, especially the cultural elite, towards these specific Mediterranean regions, thus laying the foundation for the prevailing travel stereotypes of their time.

At this point it would be remiss not to mention the names of some of the great female travellers of modern times, whose numbers increased according to historical data. These include: (i) the British writer Mariana Starke (1762–1838); (ii) the French writer Germaine de Staël (1766–1817); (iii) the British novelist Frances Trollope (1779–1863); (iv) the British banker and travel writer Charlotte Eaton (1788–1859); (v) the Anglo-Irish art historian and writer Anna Jameson (1794–1860); (vi) the French mountaineer Henriette d'Angeville (1794–1871); (vii) the American writer Harriet Beecher Stowe (1811–1896); (viii) the French explorer and writer Adèle Hommaire de Hell (1819–1883); (ix) the Belgian explorer and writer Carla Serena (1820–1884); (x) the British biologist and botanist Marianne North (1830–1890); (xi) the British writer, photographer and naturalist Isabella Lucy Bird (1831–1904); (xii) the American writer Mary French Sheldon (1847–1936); (xiii) the American mountaineer Annie Smith Peck (1850–1935); (xiv) the British ethnographer Mary Henrietta Kingsley (1862–1900); (xv) the American journalist Nellie Bly (1864–1922); and (xvi) the British archaeologist Gertrude Bell (1868–1926). These women, and many others, have gone down in history for travelling independently and to places that, at the time, were considered off the beaten track, and writing about their adventures (Monicat, 1994; Walchester, 2007; Thompson, 2011).

It is noteworthy that many of the writers and poets of the post-Napoleonic period who travelled and wrote about their journeys considered themselves 'travellers'. They arrogantly distinguished themselves in their works from the 'mere tourists', the gradually growing mass groups composed of ordinary people whom they often described disparagingly, even though they may have used the same means of transport and sometimes even similar accommodation and visited the same places as themselves (Buzard, 1993).

Be that as it may, we can only be sure that the quantity, expressiveness and literariness of travel texts, as well as the public interest in the books whose authors described travels in Europe and the world in general, defined the typical itineraries of the time. In this way, 'mere tourists' or readers who had not travelled to the places visited by all these writers and poets accepted their travel fantasies, visual habits, images of places, travel methods and patterns as they read them in books.

In view of these points, it should be made clear that literature in general had a decisive influence on the social construction of images of distant places for the masses until the beginning of the 20th century, and mostly in a mediating way (i.e. in a certain way according to the model of the 'two-phase flow').

Those who did not travel learnt about faraway places from people who had read travel literature and were able to form an opinion. In other words, the majority of the population formed the basic mental images of places beyond their direct experience from the oral accounts of a few people, usually from the educated and wealthy bourgeoisie or aristocracy, who all read these books or had the time and money to travel. These people acted as image bearers or opinion leaders because they had several advantages. First, they could travel easily because of their financial situation, and secondly, they had grown up with art and literature and had the time to read and write books, read newspapers, magazines, novels, etc., so they could tell stories and describe distant lands.[4]

## 8.6 Twentieth-Century Travel Literature and the Development of Modern Tourism

The spread of literacy among the population of the Western world in the 20th century, combined with further technological developments in the field of printing that enabled the increased and, at the same time, more economical production of books and other printed media (magazines, newspapers, etc.), provided more and more people with immediate access to travel information.

In addition, the emergence and spread of some media with dynamic content in this century, such as cinema, radio and television, which we will discuss below, gave the whole world the opportunity to have personal experiences and to come into direct contact with the 'images of places' constructed by these media. From all this, it is clear that the enormous change in the landscape of mass communication that has taken place over the last 100 years, as many more people have now entered the circle of media communication, has been crucial to the emergence of social representations of places near and far, and to the understanding of the concept of travel as a leisure activity and cultural enrichment by a wider population.

Many successful books from the first decades of the 20th century, such as the stories of the widely read American author Francis Scott Fitzgerald (1896–1940), had already begun to form a collective image of the Mediterranean landscape in the minds of Westerners. One could say that his novel *Tender is the Night* (1934), set on the Côte d'Azur, and other stories of this kind, helped to form the 'core images' for tourist representations of this region and set the framework for the later development of this place into a 'tourist myth', classifying it as one of the most attractive destinations for summer holidays.

Some of the novels by the American writer Ernest Hemingway (1899–1961) are also typical examples of the same theme associated with the construction of the tourist image of the Mediterranean. Books like Hemingway's *Death in the Afternoon* (1932) and *The Sun Also Rises* (1926) stimulated the public's imagination in a realistic way by embedding the customs and traditions of Spain in the plot, thus anticipating ideas, thoughts, judgements, expectations and knowledge about this country in general.

It is worth noting that the famous American-British writer Henry James (1843–1916) also travelled extensively throughout Europe and published numerous articles and travel books describing the geography, architecture and culture of the places he visited. Some of his books related to our discussion are *Transatlantic Sketches* (1875), *Portraits of Places* (1883) and *Italian Hours* (1909). These books are collections of essays in which he describes his extraordinarily beautiful stays in Venice, Rome and Florence, as well as trips to Siena, Assisi, Perugia, Capri, Ravenna and other Italian cities that shaped the picturesque image of Southern Europe.

Other critically acclaimed travelogues of the first half of the 20th century include T.E. Lawrence's (1888–1935) *Seven Pillars of Wisdom* (1922), D.H. Lawrence's (1885–1930) *Sea and Sardinia* (1921) and *Mornings in Mexico* (1927), as well as André Gide's (1869–1951) *Voyage au Congo* (1927). Later publications in the 1930s include Robert Byron's (1905–1941) *The Road to Oxiana* (1937), W.H. Auden's (1907–1973) and Louis MacNeice's (1907–1963) Letters from *Iceland* (1937) and W.H. Auden's and Christopher Isherwood's (1904–1986) Journey to *a War* (1939) (Thompson, 2011).

Of particular note is Rebecca West's (1892–1983) *Black Lamb and Grey Falcon: a Journey Through Yugoslavia* (1941), a famous example of travel literature enriched with Balkan history and ethnography. In her work, West describes life during a trip to Yugoslavia, which may have contributed to the country becoming a popular tourist destination a few decades later.

Jack Kerouac (1922–1969), an American writer and poet of the Beat Generation, wrote his autobiographical novel *On the Road* (1991) in 1951. He skilfully contrasted the lively authenticity of life on the open road with the banal, regimented existence in the modern city. He portrayed travel as a liberating escape from the confines of city life and the daily grind, ironizing the glamour of the American dream by suggesting that true fulfilment can lie in unconventional paths, personal experiences and the rejection of the conventional markers of success. As a result, his book came to epitomize the beatnik generation.

The main character and narrator of the novel traverse the American continent on a series of journeys, leading an itinerant life of hitchhiking, driving cars and jumping on trains, thus embodying the lifestyle of a vagabond. Their adventures resemble the lives of hobos and are essentially a constant search for self-discovery, with no fixed destination on the boundless American highways. For Kerouac, life itself is a perpetual road, a search for meaning resulting from human's inherent desire to move.

Basically, Kerouac stands for a more authentic form of tourism that recalls the almost forgotten essence of travel that we want to portray in this book – a turning away from the trappings of the tourism industry and the subordination of human life to economic interests.

As mentioned earlier, we should note here that although many female travellers have produced a considerable amount of travel literature over the centuries, only a few texts have been recognized and remembered in the formal history of travel literature compared with those written by men. Many others never found their way into print, while some were printed and then forgotten.

Since the 1970s, feminist scholars have sought to recover and re-evaluate this material produced by women, and thus a remarkable field of studies on women's travel writing has developed in recent decades. Some have raised the question of whether men and women travel differently, see things and sights differently, and therefore inherently produce different kinds of travel writers. Others have argued that one cannot speak of such a great difference and that the writing is often similar, while there are many other parameters to consider besides gender.

In this context, some questions arise that are very difficult to answer and depend on the perspective with which we encounter them. For example, are there really gender-specific styles of travel writing (i.e. more adventurous, practical and scientific descriptions, traditionally labelled masculine, or more picturesque, aesthetic and sentimental ones, labelled feminine), or does this just depend on the personality of the author? If so, was it really endogenous characteristics of male and female writers that led to different writing styles, or was it the influences of the social environment and the stereotypes of publishers and critics? Does a travel writer's gender play a more important role in our reception and understanding of the text than their personal style and content when we make assumptions about gendered styles, etc. (Robinson, 1990; Mills, 1991; Foster and Mills, 2002; Morris, 2007; Thompson, 2011)?

Be that as it may, regardless of gender, the last half of the 20th century has produced numerous commercially successful, entertaining, informative and sometimes highly innovative books on travel and walking, where the boundaries between travel journalism and travel literature are not so clear.

Some of these notable travel books that have appeared are: (i) Arthur Frommer's *Europe on 5 Dollars a Day* (1957); (ii) Dervla Murphy's (1931–2022) *Full Tilt: Ireland to India with a Bicycle* (1965); (iii) Paul Theroux's *The Great Railway Bazaar: By Train Through Asia* (1975) and *The Old Patagonian Express* (1979); (iv) Peter Matthiesen's (1927–2014) *The Snow Leopard* (1979); (v) Bruce Chatwin's (1940–1989) *In Patagonia* (1977); (vi) Robyn Davidson's *Tracks* (1980); (vii) Mary Morris' *Nothing to Declare: Memoirs of a Woman Traveling Alone* (1987); (viii) Barbara Grizzuti Harrison's (1934–2002) *Italian Days* (1989); (ix) Mary Lee Settle's (1918–2005) *Turkish Reflections: a Biography of Place* (1991); and (x) Dea Birkett's *Serpent in Paradise* (1997).

More recently, when sexuality and gender identity enjoyed relative freedom in Western countries, many people were able to free themselves from the heteronormative constraints imposed on earlier travel writers. This led to remarkable works depicting travel from their own perspective, such as Edmund White's *States of Desire: Travels in Gay America* (1980), Tobias Schneebaum's *Where the Spirits Dwell: an Odyssey in the Jungle of New Guinea* (1988), Aldo Busi's *Sodomies in Elevenpoint* (1992), Lucy Jane Bledsoe's *Lesbian Travels: a Literary Companion*(1998), Raphael Kadushin's *Wonderlands: Good Gay Travel Writing* (2004), Gillian Kendall's *Something to Declare: Good Lesbian Travel Writing* (2009), and many others.

So far, the main focus has been on literature as an autonomous factor in the production of destinations (i.e. the unintentional creation of organic touristic images resulting from the written products of artistic language). Simply put, in these cases, the authors told stories set in specific areas without the intention of promoting tourist demand for those areas. Their main purpose was to praise nature and true beauty in the context of a story they were describing; it was 'literature' that conveyed aesthetic pleasure. However, because they succeed in connecting the reader emotionally with the places of action in question, their descriptions form intense tourist representations. Even in travelogues, which are difficult to distinguish from tourism journalism, the author, who is fascinated by a destination or a journey, aims to inform through an emotional account; he or she is not usually commissioned by a travel agency, an airline, a country's tourism organization, etc.

### 8.6.1 Literary tourism

In order to look at the relationship between literature and tourism from a different angle, it is necessary to clarify the concept of so-called 'literary tourism'. Apart from the relationships between tourism and literature analysed so far, 'literary tourism' is just another form of tourism developed by the tourism industry, like all others (i.e. such as art tourism, gastronomic tourism, sports tourism and conference tourism).

In particular, literary tourism is supported by the tourism industry with promotional strategies that in some way present literature as the main motivation for visiting the destination in question. That is, it may involve the production and promotion of certain areas, museums, festivals, etc. that are directly related to literary works or the lives of writers. For example, today there are tourists who visit Florence because it is touted by the tourism industry as the city of Dante, or Prague to visit Kafka's house, which has been turned into a museum.

In the context of literary tourism, then, it is obvious that the marketing people of the tourism industry are constantly inventing new tourist products that draw their themes from literature in order to stimulate demand for established destinations as well. Planning city tours based on famous detective novels, for example, is a relatively recent invention, with the 'Sherlock Holmes Routes' in London or the 'Raymond Chandler Streets' in Los Angeles being particularly noteworthy.

## 8.7 Summary

To summarize, it can be said that ancient and modern texts have profoundly influenced the formation of cultural identities and the creation of distinctive 'images of place' that serve as the basis of travel literature. From the works of ancient Greek philosophers and historians, to the proliferation of travel literature in later eras, to the novels, poems, plays and essays of modern times, literature has awakened wanderlust and shaped cultural perceptions. It is true that all types of literature often unintentionally encourage the construction of intense tourist representations, as readers form an emotional attachment to the places where the plot of the narrative unfolds. We can therefore say that literature has, over time, crystallized our inner tendency to move through space with certain representations. Be that as it may, of particular interest to those who study the relationship between tourism and travel is the concept of literary tourism, where the tourism industry uses literary themes to promote destinations, for example by planning city breaks to places described in famous novels or to the authors' hometowns.

## Notes

[1] Homer was the greatest epic Greek poet, to whom two poems are attributed: (i) the *Iliad*; and (ii) the *Odyssey*. Although many scholars and common people believe that he was a real person, this is less strongly accepted by some contemporary scholars. If there was a historical figure named Homer, he may have lived sometime between 850 and 750 BC. Several places such as Chios, Smyrna, Athens, Egypt and Rome claimed to be Homer's birthplace. It is likely that Homer lived in the Greek region of Ionia on the west coast of Asia Minor, as the Ionian dialect is prevalent in his poems and this place was the most culturally developed part of the Greek world at the time. His second poem, the *Odyssey*, describes the journey of the cunning and resourceful Greek hero Odysseus, the king of Ithaca, on his way home to his country, accompanied by supernatural adventures. Many believe that Homer, who was blind, was a sailor before he went blind, and that for this reason he possessed a great knowledge of distant places as well as nautical skills. It should be noted at this point that epic poetry has its roots in oral rather than written tradition and was therefore part of the communicative cultural memory of the ancient Greeks. Originally, these texts were handed down by bards or professional poets who earned their living by reciting folk tales and epic poems to an audience. Homer's poems could therefore also be categorized as media with dynamic messages. Later, however, in the sixth century BC, during the tyranny of Peisistratos, it was decided to write down the Homeric epics in a definitive version for the first time, making them static (Sacks, 2014).

[2] As early as 600 BC, numerous Greek settlements, often referred to as 'colonies', were founded on the coasts of the Mediterranean and neighbouring seas. These communities were spread across various regions, including the coasts of southern Italy (such as Naples and Taranto), Sicily (with notable centres such as Leontine and Syracuse), the southern French coast (including Marseille and Nice), Catalonia (especially Empúries), North Africa (including Cyrene and Naucratis) and the Black Sea region (with settlements such as Cyzicus and Sinope). This extensive network of Greek colonies served as important intermediaries and facilitated the spread of Greek goods and culture throughout the then-known world (Beaton, 2021).

[3] The Chinese knew this printing technique many centuries earlier, but this book is limited to the development of Western travel culture, so Chinese media on tourism will not be examined. We can, however, mention that the technique of printing with movable type was introduced in China around 1040 AD during the Song Dynasty (Needham, 1974).

[4] In this context, however, it should also be mentioned that with the progress of printing technology and the widespread distribution of printed media, reading aloud was also common, so that the texts were not only reached by those who could read, but were often read aloud in the family, in taverns or other places, so that the surrounding community also benefitted (Burke, 2003).

## References

Adams, P. (1978) Travel literature of the seventeenth and eighteenth centuries: a review of recent approaches. *Texas Studies in Literature and Language* 20(3), 488–515.

Akbari, S. (2009) *Idols in the East: European Representations of Islam and the Orient 1100–1450*. Cornell University Press, Ithaca, New York.

Atwood, M. (2005) *The Penelopiad*. Canongate, Edinburgh, UK.

Auden, W.H. and Isherwood, C. (1939) *Journey to a War*. Random House, New York.

Auden, W.H. and MacNeice, L. (1937) *Letters from Iceland*. Faber and Faber, London.

Aujac, G., Harley, J.B. and Woodward, D. (1987) The foundations of theoretical cartography in archaic and classical Greece. In: Harley, J.B. and Woodward, D. (eds) *The History of Cartography,* 1. The University of Chicago Press, Chicago, Illinois, pp. 130–147.

Bacon, F. (1625 [1971]) *Essays: Of Travel*. The Scolar Press, Menston, UK.

Barker, P. (2018) *Silence of the Girls*. Hamish Hamilton, London.

Beaton, R. (2021) *The Greeks: a Global History*. Basic Books, New York.

Birkett, D. (1997) *Serpent in Paradise*. Picador Publishing, London.

Blackmore, N. (2020) Top 10 books about the Grand Tour. *The Guardian* 12 Aug 2020. Available at: https://www.theguardian.com/books/2020/aug/12/top-10-books-about-the-grand-tour-neil-blackmore-the-intoxicating-mr-lavelle (accessed 12 May 2021).

Bled, J.-P. (2003) European cultures or European culture? In: Ahrweiler, H. and Aymard, M. (eds) *The Europeans, Volume II*. Savallas, Athens. (in Greek), pp. 313–343.

Bledsoe, L.J. (1998) *Lesbian Travels: a Literary Companion*. Whereabouts Press, San Francisco, California.

Bowers, T.N. (2000) Margery Kempe as traveler. *Studies in Philology* 97(1), 1–28.

Brown, P. (2013) *Augustine of Hippo: a Biography*. University of California Press, Los Angeles, California.

Burke, P. (2003) Unity and diversity of European culture. In: Ahrweiler, H. and Aymard, M. (eds) *The Europeans, Volume I*. Savallas, Athens. (in Greek).

Busi, A. (1992) *Sodomies in Elevenpoint*. Faber Faber, London.

Buzard, J. (1993) *The Beaten Track: European Tourism, Literature, and the Ways to Culture, 1800–1918*. Clarendon Press, Oxford, UK. DOI: 10.1093/acprof:oso/9780198122760.001.0001.

Byron, R. (1937) *The Road to Oxiana*. MacMillan & Co, New York.

Campbell, M.B. (1988) *The Witness and the Other World: Exotic European Travel Writing, 400–1600*. Cornell University Press, Ithaca, New York.

Casson, L. (1994) *Travel in the Ancient World*. Johns Hopkins University Press, Baltimore, Maryland. DOI: 10.56021/9780801848087.

Chaney, E. (1985) *The Grand Tour and the Great Rebellion: Richard Lassels and the Voyage of Italy in the Seventeenth Century*. Slatkine, Geneva, Switzerland.

Chatwin, B. (1977) *In Patagonia*. Jonathan Cape, London.

Coryat, T. (1611) *Coryat's Crudities*. William Stansby, London.

Coulet, C. (1996) *Communiquer en Grèce ancienne: écrits, discours, information, voyages*. Les Belles Lettres, Paris. (in French).

Davidson, R. (1980) *Tracks*. Pantheon Books, New York.

Equiano, O. (1789) *The Interesting Narrative of the Life of Olaudah Equiano, or Gustavus Vassa, the African*. Self-published by Equiano, London.

Foster, S. and Mills, S. (eds) (2002) *An Anthology of Women's Travel Writing*. Manchester University Press, Manchester, UK.

Frommer, A. (1957) *Europe on 5 Dollars a Day*. Prentice Hall, Englewood Cliffs, New Jersey.

Gide, A. (1927) *Voyage au Congo*. Éditions Gallimard, Paris.

Grizzuti Harrison, B. (1989) *Italian Days*. Atlantic Monthly Press, New York.

Hemingway, E. (1926) *The Sun Also Rises*. Charles Scribner's Sons, New York.

Hemingway, E. (1932) *Death in the Afternoon*. Charles Scribner's Sons, New York.

James, H. (1875) *Transatlantic Sketches*. James R. Osgood and Co., Boston, Massachusetts.

James, H. (1883) *Portraits of Places*. Macmillan and Co, London.

James, H. (1909) *Italian Hours*. William Heinemann, London.

Joyce, J. (1922) *Ulysses*. Shakespeare and Company, Paris.

Kadushin, R. (ed.) (2004) *Wonderlands: Good Gay Travel Writing*. University of Wisconsin Press, Madison, Wisconsin.

Kakridis, F.I. (2006) *Ancient Greek Grammar*. Institute of Modern Greek Studies. Manolis Triantafyllidis Foundation, Thessaloniki, Greece. (in Greek).

Kendall, G. (ed.) (2009) *Something to Declare: Good Lesbian Travel Writing*. University of Wisconsin Press, Madison, Wisconsin.

Kerouac, J. (1991) *On the Road*. Viking Penguin, New York.

Khanmohamadi, S.A. (2014) *In Light of Another's Word: European Ethnography in the Middle Ages*. University of Pennsylvania Press, Philadelphia, Pennsylvania. DOI: 10.9783/9780812208979.

Lassels, R. (1670) *The Voyage of Italy*. John Starkey, Vincent du Moutier, M. DC. LXX, Paris/London.

Lawrence, D.H. (1921) *Sea and Sardinia*. Thomas Seltzer, New York.

Lawrence, D.H. (1927) *Mornings in Mexico*. Martin Secker, London.

Lawrence, T.E (1922 [1997]) *Seven Pillars of Wisdom, a Triumph*. Castle Hill Press, Salisbury, UK.

Legassie, S.A. (2017) *The Medieval Invention of Travel*. University of Chicago Press, Chicago, Illinois. DOI: 10.7208/chicago/9780226442730.001.0001.

Le Goff, J. (1992) *History and Memory*. Columbia University Press, New York.

Matthiesen, P. (1979) *The Snow Leopard*. Chatto and Windus, London.

McGowan, A. and Bradshaw, P.F. (2018) *The Pilgrimage of Egeria: a New Translation of the Itinerarium Egeriae with Introduction and Commentary*. Liturgical Press, Collegeville, Minnesota.

Mills, S. (1991) *Discourses of Difference: an Analysis of Women's Travel Writing and Colonialism*. Routledge, London.

Milton, J. (1667) *Paradise Lost*. Samuel Simmons, London.

Monicat, B. (1994) Autobiography and women's travel writings in nineteenth-century France: Journeys through self-representation. *Gender, Place & Culture* 1(1), 61–70. DOI: 10.1080/09663699408721201.

Morris, M. (1987) *Nothing to Declare: Memoirs of a Woman Traveling Alone*. New York Times Company, New York.

Morris, M. (ed.) (2007) *The Illustrated Virago Book of Women Travellers*. Virago, London.

Murphy, D. (1965) *Full Tilt: Ireland to India with a Bicycle*. John Murray, London.

Needham, J. (1974) *Science and Civilisation in China*, Vol. 5. Cambridge University Press, Cambridge, UK.

Newby, P.T. (1981 [2014]) Literature and the fashioning of tourist taste. In: Pocock, D.C.D. (ed.) *Humanistic Geography and Literature*. Routledge, London, pp. 130–141.

Oliva, M. (2020) *The Odyssey: Told by Penelope, Circe, Calypso and the Others*. Solferino, Milano, Italy. (in Italian).

Parr, A. (1992) Thomas Coryat and the dIscovery of Europe. *Huntington Library Quarterly* 55(4), 579–602. DOI: 10.2307/3817634.

Phillips, K.M. (2014) *Before Orientalism: Asian Peoples and Cultures in European Travel Writing, 1245–1510*. University of Pennsylvania Press, Philadelphia, Pennsylvania. DOI: 10.9783/9780812208948.

Putnam, G.P. (1838) *The Tourist in Europe: Or, a Concise Summary of the Various Routes, Objects of Interest, &c in Great Britain, France, Switzerland, Italy, Germany, Belgium, and Holland; with Hints on Time, Expenses, Hotels, Conveyances, Passports, Coins, &c*. Wiley & Putnam, New York.

Robinson, J. (1990) *Wayward Women: a Guide to Women Travellers*. Oxford University Press, Oxford, UK.

Sacks, D. (2014) *Encyclopedia of the Ancient Greek World*. Infobase Publishing, New York.

Said, E. (1978) *Orientalism: Western Concepts of the Orient*. Pantheon, New York.

Schneebaum, T. (1988) *Where the Spirits Dwell: an Odyssey in the Jungle of New Guinea*. Weidenfeld and Nicolson, London.

Scott Fitzgerald, F. (1934) *Tender is the Night*. Charles Scribner's Sons, New York.

Settle, M.L. (1991) *Turkish Reflections: a Biography of a Place*. Simon and Schuster, New York.

Shelley, M. (1817) *History of a Six Weeks' Tour*. T. Hookham, Jr, London.

Simopoulos, K. (2003) *Foreign Travelers in Greece: Public and Private Life, Popular Culture, Church and Economic Life, from Travel Chronicles*, Vol. 3. Stachy, Athens. (in Greek).

Smollett, T. (1751) *The Adventures of Peregrine Pickle*. D. Wilson, London.

Smollett, T. (1766) *Travels Through France and Italy*. R. Baldwin, London. DOI: 10.1093/actrade/9780198126119.book.1.

St Augustine (2017) *Confessions*. Word on Fire, Los Angeles, California.

Sterne, L. (1768) *A Sentimental Journey Through France and Italy*. T. Becket and P.A. de Hondt, London. DOI: 10.1093/actrade/9780813017716.book.1.

Themelis, P. (2016) Pausanias and travels in antiquity. In: Zampas, K., Lambrinoudakis, V., Simantoni-Bournia, E. and Ohnesorg, A. (eds) *Architect, Honorary Volume for Professor Manolis Korre*. Melissa Publishing House, Athens. (in Greek), pp. 637–644.

Theroux, P. (1975) *The Great Railway Bazaar: By Train Through Asia*. Houghton Mifflin, New York.

Theroux, P. (1979) *The Old Patagonian Express*. Houghton Mifflin, New York.

Thompson, C. (2011) *Travel Writing*. Routledge, Abingdon, UK. DOI: 10.4324/9780203816240.

von Goethe, J.W. (1816 [1982]) *Italian Journey, 1786–1788*. Penguin Classics, London.

Walchester, K. (2007) *Our Own Fair Italy: Nineteenth Century Women's Travel Writing and Italy 1800–1844*. Peter Lang, Oxford, UK.

West, R. (1941) *Black Lamb and Grey Falcon: a Journey Through Yugoslavia*. Macmillan, London.

White, E. (1980) *States of Desire: Travels in Gay America*. EP Dutton, New York.

# 9 Tourism, Photography, Press, Posters and Other Media

**Abstract**

The advent of photography, which coincided with the modern tourism industry, triggered a transformative convergence and changed the way we perceive and experience travel. Examining the construction of tourism images by the press and tourism journalism reveals the synergy of words and images that create evocative representations of distant worlds that capture the imagination of readers. The investigation in this chapter also extends to the influential role of tourism posters as artistic representations central to the marketing of the dream of travel. Postcards prove to be powerful mediators, shaping the image of tourists and spreading travel myths. A brief but insightful look at brochures, guidebooks and maps emphasizes their indispensable role in shaping the tourism industry. In short, this chapter shows how photographs, in combination with various other media, convey important information about people, local culture and attractions and play a crucial role in shaping the overall image of tourism.

Since the beginning of the modern tourism industry, various media such as photographs, advertisements and articles in the daily press, posters, postcards, travel brochures, guidebooks and maps have played a crucial role in creating tourism images. Most of these media rely primarily on textual narratives and visuals to disseminate information with static messages and to capture the interest of potential travellers. Through the strategic use of still images, illustrations, textual content and a range of graphic elements, sometimes artistically designed, destinations, attractions and experiences are effectively presented and promoted, stimulating demand for tourism.

All these media usually use their static messages to convey important information and symbolic associations about destinations. This includes details about local culture, attractions, accommodation, transport, activities, directions, romance and nostalgic colours.

With the advent of digital technology and the Internet, many of these traditional forms of tourism media have transformed into digital formats, making them accessible to a global audience through websites, mobile apps and social media.

In the following, we will address each of these elements separately, but they complement each other to create a holistic travel experience for tourists and travellers.

## 9.1 Photography and Tourism

The coordination and organization of the tourism industry has been linked by many researchers to the discovery of photography (Chalfen, 1979; Albers and James, 1983; Urry, 1990; Cohen *et al.*, 1992; Crang, 1997; Bonarou, 2012).

Indeed, photography is now a necessary ingredient for the creation of social representations in Western culture, and it would not be an exaggeration to say that photographs (including their present digital form) are perhaps the most important means of communication related to the promotion of the tourism industry, but also to the preservation of memories on the part of tourists (Albers and James, 1983).

The term ‘photography’ comes from the fusion of the Greek elements ‘photo’ (derived from *phos*, meaning light) and *graphein*, meaning



DOI: 10.1079/9781800626034.0009

writing (Encyclopædia Britannica, 2006; Watson and Hill, 2015).

We can define photography as the art or method of recording permanent images through the action of light projected through a lens in a camera onto film or other light-sensitive material, but when we speak of photography today we also include analogue and digital video technologies for capturing images.

The history of the discovery of photographic imaging technology goes back a long way, but suffice it to say that the crucial development began with Joseph Nicéphore Nièpce (1765–1833) and his brother Claude (1764–1828), who were the first to fix camera obscura images by chemical means in 1793.

It is generally believed that the first photographic image was taken by Joseph Nièpce in 1826, by imaging a roofscape on a tin plate, which he called heliography. In 1830, he teamed up with painter and theatre designer Louis Daguerre (1789–1851) and the two invented the first commercially successful photographic process, the daguerreotype (1837).

In addition, the two British inventors Thomas Wedgwood (1771–1805) and William Henry Fox Talbot (1800–1877) patented the negative-positive calotype process, which became the forerunner of modern photographic technology.

With technological advances in the improvement of photographic equipment, materials and techniques, the scope of photography has expanded enormously.

The invention of photography was officially presented to the public at a meeting of the French Academy of Sciences in Paris in 1839, and more or less at the same time the idea of some resourceful businessmen to turn tourism into an industry was born. It is therefore instructive to emphasize that photography and the modern tourism industry emerged together and have a close connection.

In fact, the relationship between photography and tourism is symbiotic and mutually beneficial. Photography plays an important role in promoting destinations and experiences, while tourism offers photographers numerous opportunities to capture and present captivating images.

If the photos are appealing, they can arouse curiosity and a desire to visit and take a closer look at the places depicted. This can happen in an organic or induced way (i.e. either through a random photo that has nothing to do with tourism advertising or through those produced by the tourism industry). So not only tourism advertisements, but all kinds of pictures in magazines, newspapers, art photos, photos taken by our friends as tourists, etc. can create a motivation for tourism. Every beautiful photo can awaken dreams and longings and serve as an invitation to a place, and that is just as good as, and maybe even better than, official tourism advertising.

Since the introduction of Kodak's first simple and portable camera at the end of the 19th century, which was easy to use, photography became a part of typical tourist activities.

Many scholars believe that the tourist gaze would look very different today had it not been for the invention of the cheap camera, which led to a broader and more homogeneous form of ways of seeing in tourism (Urry, 2005).

Since then, photography has become so ingrained in the minds of tourists that it is now unthinkable to return home from the places we have visited without posting photos of them on social media to prove we were there.

In the past, photos were printed out in photo labs and pasted into personal albums. It would not be an exaggeration to say that if we look at old photo albums or photo archives, we are likely to find that, if not most of them, a large proportion of them are about places their owners have visited or are related to tourism for one reason or another.

As we have already mentioned, in order to retrieve or reconstruct the mental images of the places we have visited in our imagination, we necessarily need the evidence of visual images, and in the best case, photographs.

However, tourism photography does not show the most realistic image of a place, it only selects elements and visual frames for the tourist's reality, it shows what the tourist wants to see and remember. Therefore, in the end, tourist photos are not exclusively a proof of what is really present at the destination, but of what the tourist sees. In most cases, tourists want to create beautiful memories with their photos, and the easiest way to do that is to follow the semiotic markers that the tourism industry has set.

The possibilities of digital technology and mobile communication have decisively changed the practice of photographic imaging. Unlimited shots are now possible as high-capacity memory cards have replaced expensive film and later prints in photo shops. In addition, new technologies have made it easier and faster to take, instantly view and delete images. Moreover, tourists are no longer used to printing their photos in photo labs. They prefer to view them on their mobile phone or computer screens and share them on social media.

On the other hand, photography as an instrument of the tourism industry plays a central role in the promotion of travel destinations. For this reason, various media channels are used in which photos are embedded, from posters and postcards to brochures, TV spots, travel blogs by influencers, online tourism platforms, virtual reality (VR) and augmented reality (AR) experiences, etc.

The tourism industry strives to ensure that these photos reflect what travellers want and show the kind of sights they find captivating, elusive, humorous and entertaining.

By and large, photography is an indispensable tool that actively shapes travellers' perceptions and whets their appetite to explore specific destinations, thereby influencing their travel decisions. In a highly competitive global market where destinations vie for travellers' attention, professional and captivating images are a necessity.

Given the profound role of photography in shaping social representations, we could shed light on many aspects of the relationship between photography and tourism, but we will not do so here, as this relationship is found in most of the media we will examine in this book.

## 9.2 Press and Travel Journalism

Tourism journalism is another form of discourse that is reproduced by the print media, but can also be found in other media with dynamic messages such as radio, television and the Internet.

There is, of course, no doubt that both travel literature and travel journalism broadly describe travel and faraway places to the reader. However, there are significant differences between these two types of writing because of the way the subject is approached. We can say that travel literature describes the area in question in an elegant way and appeals to the reader emotionally, while travel journalism focuses on the practical side of travel and provides useful information and guidance necessary for an enjoyable trip.

If we want to examine the differences between travel literature and travel journalism in more detail, we can say that they primarily lie: (i) in the objective (i.e. personal expression on the one hand and information on the other); (ii) in the perspective taken (entertaining the reader vs providing useful and relevant details for potential travellers); (iii) in the medium of publication (typically books, essays and memoirs vs newspapers, magazines, television programmes and online travel websites); and (iv) in the accuracy of the approach (subjective experiences and perceptions of the author vs verified information from more than one source).

In short, travel literature refers to the author's subjective but sometimes imaginative descriptions of the place visited as an expression of his or her inner creativity, while travel journalism is a tool created by a journalist to provide practical advice on how to have a better time visiting a destination (Aslanidou, 2017).

It is generally recognized that the press, together with literature, had a major influence on the development of tourism early on.

From the end of the 15th century, periodical publications of a few pages appeared irregularly at first, but then more and more frequently, such as the *occasionnels* and later the *canards* in France, and the 'gazettes' in Italy. These publications gradually spread, disseminating news and real or fictional stories that took place in regions near or far, and generally stimulating the public's imagination, sensibility and interest (Jeanneney, 1996).

It is worth recalling at this point that, as argued elsewhere in this book, the unintentional mention of places by the mass media can create future destinations, so it can be said that these first newspapers contributed to building the image of places from the very beginning, even before the emergence of pure tourism journalism.

The early forms of newspapers were initially loose and inexperienced, but had some potential to convey information based on the author's analysis of facts and opinions to a relatively wide audience, rather than focusing on the immediacy of the news. After some time, however, better organized and more sophisticated periodical newspapers appeared.

Finally, in the late 19th century, changes and advances in printing technology led to a real boom in the circulation of the press. It was also at this time that the cheap daily newspaper appeared, aimed at the general public.

The phenomenon of the 'high-circulation newspaper' began to spread throughout the Western world. Scandal sheets with a circulation of hundreds of thousands, which could be distributed to every corner of every country, contributed greatly to shaping the image of faraway places and the image of countries in general, in addition to all the other effects on society.

Thus, a new power was born: the fourth estate, which was directly linked to public opinion. Very quickly, the press magnates learned to play skilfully with people's emotions, to spread rumours, to promote or conceal facts, to create, promote and exploit stars and political personalities and, most importantly for our case, to construct images of exaggeratedly beautiful places and leisure habits. During this period, theories were developed that mass media messages can have a powerful effect on people (the 'hypodermic needle model' and 'magic bullet theory' referred to in Chapter 5, this volume), while the idea of mass tourism gradually took hold.

According to Jeanneney (1996), in those years the Western world experienced the 'golden age' of the written press, which lasted from the early 1870s until the First World War, when newspapers consolidated their power and had no competition from radio, or television that came much later.

A newspaper is characterized by the fact that: (i) it usually has a commercial form; (ii) it is published regularly (daily, weekly, etc.); (iii) its main purpose is to inform; (iv) it has relative freedom to intervene in the public sphere; and (v) it usually addresses an educated audience of urban and popular origin (McQuail, 2010).

We would even go so far as to say that newspapers are different from other media in that their main purpose is usually to report news, that they employ journalists and reporters to research, collect and verify information in order to write news, and that they are usually divided into different sections (e.g. local news, world news, journalistic analysis, sports, business, entertainment, lifestyle and travel). Also they use headlines, have a specific print format, layout and design, and usually they contain advertising and sometimes rely on advertising revenue to finance their operation.[1]

Thus, in addition to political analyses, news and social columns, a number of newspaper articles were published over time that contained valuable information and advice, particularly on travel and tourism practices. It can be argued that these early articles were important factors in the formation of tourism consumption patterns among the general public and contributed to the development of societal ideas about tourism in general. In this context, it could also be said that the articles in the society columns, which often showed how various public figures, such as actors, politicians and rich businessmen, spent their holidays, encouraged the public's tendency towards conspicuous consumption. Influenced by what they read in the feature pages, ordinary people tried to improve their social standing by, among other things, imitating the tourist choices of the elite.

Often newspaper correspondents or publicists who travelled to foreign countries in the course of their work presented distant cities, the modern or traditional behaviour and lifestyle of the inhabitants, places worth seeing, cuisine and so on. In this way, travel journalism emerged gradually but experienced significant growth, especially from the end of the 20th century, along with the dynamic development of other media forms that reinforced 'armchair tourism', such as the international TV channels (satellite, cable, etc.) and, later, tourism platforms, social media, and so on (Waade, 2006; Abram and Norum, 2016).

As we mentioned earlier, and as it is relevant to tourism journalism, 'armchair tourism' is practised from the comfort of one's own living room through the media (i.e. reading books about travel, watching travel programmes on TV or visiting websites about faraway places). It essentially includes 'imaginative travel', 'virtual travel' and 'communicative travel' (i.e. three of the five forms of mobility defined by Urry (2016), where people use media to discover distant places and destinations).

Travel journalism is essentially a specific field of journalism that deals with travel and tourism information, but also aims to provide useful tips and practical guidance for travelling to specific places. It should be noted here that travel articles, as part of the general social function of journalism, aim for accuracy, clarity and truth, and should be characterized by objectivity and impartiality. However, many publications in this field are subsidized by destination management organizations, which moves travel journalism out of the realm of objective journalism and into the realm of advertising or public relations. Thus, tourism journalism is now essentially a business and a product of the mass media. The media invest time and money to send journalists, photographers, cameramen, etc. to a destination and in return receive some form of profit (i.e. facilities and production subsidies from a country's tour operators, or the attention of the public interested in a particular fashionable destination, which in turn attracts advertising money, and so on).

We must not forget that journalism in general has moved away from its traditional meaning, as the general field of mass media follows the 'infotainment' model mentioned earlier.

Thus, the topics often covered in travel journalism can be part of a promotional strategy of tour operators or destination management organizations, covering everything related to the organization, planning and execution of a trip to the destination. At the same time, they also provide information about the history, geography, culture, cuisine, art and daily life in these destinations.

It would be remiss not to recognize print travel journalism as a crucial factor in shaping the contemporary model of the tourist and promoting today's 'travel culture'. However, with the development of communication technology, travel journalism has outgrown its traditional role and has become a cornerstone of tourism promotion. As mentioned earlier, travel journalism has seamlessly transitioned to television and has subsequently found a prominent place on new media platforms. More recently, it can also be practised by individuals who are not professional journalists but disseminate information via websites, blogs and social media.

## 9.3 Travel Posters

It is no exaggeration to say that today's cityscape is saturated, if not inundated, with advertising signs. Wherever we turn our gaze, we encounter large billboards, printed posters and a variety of other advertising media, alongside street signs, outdoor retail banners and signs for markets, shopping centres and the like. In this patchwork of commercial imagery prevalent in today's society, advertisers work ceaselessly with graphic design, illustration, photography, sketching and typography to create posters that are reproduced by the thousands on large billboards and other surfaces to stimulate consumption.

Traditional promotional printed posters are eye-catching prints on paper, and although they are spatially limited because they are pasted on a surface, they are a widely used visual means of communicating static messages to a relatively large audience. They can be printed in large numbers and placed on large billboards, city walls, or other surfaces in public spaces to reach their audience, and are typically used to advertise a product, promote a political issue or a cultural event.

On the other hand, due to their particular artistic form and limited circulation, some advertising posters may increase in value over time, and when their function as advertising media diminishes, they may be considered as works of art by collectors.

Advertisers try to combine art and psychology on their posters to evoke emotional responses and leave a memorable impression on the audience. This process of appropriately presenting products and services to promote sales and the fact that they have to take into account customer preferences and brand guidelines limits the creative freedom of their artistic expression.

On closer inspection, it becomes clear that visual aesthetics such as advertising posters, even if they occasionally have impressive aesthetic elements, are something different from art. This difference arises not from their medium, but from their actual purpose. They are mass-produced for commercial purposes, for the sale of goods and the promotion of services, and are therefore socially categorized as such, which distinguishes them from the field of fine art. Be that as it may, in some cases, the commercial images created by famous painters are also collected and appreciated in the art world, similar to other forms of artistic expression (Fig. 9.1). The overlap between commercial work and fine art can be complex, but it is not uncommon.[2]

Anyway, posters often serve the purpose of reproducing original works of art and decorating the walls of our homes. In this context, posters become a means to make art more accessible and allow individuals to enjoy iconic or admired works without the need for an original artwork.

Travel posters can appeal to the viewer not only with information but also with emotion. The impressive power of their images combined with the usually bold text and sometimes their artistic synthesis produce powerful, dramatic, moving, romantic or humorous results that often create a special bond with the viewer.

According to marketers, posters can promote tourism at low cost compared to other media. Since they are placed in different locations in cities, they can be targeted to a specific audience. Depending on their placement, shape and quality, they sometimes have a larger audience and a longer lifespan (e.g. metal posters on public transport buses).

Printed posters were effectively used by the tourism industry very early on to create 'tourist images' and to promote transportation services, events and tourism products in general.

By the mid-19th century, the widespread availability of colour lithographs made it easy to disseminate static messages about rail travel, cruises and popular destinations to a wide audience. Therefore, tourist boards and companies in the tourism industry began to use this medium systematically at an early stage.

It seems that the travel poster for Thomas Cook's first publicly advertised trip in 1841 was very simple, containing only the following text:

> Public notice: on Wednesday, the 5th of July, Mr. Thomas Cook, with the Midland Railway Company, will take an excursion by train from Leicester to Loughborough, at the rate of one shilling per head for double the distance. All are welcome.
>
> (Prescott and Gillett, 2019)

But soon the group tours and excursions organized by the innovative British tour operator were advertised with excellent posters that attracted public interest and are still appreciated for their artistic value. Very characteristic is a poster produced in 1910 for the Southeastern and Chatham Railway (SE & CR) to advertise the train connections with the world travel tickets offered by the tour operator Thomas Cook & Son. It shows a woman in Edwardian dress holding a walking stick and walking towards a globe on a yellow background, while the caption reads 'A Cook's Ticket Will Take You Anywhere on the Southeastern and Chatham Railway or Around the World'.

Although the woman is depicted as an elegant and dainty figure in the fashionable dress of the time, she occupies the position of being first in the picture, followed by two men standing much further back. This contradicts the usual depiction in tourism posters of the time (or even our time), which show the woman with limited independence in a second role in the company of male family members, and generally in a more controlled and safe environment. Perhaps this poster can be considered another breakthrough among the many other traditional posters Cook published, for he was always ahead of his time. In any case, this poster can be considered an innovation and milestone in the history of travel advertising, as it shows a woman in a leading position since those early years.[3]

All these travel posters were usually accompanied by meaningful texts that informed the viewer, but it was the pictures that attracted attention and promoted tourism most effectively. Thus, the travel poster has been directly linked to artistic creation since the early years of its development.

It should be emphasized that posters with drawings and sketches were for a long time the most common means of informing the public about topics such as means of transport, hotels, destinations and their history, as well as cultural events (e.g. concerts, sporting activities, festivals and celebrations). Later, when air travel became popular all over the world, airline posters were also widely used.

Thus, together with the press, the poster established itself for many years as the basic advertising medium for travel and tourism, adapting to the fashions and trends of each era, such as the art movements of Art Nouveau, Art Deco, Futurism and Constructivism (Sampson, 1874; Victoria and Albert Museum, 2021).

**Fig. 9.1.** Greek National Tourism Organisation (GNTO) poster, *Hydra* (1956), created by Yiannis Moralis (1916–2009). While there is no doubt that there are differences between art and aesthetic advertising, as they differ in their intention, context and purpose, when it comes to famous artists, the boundaries are sometimes blurred. No one can deny that artists can create commercial works while maintaining their artistic integrity by challenging the traditional boundaries between the two fields. Image used with permission from the GNTO.

It is important to stress that the promotion of tourism in many countries, including Greece, until recently relied heavily on the production of tourism posters. For example, the Greek National Tourism Organisation (GNTO) is said to have created more than a thousand posters throughout its history (Katsigiannis, 2017).

Some early travel posters, influenced by the concept of 'conspicuous consumption', fancifully promised social advancement. Sometimes they reinforced cultural stereotypes that suggested superiority over locals at the advertised destination, or they omitted elements of the place visited that were less enticing to a tourist mindset (O'Dowd, 2013).

In the context of our discussion, we can also draw parallels with colonial exhibition posters from the late 19th century. These posters aimed to gain support for the colonial possessions of great powers such as Britain, France and the Netherlands. They used similar techniques to travel posters and depicted the culture, resources and products of their colonies in vivid images. These depictions presented the colonies as exotic and enticing places full of adventure and riches to attract visitors, investors, traders and consumers. They also served as a means to legitimize colonial rule by portraying the colonies as places in need of European intervention and civilizing influence.

However, it is important to note that these posters and exhibitions reflected the prevailing attitudes towards imperialism, racial hierarchy and exploitation at their time. Examining the Orientalist aesthetics of these early posters shows how politics can influence artistic styles, how Europeans portrayed cultural differences and how these posters influenced the mentality of Western tourists (Benjamin, 2003).

In the early 1900s, well-known graphic designers and poster artists such as the Frenchman Hugo d'Alesi (1849–1906), the Swiss Emile Cardinaux (1877–1936) and the German Adolfo Hohenstein (1854–1928) combined bold lettering with very effective artistic images to motivate people to travel and explore the world. Their posters were hung in various places in cities, such as kiosks, train stations, travel agencies, ports and warehouses, and later, as aviation developed commercially, at airports to increase tourist demand. A few decades later, in the 1920s, visual artists such as the Frenchman A.M. Cassandre (1901–1968), the British Tom Purvis (1888–1959) and Frank Newbould (1887–1951) and others, succeeded in transforming the travel poster from an illustration into a mass-reproduced quasi-artwork (Novin, n.d.).

In the 20th century, travel posters retained many of the themes and techniques already used, but also utilized photography as a central design element. In this way, photography was able to convey a more realistic impression of the destinations.

As photography captures the subject more accurately, the image appears as a more reliable and realistic representation of the destination. In addition, recent technological developments have enabled advances in billboard advertising that allow the production of huge billboards that would not have been profitable in the past (Morgan and Pritchard, 2012).

In 2020, for example, motorists on the Sydenham bypass in London were confronted daily for several months with a giant billboard containing 336 sheets of paper inviting them in large letters to 'Explore Northern Ireland with confidence' (McGowans Digital Print, 2020).

In the digital age, billboards have evolved from physical printed materials to digital formats used on large digital screens in cities. They offer more flexible and eye-catching ways to reach their target audience as content can be easily updated and tailored to specific times of day or events. Unlike static printed billboards, digital billboards can include animations, videos, slideshows, live data feeds and interactive elements that allow advertisers and businesses to create more engaging and attention-grabbing content that effectively captures the attention of passers-by. However, these billboards are part of dynamic media.

Finally, it is important to refer once again to the results of the semantic analyses of various scholars on the subject of tourism posters. Most of these studies come to the conclusion that travel posters, like all advertising, reflect the prejudices prevalent in the society from which they originate. They reinforce inequalities related to gender, race, ethnicity and more. Academic work such as that of Scott (2021) and Iskin (2003) particularly emphasize that some tourism posters also reflect the indifference of the tourism industry to environmental concerns.

Overall, however, it can be said that printed travel posters remain an effective and engaging medium for tourism marketing, bridging the gap between the attractiveness of a destination and the curiosity of travellers; they invite people to discover new places and are therefore an essential part of the tourism industry's communication and promotion strategies.

## 9.4 Other Media (Postcards, Travel Brochures, Guidebooks and Maps)

In addition to the print media already mentioned, there are other forms of communication that used to play an important role in promoting tourism but have since faded into the background. This decline can be attributed to the emergence of more efficient and cost-effective new media or to changing fashion trends.

However, some of these forgotten means of communication are still important for tourism marketing, while others serve as repositories for the sentimental nostalgia of bygone eras. Tourists often view them as romantic mementos of the place they visited.

In this respect, printed postcards, printed travel brochures, printed travel guides and maps, despite their outdated nature, continue to have an impact on destination promotion and the tourism industry in general. They have become retro items that tourists value as souvenirs.

These outdated tools, which shaped tourists' perceptions of destinations before the advent of new media, have played a crucial role in the creation and perpetuation of various tourism myths. This was achieved, among other things, by presenting idealized images, offering selective information and often including romantic narratives.

A postcard is a small, stiff card sent by traditional mail with a picture on one side and a blank side for the sender's message and the recipient's name and address on the other.

According to Bonarou (2012), postcards are meaningful social signs with diverse and complex connotations that reflect and characterize the identity of a place and the identity of the tourist, while at the same time representing an important component of tourism marketing for the destination in question.

In the late 19th century, but especially at the beginning of the 20th century, postcard production experienced a significant boom. The popularity of postcards was directly linked to the development of organized mass tourism, and it is no coincidence that companies such as Thomas Cook & Son, who are often credited with inventing modern tourism, frequently printed postcards.

However, the postcard industry as a whole developed independently as a form of communication and visual representation of travel destinations. Postcards were initially produced by various publishers, photographers and printers. They became a popular way for travellers to share scenic views and experiences with friends and family.

In fact, for a long time and until the development of social media, postcards were one of the most typical ways for people to share their feelings with each other and maintain their social relationships. Tourist postcards could show landmarks, beaches, landscapes, architecture, delicacies, local clothing, tourist maps and anything else that had to do with a place. They were a kind of 'certificate of authenticity', attesting to the genuineness of the travel experience and enabling the recipient to recognize and appreciate the sender's preferences and choice of destination.

In this way, postcards functioned both as shapers of the tourist social image and as advertisements for destinations, as a means of promoting and expediting travel for those left behind. For many years they enjoyed great popularity and were an integral part of tourist activities (Smithsonian Institution Archives, n.d.; Bonarou, 2012).

Since postcards could be reproduced in large numbers and were sent by tourists to all parts of the world, they helped to shape the image of many countries and cities internationally. Above all, they could identify a destination with a landmark and thus create international characteristic tourist symbols. Thousands of postcards featuring Big Ben sent by tourists made it an iconic symbol for London, just as the Eiffel Tower on postcards became a synecdoche for Paris and the Acropolis for Athens. You could even say that postcards were a kind of tourist ritual that everyone had to follow.

Nevertheless, in the 20th century it was almost unthinkable to travel without sending a postcard to one's loved ones. Therefore, postcards, with their symbolic, iconic images and personal sentiments, played a multiple role: (i) they advertised places; (ii) preserved cherished memories; (iii) significantly influenced global perceptions of destinations; and (iv) fostered interpersonal bonds.

A brochure is a pamphlet or small booklet distributed free of charge containing promotional, pricing and analytical information about products or services.

Tourism brochures, travel agency magazines and advertisements in the press have long been the tourism industry's most important information channels for tourism programmes and prices. It is worth noting that many graphic design schools had guidelines for the graphic design of printed tourist advertising brochures and for the marketing methods for such publications.

Most of these printed publications were distributed directly to potential tourists, sent out by travel agencies or published as inserts in newspapers, magazines, etc. However, with the development of the Internet and the emergence of tourism platforms, this situation started to change. Some of these methods are still used, but most tourism businesses publish e-brochures online or send promotional material by e-mail.

Travel guides or handbooks are books in the traditional format that contain typical information about a place; they are simple manuals specifically designed for use by tourists. They usually contain catalogues and directions to hotels, accommodation, sights, restaurants, transport and all kinds of tourist activities; some also contain a map. Very often they also contain advertisements for local tourism businesses.[4]

Travel guides originated in historical and geographical texts of antiquity, but in modern times the advent of the railways led to an explosion of these manuals, which contain valuable information and recommendations for tourists (Zuelow, 2015).

Today, the best-known and most recognized guidebook series include Lonely Planet, Fodor's, DK Eyewitness Travel, Rick Steves' Europe, Insight Guides, Bradt Travel Guides, Michelin Green Guides, Blue Guides, Time Out, Footprint, Cicerone and Marco Polo.

The truth is that it can sometimes be difficult to distinguish between travel literature, travel journalism and travel guides, or that sometimes studies on travel literature lead to reports on travel journalism or travel guides, or vice versa. Nevertheless, we have talked above about well-known houses that published travel guides in the 19th century, such as those of Baedeker and Murray, although there were even earlier examples, at the beginning of the 19th century, such as the *Guide des Voyageurs en Europe* (1805) by Heinrich August Ottokar Reichard (1751–1828).

Be that as it may, according to many scholars, modern guidebooks have drawn heavily on the publications of the English writer Mariana Starke (1762–1838), written for British continental travellers of the early 19th century (Buzard, 1993; Walchester, 2007; Sørensen and Peel, 2016).

Starke was born in Surrey in the UK and then grew up in India, where her father was Governor of Fort St George Madras. When she returned to Britain, she began a career as a playwright. After a visit to Italy, she decided to write about her travels and the countries she visited.

In 1802, she published *Travels in Italy between the years 1792 and 1798*, consisting of 25 letters in two volumes, in which she described the political situation in Italy after the first Napoleonic campaigns and gave practical advice to tourists visiting the cultural sites of Italy (Starke, 2010).

However, it is claimed that another book by Starke, *Information and Directions for Travellers on the Continent* (1828), was one of the most valuable works for travellers of the time.

It is a widely held opinion that her books set the standard for travel guides as they introduced two innovative elements: (i) planned itineraries; and (ii) rating systems for tourist products and sights. They also contained, for the first time, very specific information such as a wealth of advice on luggage, how to obtain passports, and the exact cost of accommodation and food in each city, and contributed in some way to the theoretical substance of the new concept of tourism that was developing at the time (Walchester, 2007).

It is undisputed that guidebooks contributed to the spread of tourism and the development of an extensive tourism industry, but they were also criticized early on, for example, for setting rules for likes and dislikes and shaping the tastes and preferences of the masses, while many claimed that tourists ultimately visit sights only to validate the guidebook (Buzard, 1993).

Tourist maps serve as optimized topographical guides that are enriched with additional information tailored to the traveller. These details include places of interest, hiking, skiing or cycling trails, mountain huts, water sports facilities and more. Often these maps are accompanied by photos or illustrations showing archaeological monuments, villas, beaches and amusement parks, and they also include advertisements for restaurants, hotels and other amenities.

Before the advent of tourist software, GPS (Global Positioning System) and route planners, printed tourist maps played an indispensable role, especially when travelling by car through rural areas or exploring urban landscapes on foot. They facilitated orientation and helped travellers to find roads, important sights and other key points of interest.

In general, it is noteworthy that certain media products designed for functions such as maintaining friendships (postcards) or topographical orientation for tourists (maps) also functioned as advertising media. These products, often printed by major players in the tourism industry such as tour operators and hotel chains, were sometimes distributed either free of charge or at very favourable prices. In addition to their main function, these products also played a dual role in the subtle promotion of various tourist offers.

Another particularly important point we would like to emphasize here is that printed representations do not always convey messages as clearly as many of us believe. In most cases, they conceal symbolism that individuals absorb unconsciously, but which can be deciphered by the researcher through semiotic analysis.

In this context, it is believed that tourist prints can use artistic means of expression (paintings, art photographs, etc.) to cleverly transform a destination into a myth, depending on the points the tourism industry wants to promote. Postcards, for example, can serve both as a social link and as proof to the sender that their specific tourist experience is valid. They can also be seen as virtual texts that create tourist symbols, as we saw above, but they can furthermore combine symbolic connotations to create tourist myths that naturalize certain forms of tourism.

In this sense, mass tourism, characterized by large numbers of travellers to well-known destinations, owes its existence and success not only to certain socio-economic reasons and other anthropological, technological, etc. factors but also to the skilful orchestration of marketing strategies. Posters, guidebooks, maps and a series of brochures have helped to imprint destinations on travellers' memories and prescribe the route of their tourist odyssey, forever linking the past to the present.

## 9.5 Summary

To summarize, the modern tourism industry relies heavily on various media, including photography and the press, to create vivid tourist images and promote destinations, with photography capturing the essence of places and the press providing practical guidance and information that shapes contemporary travel culture. Alongside the central role of photography and journalism in new media and its applications, the cityscape today is awash with advertising signage that serves as a patchwork of commercial imagery, combining art and psychology to stimulate consumption. However, traditional travel posters, postcards, brochures, guidebooks and maps continue to shape tourists' perceptions as souvenirs from another era, promoting destinations through symbolic representations and promotional strategies.

## Notes

[1] Tourism journalism also falls into the realm of magazines, a category we will not go into in detail. Newspapers and magazines, which both fall under the term periodicals, have specific characteristics that distinguish them from each other. Key differentiators include frequency of publication, thematic focus, occasional variations in writing style and different approaches to design and layout. Although these

distinctions provide a broad perspective, it is important to recognize that there are nuances within each category and that certain publications can defy clear categorization and blur the lines between traditional newspapers and magazines.

[2] Ultimately, however, one could argue that whether advertisers are considered 'true artists' depends on one's perspective and the criteria used to define artistry. Some may argue that advertising is different from traditional art forms because of its commercial nature, while others appreciate the creativity and ingenuity of advertisers in creating compelling messages and visuals. We should not forget that many great artists such as Henri Toulouse-Lautrec, J.C. Leyendecker (1874–1951), Edward Hopper (1882–1967), Norman Rockwell (1894–1978), Roy Lichtenstein (1923–1997), Andy Warhol (1928–1987), and many others worked for the advertising industry for a time.

[3] The stereotypical and prejudiced position of women as portrayed in advertising has been the subject of much debate and research. Many studies have examined how advertising can reinforce, perpetuate or challenge gender roles and expectations. Research in this area can involve a variety of methods, including content analysis of advertising, semiotic analysis, consumer perception surveys, focus groups to examine audience responses, and experimental studies to measure the effects of gender representations on attitudes and behaviour (Goffman, 1979; Courtney and Whipple, 1983; Kilbourne, 1990; Nina–Pazarzi and Tsangaris, 2008). Similar research on the portrayal of gender in advertising has also been conducted in the field of tourism advertising (Heatwole, 1989; Sirakaya and Sonmez, 2000; Khan, 2011).

[4] It is said that the term 'handbook' was coined by John Murray II (1778–1843) for these books in conventional format were easy for tourists to handle. However, the idea for the series came from his son John Murray III (1808–1892), who wrote the first volumes himself, and Karl Baedeker adopted the same word into German. Later the term referred more generally to collected works on a subject or other summaries of information, or to handbooks intended for quick reference (Lister, 1993).

## References

Abram, S. and Norum, R. (2016) An archaeology of Arctic travel journalism. *Studies in Travel Writing* 20(3), 272–284. DOI: 10.1080/13645145.2016.1230292.

Albers, P. and James, W. (1983) Tourism and the changing photographic image of the Great Lakes Indians. *Annals of Tourism Research* 10(1), 123–148. DOI: 10.1016/0160-7383(83)90053-1.

Aslanidou, A. (2017) Travel journalism: from Herodotus on social media. In: Iordanidou, S. (ed.) *On the Occasion of Crete: From Place Branding to Travel Journalism*. Metamesonikties Publications, Athens. (in Greek), pp. 29–52.

Benjamin, R. (2003) *Orientalist Aesthetics*. University of California Press, Berkeley, California. DOI: 10.1525/9780520924406.

Bonarou, C. (2012) *Visual Culture and Tourism. Representations of Greece on Tourist Postcards*. Papazisi Publications, Athens. (in Greek).

Buzard, J. (1993) *The Beaten Track. European Tourism, Literature, and the Ways to Culture, 1800–1918*. Clarendon Press, Oxford, UK. DOI: 10.1093/acprof:oso/9780198122760.001.0001.

Chalfen, R. (1979) Photography's role in tourism – some unexplored relationships. *Annals of Tourism Research* 6(4), 435–447. DOI: 10.1016/0160-7383(79)90006-9.

Cohen, E., Nir, Y. and Almagor, U. (1992) Stranger-local interaction in photography. *Annals of Tourism Research* 19(2), 213–233. DOI: 10.1016/0160-7383(92)90078-4.

Courtney, A. and Whipple, T. (1983) *Sex Stereotyping in Advertising*. Lexington Books, Lexington, Kentucky.

Crang, M. (1997) Picturing practices: research through the tourist gaze. *Progress in Human Geography* 21(3), 359–373. DOI: 10.1191/030913297669603510.

Encyclopædia Britannica (2006) *Britannica Concise Encyclopedia*. Encyclopædia Britannica, Inc, London.

Goffman, E. (1979) *Gender Advertisements*. Harper and Row, New York.

Heatwole, C.A. (1989) Body shots: women in tourism-related advertisements. *Focus* 39(4), 7–11.

Iskin, R. (2003) The pan-European flâneuse in fin-de-siècle posters: advertising modern women in the city. *Nineteenth-Century Contexts* 25(4), 333–356. DOI: 10.1080/0890549032000167844.

Jeanneney, J.N. (1996) *Une Histoire des Medias: Des Origines à Nos Jours*. Seuil, Paris. (in French).

Katsigiannis, K.X. (2017) *Hellenic Tourism Organization: Travel Through Time*. Private Edition, Athens. (in Greek).

Khan, S. (2011) Gendered advertising in tourism: an assessment of tourists' perceptions. *South Asian Journal of Tourism and Heritage* 4(2), 54–63.

Kilbourne, J. (1990) Female stereotyping in advertising: an experiment on male-female perceptions of leadership. *Journalism Quarterly* 67(1), 25–31. DOI: 10.1177/107769909006700105.

Lister, J. (1993) *Murray's Handbooks for Travellers*. University Publications of America, Lanham, Maryland.

McGowans Digital Print (2020) Facebook, 3 September 2020. Available at: https://www.facebook.com/McGowansPrint/posts/the-largest-billboard-poster-weve-printed-so-far-a-336-sheet-on-sydenham-by-pass/10158030995068052/ (accessed 12 May 2022).

McQuail, D. (2010) *McQuail's Mass Communication Theory*. Sage Publications, London.

Morgan, N. and Pritchard, A. (2012) *Advertising in Tourism and Leisure*. Routledge, Oxford, UK. DOI: 10.4324/9780080498898.

Nina–Pazarzi, E. and Tsangaris, M. (2008) Constructing women's image in TV commercials: the Greek case. *Indian Journal of Gender Studies* 15(1), 29–50. DOI: 10.1177/097152150701500102.

Novin, G. (n.d.) Chapter 89: a history of travel posters. In: *A History of Graphic Design*. The Online Textbook. Available at: http://guity-novin.blogspot.com/2018/03/a-history-of-travel-posters.html (accessed 6 August 2021).

O'Dowd, S. (2013) *Imagining Destinations: Art Posters and the Promotion of Tourism*. Arizona State University, Tempe, Arizona.

Prescott, K. and Gillett, F. (2019) Thomas Cook: the much-loved travel brand with humble roots. *BBC News*, 23 September 2019. Available at: https://www.bbc.com/news/business-49789073 (accessed 10 May 2021).

Reichard, H.A.O. (1805) *Guide des Voyageurs en Europe*. Bureau de l'Industrie, Weimar, Germany.

Sampson, H. (1874) *A History of Advertising From the Earliest Times: Illustrated by Anecdotes, Curious Specimens and Biographical Notes*. Chatto and Windus, London.

Scott, D. (2021) *Design in Airline Travel Posters 1920–1970: a Semiology and Sociocultural History*. Anthem Press, London. DOI: 10.2307/j.ctv1dv0smw.

Sirakaya, E. and Sonmez, S. (2000) Gender images in state tourism brochures: an overlooked area in socially responsible tourism marketing. *Journal of Travel Research* 38(4), 353–362. DOI: 10.1177/004728750003800403.

Smithsonian Institution Archives (n.d.) 'Greetings from the Smithsonian A Postcard History'. Available at: https://web.archive.org/web/20181123195447/https://siarchives.si.edu/history/featured-topics/postcard/postcard-history (accessed 10 October 2023).

Sørensen, A. and Peel, A. (2016) Guidebooks. In: Jafari, J. and Xiao, H. (eds) *Encyclopedia of Tourism*. Springer, Cham, Switzerland, pp. 410–411. DOI: 10.1007/978-3-319-01384-8.

Starke, M. (1828) *Information and Directions for Travellers on the Continent*. A. and W. Galignani, Paris.

Starke, M. (2010) *Travels in Italy,between the Years 1792 and 1798, Containing a View of the Late Revolutions in that Country*. Cambridge University Press, Cambridge, UK. DOI: 10.1017/CBO9780511732119.

Urry, J. (1990) *The Tourist Gaze: Leisure and Travel in Contemporary Societies*. Sage, London.

Urry, J. (2005) Gazing on history. In: Kleniewski, N. (ed.) *Cities and Society*. Wiley, Oxford, UK, pp. 195–214. DOI: 10.1002/9780470752814.

Urry, J. (2016) Does mobility have a future? In: Griego, M. and Urry, J. (eds) *Mobilities: New Perspectives on Transport and Society*. Routledge, Abingdon, UK. DOI: 10.4324/9781315595733.

Victoria and Albert Museum (2021) A short history of the poster. Available at: https://www.vam.ac.uk/articles/a-shorthistory-of-the-poster (accessed 7 October 2021).

Waade, A.M. (2006) Armchair travelling with pilot guides: cartographic and sensuous strategies. In: Jansson, A. and Falkheimer, J. (eds) *Geographies of Communication: the Spatial Turn in Media Studies*. Nordicom, University of Gothenburg, Gothenburg, Sweden, pp. 155–170.

Walchester, K. (2007) *Our Own Fair Italy: Nineteenth Century Women's Travel Writing and Italy 1800–1844*. Peter Lang, Oxford, UK.

Watson, J. and Hill, A. (2015) *Dictionary of Media and Communication Studies*. Bloomsbury Publishing, London. DOI: 10.5040/9781501304712.

Zuelow, E. (2015) *A History of Modern Tourism*. Bloomsbury Publishing, London. DOI: 10.1007/978-0-230-36966-5.

# 10 Media with Dynamic Messages

**Abstract**

The central role of shaping the 'destination image' in the promotion of tourism is analysed in this chapter based on the influence of modern mass media with their dynamic messages. Starting with cinema, emotionally charged moving images have a great impact. They tell captivating stories that evoke excitement and inspiration and fundamentally shape the perception of the destination. Radio also transports distant events and places directly into living rooms through captivating audio stories, awakening wanderlust and motivating people to travel. Television, with its transformative audio-visual possibilities, continuous narrative flow and wide reach, has revolutionized the way in which the image of a destination is built. Advertising on international television channels is shaping travel trends and contributing significantly to the expansive growth of global tourism. Through these media, the intricate threads that weave together the perception of destinations and the evolving image of tourism are unravelled.

Often the signifiers of messages conveyed through the media do not remain constant, as in the case of illustrations on posters, newspaper titles or brochures, but change and disappear during the communication process, so that the audience must be involved throughout the transmission of the message in order to grasp the full meaning.

In this sense, the flow of sounds and any kind of image movement can be seen as a form of communication that is transmitted dynamically in terms of continuity. The vitality of messages can thus range from an audio or visual flow such as a speech or a play to advanced radio streams, films, television, interactive video games, computer graphics animation and multimedia applications.

Static messages, then, are like snapshots in time, unchanging and self-contained, while dynamic messages with a flow are like live streams that are constantly evolving and require the audience's constant attention to understand the whole message.

This distinction is particularly important when it comes to audience engagement and how different media convey information or stories.

Bypassing non-mediated communication (i.e. face-to-face conversation, rhetoric and direct address to an audience, theatre, etc.), in the following we will refer only to the basic modern mass media that convey dynamic messages, namely cinema, radio and television, always in relation to tourism.

## 10.1 Cinema and Tourism

As in the case of tourism, it is often said that cinema is a cultural industry, but also an art. Indeed, it is hard to forget that cinema is a profitable business based on the production and distribution of films and box office revenues. In this context, many researchers have labelled cinema as a dream industry because they believed that it has the potential to control people's imagination (Marsh, 1992; Roberts, 2007).

Be that as it may, the Italian art theorist Ricciotto Canudo (1877–1923) considered cinema as the seventh art in an article as early as 1923, acknowledging that it can be established as an artistic creation by combining technology and science with the other six arts (Canudo, 1988).

In general, it can be said that cinema is a social phenomenon that can be analysed from different scientific and philosophical points of


DOI: 10.1079/9781800626034.0010

view (e.g. aesthetic, economic, social, political, cultural and ethical), in different combinations, leading to very interesting multi-perspective considerations.

There is a widespread view among modern social scientists that film is a collective work, an artistic product to which many factors contribute, such as the producer, the scriptwriter, the director, the actors and the various film crews. Each of these factors can, to varying degrees, influence and control part or most of the production process of a film, but above all they must realize their own project with certain goals in mind.

The screenwriter's goal, for example, is to write or edit a story to best fit the narrative and expressive devices of cinema. The actor endeavours to embody a character and accurately portray their emotions and character traits. The director's intention is to portray a certain scenario as well as possible and according to his/her personal, artistic and also philosophical positions. The director's goal can be the commercial success of the film, but he/she can also be guided by art without regard to profit. As one of the main factors in the production, the director therefore has the task of controlling, coordinating and regulating the collective work that everyone else (actors, technicians, engineers, etc.) must do to make a film.

The area in which a film is to be shot is either inevitably determined by the script or, if necessary, is selected by the production management and the respective location team in collaboration with the director. It is therefore easy to understand that the main objective of the production staff when selecting a location is the appropriateness of the space in relation to the script. At the same time, however, many other factors must also be taken into account, such as lighting, accessibility, cost, possibilities of the location and infrastructure. It would not be an exaggeration to say that initially and for several years, film producers and directors did not think at all about the impact of their work on location tourism, just like painters or poets. However, as we will show below, this situation has changed.

On the other hand, it would be wrong to claim that the influence of film on tourism is something new that only emerged in the course of the dynamic development of tourism in the last half of the 20th century. Film has promoted tourism since its beginnings, even without the mediation of the tourism industry. As early as 1895, when ten short films by the Lumière brothers were shown for the first time at the Grand Café in Paris, the cinematic narratives awakened a longing in viewers for the places shown. As examples, we can also cite classical themes such as the representation of characters from Greek mythology and the epics that have appeared since the dawn of cinema, such as in Méliès' *Pygmalion and Galatea* (1898) and *Ulysses and the Giant Polyphemus* (1905), which undoubtedly evoked in the viewer the region in which Greek civilization developed.

Admittedly, the images of cities or landscapes projected in artistic works are fundamental elements for the creation of social representations of places. It is generally accepted that the grandiose and evocative world of the spectacle, together with the brilliance of the actors, has the potential to have an emotional impact on people who are overwhelmed by everyday life.

The screen frames the world in specific aesthetic images and encourages viewers to enliven their imagination and engage dreamily with the cinematic narrative. When cinematic films, as works of art in their own right, depict cities or outdoor spaces where the action of a story takes place, viewers develop an emotional attachment to these places, which become imprinted in their memory in a special way. Consequently, cinematic narratives creatively contribute to composing central mythic images and constructing the viewer's touristic conceptions of the places shown on the screen. This process is reinforced by similar representations projected through other communication systems, ultimately forming the integrated image of the destination. It is clear, then, that the power of the screen entices people to travel dreamily to cinematic places and reshape real spaces into 'imaginary geographies' (Bronfen, 2004; Urry and Larsen, 2011).

At this point, it should be emphasized that a film, as an independent work with artistic value, can reflect the image of a destination and attract visitors for a long time (i.e. as long as it is shown on the big screen). Compared with the past this length of time is now extended as with the various online archives and on-demand digital media services, films can be accessed at any time, which enhances the power of a particular film to attract tourists to a particular place.[1]

One of the most interesting aspects of cinema's special relationship to tourism is that it literally captures and consolidates tourist representations. Although social representations have been defined as dynamic cognitive systems that can change over time, we cannot deny that certain core images persist. In this respect, many landmarks of world cinema lend stability to some of their elements and have formed the core images for the social representations of many large and well-known cities.

Films shot in Paris such as *Les Cousins* (1959) by Claude Chabrol, *Paris Nous Appartient* (1960) by Jacques Rivette, *À bout de souffle* (1960) by Jean-Luc Godard, *Zazie Dans le Métro* (1960) by Louis Malle and *Cléo de 5 à 7* (1962) by Agnès Varda have formed stable key images in the collective perception of the city in the world. This applies to other cities too, for example films shot in Rome, including *Bicycle Thieves* (1948) by Vittorio De Sica, *Roma città aperta* (1945) by Roberto Rossellini, *Mamma Roma* (1962) by Pier Paolo Pasolini and *La Dolce Vita* (1960) by Federico Fellini.

Indeed, one could argue inductively that the many films of the French Nouvelle Vague have formed stable nostalgic key images in our collective perception of Paris, and that many films of Italian neo-realism have done the same for Rome or Milan, and so on.

So we can say that old and qualitative films have value for tourism promotion because they can timelessly reproduce the nostalgic, artistic image of a place that arouses the romantic interest of the audience by recalling past eras.

New films, on the other hand, are agents of change and can offer progressive directions for the development of contemporary elements in the core images of tourist representations.

Be that as it may, it has been argued by many researchers that promoting a place through a film can be much more effective in the long run than promoting tourism through advertising, which has a limited duration (Dore and Crouch, 2003).

In this context, however, it should be emphasized that cinema in our time is increasingly becoming a spectacle on the small screen. Admittedly, a large part of the population watches cinema films on television, computer or cell-phone screens, which compromises their original charisma and artistic autonomy as works of art, degrading them to parts of a continuous, fragmented digital stream.

A derivative of the relationship between cinema and tourism is 'film-induced tourism'. As with literary tourism, this form of tourism refers to the tourist activities triggered by the production and screening of cinema films in collaboration with the tourism industry.

In other words, 'film-induced tourism' or 'cinema tourism' is directly supported by the tourism industry with promotional strategies that present a film or cinema in general as the main motive for visiting the destination in question. However, many tourism researchers use this term in a way that extends it to television, video and DVD productions (i.e. they arbitrarily rename all forms of moving images as cinema or film).[2]

The USA, Canada, Australia, the UK, Switzerland, the Bahamas and several other countries developed film tourism at a very early stage. As a result, destination management organizations in these countries have extensive experience and long-standing programmes to systematically promote the territories they manage to international film producers (Morgan and Pritchard, 1998; Hudson and Ritchie, 2006).

It is a fact that destination management organizations, national film production support centres and film producers are now increasingly seeking to build relationships with each other on a global scale to ensure mutual benefit. A simple search of the international literature can reveal the universal dimension of this phenomenon, as there are hundreds of research publications in the field of tourism studies dealing with film productions that have resulted from such collaborations.

Cooperation between the film and tourism industries can be based either on direct financial support for film production by destination management organizations in the form of grants and tax exemptions, or on some other form of financial facilitation.

In addition, following a film production, businesses and local authorities may jointly or individually propose the printing of promotional cards or the creation of websites leading to filming locations to jointly promote the film with the region and businesses. It is also possible for local tourism businesses as well as the major film producers to organize tourist tours that follow the plot of the film. The Backlot Tour, the Harry Potter Warner Bros Studio Tour, the Cinecitta Studio Tour, etc. can be mentioned here as examples.

Numerous studies conducted in different countries confirm the close relationship between cinema and the world population's preferences for travel destinations. For example, a study conducted in the first decade of the new millennium on the economic contribution of the film industry in the UK found that a relatively large part of the increase in annual tourist visits in the UK was in some way related to British cinema (Oxford Economics, 2010).

Other surveys conducted in the UK have shown that residents taking a holiday abroad visit at least one place where a film or television production has taken place during their vacation, and some of them indicated that this was the main reason for choosing their destination (Statista Research Department, 2020).

In addition, several studies conducted in different countries have shown that a particular film can bring positive publicity in the form of tourist visits to the regions where it was filmed (Tooke and Baker, 1996; Busby *et al.*, 2003; Croy and Walker, 2003; Grihault, 2003).

More specifically, research has shown that many successful fantasy films, such as the *Harry Potter* film series and *The Lord of the Rings* and *The Hobbit* trilogies, have boosted domestic and international tourism in the places where they were filmed (i.e. the UK and New Zealand, respectively) (Lee, 2012; Li *et al.*, 2017).[3]

Other studies have shown a direct link between certain films depicting real historical events and an increase in tourist appeal. Notable examples include: (i) Mel Gibson's *Braveheart* (1995), which was filmed in Scotland; (ii) John Madden's *Captain Corelli's Mandolin* (2001), which was filmed on Kefalonia in Greece; and (iii) Gregor Jordan's *Ned Kelly* (2003), which was filmed in the Australian state of Victoria (Edensor, 2005; Frost, 2006; Hudson and Ritchie, 2006).

## 10.2 Radio and Tourism

The discovery of the ability of radio waves to transmit the human voice and sounds, thanks to scientists such as James Clark Maxwell (1831–1879), Heinrich Hertz (1857–1894), Eduard Branly (1844–1940), Lee De Forest (1873–1961) and above all Guglielmo Marconi (1874–1937), was followed by a phase of research into practical applications. This era culminated in the dominance of the one-way radio model, in which a single transmitter communicated with numerous receivers. The dominance of this model was determined by a combination of political, technical and commercial considerations (Jeanneney, 1996).[4]

It was undoubtedly possible to use radio waves to develop communication models close to the many-to-many model of today's Internet, but one does not have to be particularly inventive to realize that anyone able to transmit audio messages to a large audience using cheap and easy-to-use technology could contribute to the formation of collective representations. However, collective representations that stimulate the imagination can have some influence on political thinking, they can motivate people to act, they can promote the sale of products and – in the context of our topic – they can create attractive images for mass tourism destinations. So, the model that was considered appropriate for the socio-economic conditions of the time was the model of one-way communication from one sender to many receivers.[5]

Thus, the first commercial radio broadcasts began in the 1920s in both America and in Europe, and gradually more and more radio stations appeared in cities throughout the Western world.

America was a pioneer: as early as 1920, various stations were broadcasting regularly to an unspecified audience; by 1921, there were 50,000 receivers in the USA; by 1929, there were 10 million; and by 1939, 31 million. In the USA, economic liberalism led to a model in which commercial broadcasters were free and programmes were allowed to advertise any product or service. In the UK, the model of state-controlled broadcasting eventually prevailed; France chose a middle course, while all other countries followed one of these lines (Jeanneney, 1996).

The invention and introduction of the radio was an innovation that greatly changed people's entertainment and the state of the culture industry, while ushering in a new era of communication. It was received with great enthusiasm and devotion by the inhabitants of Western countries as they craved communication and entertainment and loved travelling to distant lands through vivid stories.

As mentioned in the chapter on media theories (Chapter 5, this volume), in the early years of media research, the media's ability to motivate the masses was overestimated, to which the then emerging radio contributed to some extent. In any case, radio's ability to arouse emotions and stimulate the imagination cannot be denied. Radio invaded the private space of the family with a variety of informative, educational and entertaining broadcasts. Many of them reported on distant countries or played foreign top hits and exotic rhythms.

The most important characteristics of radio are its immediacy and its continuous, one-way broadcast flow. In other words, radio differs from the press and cinema because it is able to broadcast news live, one-way, in real time and without interruption to its listeners. However, radio stations can also broadcast pre-recorded programmes, and involve their listeners in programming to a limited extent by telephone.

While the concept of creating illusions about reality had already been largely realized by cinema and would be reinforced with the advent of television and the Internet, radio succeeded in giving its listeners a particularly powerful illusion. It gave them the feeling of being transported to faraway places, while at the same time convincing them that they were at the epicentre of the world, witnessing events in real time. In both cases, this illusion fostered the belief that they were in control of space and time.

Radio was thus able to capture the listener's attention in the space in which an event took place and to stimulate their imagination without visual constraints. In addition, the intangible nature of radio waves allowed messages to be transmitted over very large areas without the restriction of national borders. Radio listeners could travel by ear directly to music concerts, film festivals and other famous events taking place in remote areas, guided by both the presenter and their imagination in their living room. In other words, they could be both present and absent at the event. The concept of 'imaginary travel', one of the basic forms of mobility defined by Urry (2016), has thus taken on other dimensions.

It can be said that radio gave an impetus to the process of constructing tourist representations through mass media, which for a long time relied only on prefabricated visual and written communication, by conveying immediacy through sounds, music and the spoken word. In this way, the audience's auditory impressions were enriched and their imaginative involvement in the action space was enhanced.

The continuous stream of radio includes various types of programmes such as news, information, sports, music, education, radio magazines, talk shows and drama. Unless the radio stations intend to promote a destination, all these types of programmes are just independent factors that can shape the image of a destination. In other words, since the programmes mostly play in or describe a place, they can promote tourism demand for these destinations without being part of the tourism industry's intended promotion.

Music communicates directly with our emotions and can therefore express and evoke our feelings and influence our mood on certain subjects in a way that rational thoughts and words cannot. The various pieces of music broadcast by a radio station can also arouse a longing for contact with other cultures and a desire to travel, making certain places tourist destinations.

For example, romantic Italian music has been instrumental in shaping the image of Italy as an attractive tourist destination. In the same context, British pop and alternative music has shaped the image of the UK, and London in particular, as one of the most popular destinations for young people. The relationship between music and tourism is a very broad topic that has been thoroughly researched by media, cultural and tourism scholars in recent years[6] (Cohen, 1991; Waitt and Duffy, 2010; Kruger and Trandafoiu, 2013; Lashua *et al.*, 2014).

In any case, radio is directly and systematically used by the tourism industry to enhance the image of tourism products and to promote tourism in general through advertising and various travel programmes. Advertising directly informs potential travellers about offers, while travel programmes recommend destinations, describe places, share experiences and give expert advice.

For advertisers, radio is still an important tool of influence today, as it is an integral part of our daily lives and can reach a large and often specialized audience (motorists, visitors to large shopping centres, etc.).

Tourism advertising on the radio offers many advantages as it can be used with high frequency, at very low cost and for a controlled audience, depending on the type of programme or radio station (young people, older people, islanders, etc.). Moreover, this type of advertising is considered to have great emotional power, as it can easily stimulate the imagination through appropriate manipulations of sounds, voice and music.

It is worth noting that many countries and municipalities have set up special radio stations aimed exclusively at tourists (e.g. Radio Riviera, Radio Caribbean International, Radio Miami International and Irish Tourist Radio). Tourist radios usually use the English language to provide tourists with a variety of information about the region they are in, and regularly inform them about the sights and history of the place. But they also provide information about special events, seasonal festivals and celebrations, tourist offers taking place at that time, etc. and are one of the best ways for advertisers to reach tourists in the region.

By 1995 it was possible to broadcast live radio over the airwaves as well as the Internet. Today, it is generally accepted that the integration of digital and network technologies has opened up new possibilities and perspectives for radio. Digital radio not only improves sound quality, but also offers a wealth of new services. These services include displaying detailed information about the track currently playing on the digital screen and providing real-time updates on elements closely related to tourism, such as weather conditions, traffic reports for roads and motorways and the occupancy status of beaches.

Internet radio stations can now reach an international audience very cost-effectively, without the associated obstacles that traditional radio signal transmission has faced in the past, related to fixed frequencies, transmission power and repeaters. Most travel radio stations now broadcast around the clock to the whole world via the Internet.

Finally, it should be emphasized that radio, television and tourism have a special relationship in Greece, as both Greek radio and Greek television were officially launched in 1926 and 1960, respectively, in the framework of the Thessaloniki International Fair (HELEXPO), an institution that is linked to Greek tourism in various ways.[7]

## 10.3 Television and Tourism

Television adopted parallel broadcasting models to radio, with the private sector taking precedence in the USA. Radio Corporation of America (RCA) set up a transmitter on the Empire State Building in 1931, reflecting the early dominance of private initiatives. On the other side of the Atlantic, the British Broadcasting Corporation (BBC) began broadcasting television programmes from Alexandra Palace in London as early as 1936. In France, René Barthélemy conducted experimental 1 hour weekly broadcasts in 1932, and in 1939 the Eiffel Tower served as an antenna, enabling 15 hours of weekly broadcasts from the studio in Rue Grenelle. Remarkably, Germany had already introduced television in 1928 and was thus an example of the early use of this transformative medium (Jeanneney, 1996).

However, the great development of television in the Western world began after the Second World War, which finally brought audiovisual communication massively into family households.[8]

The spread of a one-way system of audio and video transmission from the television transmitter to people's homes led to an enormous change in the daily lives of individuals, but also in Western culture in general. The immediacy that the radio model had offered in the years before now became even more realistic (i.e. it now took on audio-visual dimensions that more effectively reinforced the viewer's false sense of really being part of the action).

With television, the viewer imaginatively travels and lives everywhere and at all times (past, present, future), regardless of whether the event is actually taking place outside his or her window or on the other side of the world; the illusion of participation knows no spatial or temporal distance (Koronaiou, 1998).

Television viewers were informed, entertained, instructed and manipulated on the basis of a continuous, one-way flow of communication that included various forms of television productions, either real-time or recorded, as well as old cinema films. The great advantage of television was the combination of sound and visual elements, which was of course also true of cinema, except that with television the messages were now transmitted into the home in a continuous stream, sometimes live, and intervened decisively in daily life.[9]

People now had the opportunity to see dramaturgically constructed images of places (beaches, forests, snowy mountains, etc.) from the comfort of their own homes, to have long conversations about them with friendly faces, and to dream that they were there.

The audio-visual messages of television are fleeting, fragmentary, specific and also dramatic, giving the false impression of capturing temporal developments and geographical, qualitative and quantitative characteristics of a place (Fiske and Hartley, 1978).

Since the places shown on television are performatively constructed as images before the viewer's eyes, they become imprinted in the viewer's memory and remain preferentially retrievable when tourist decisions are made.

Any form of television programming (i.e. commercials, TV series, documentaries, news, telefilms, cinema films) are therefore autonomous factors in the construction of tourist images that offer viewers TV favourable future trips, even without the contribution of the tourism industry.

As in the case of cinema, tourism industry players soon recognized the power of television to effectively shape the image of tourism. In addition to the obvious promotion of tourism products through advertising, tourism stakeholders took advantage of television by inventing special forms of tourism motivation. Among others, these included: (i) travel television shows; (ii) subsidized television series and documentaries produced by destination management organizations; and (iii) television presentations of international gastronomic tours etc.

With the development of satellite and cable television in the late 1980s, the increasing broadcasting of tourism advertisements and travel shows on international channels (CNN, BBC, Al Jazeera, etc.) and the creation of tourism-only channels (Travel Channel), tourism television advertising has taken on a universal dimension by shaping global tourism trends.

Originally intended to guide people's buying behaviour, television advertising has now become a dominant form of discourse that can influence a society's lifestyle, gender identity, attitudes towards nature, ethical and traditional values, and thus culture in general.

In view of this, it is not surprising that one of the most common advertising practices of the various players in the tourism industry is television advertising. Particularly in the programmes of international satellite and cable channels, the insertion of advertisements for airlines, ferry companies, large hotel chains, car rental companies, special destinations, etc. has become commonplace and takes up most of the advertising.

Tourism television advertising has become a central topic in tourism research. Accordingly, numerous studies have been conducted on various aspects, such as demonstrating that it increases awareness of tourism products, examining how tourism images are communicated, and competitive analyses investigating how destinations are differentiated through advertising (Kim *et al.*, 2005).

As travel and tourism had become one of the most attractive topics of conversation worldwide in the last decades of the 20th century, travel programmes became a sought-after product for television. Various studies have shown that newer tourism programmes broadcast around the world strive to be as 'nationally neutral' as possible, but in the context of Western culture typically emphasize individual pleasure and personal travel, ultimately prescribing and packaging culture as a commodity, much as mass tourism has done in the past (Fürsich, 2002).

Most of them are a mixture of entertainment programmes, investigative journalism, anthropological and geographical documentaries, educational television, and television advertising. Depending on the objective of the travel programme and the target audience, each of these programme types is given appropriate importance.

Tourist programmes usually present places, sights, cultures and lifestyles in an engaging way (i.e. topics related to a region's history, architecture, town planning, interior design, fashion, music scene, food, clubbing, gardening, traditions and daily habits of the inhabitants). The aim is to provide viewers with a variety of entertaining images, regardless of whether they intend to travel to the place or not, and to turn them into 'fantasy travellers' and hopefully future tourists. The complex news and entertainment format of these programmes thus increases ratings and turns viewers into potential tourists in the context of consumer culture.

On the other hand, tourism's special relationship with television is once again demonstrated by the fact that the tourism industry uses television programmes, such as television series or even television game shows, to produce tourism products. In this case, so-called 'television tourism' (television-induced tourism) is something similar to literary and cinema tourism (i.e. it is tourism activities that are created with the cooperation of the tourism industry, mainly through the production of television series, but without excluding the other forms of television programming).

For example, there are many tourism companies and travel agencies that actively offer and organize tours to the filming locations of *Game of Thrones* (2011–2019, D. Benioff and D.B. Weiss). Another example is a restaurant that bases the design and structure of its menu on a television game show.

Many researchers have argued that certain destinations have experienced a significant increase in tourist attraction because long-running and popular television series have been filmed in these locations. These include the British soap opera *Coronation Street* (1960 to present, T. Warren), the American police series *Hawaii Five-O* (1968–1980, L. Freeman) and others (Couldry, 1998; Rhodes, 2011).

Television series, of course, do not have the artistic substance that distinguishes a literary work or a motion picture, which, as artistic products in their own right, leave indelible impressions on the memory and approach the vision of eternity. However, some successful series enjoy great popularity among viewers, which leads television channels to maintain their production for very long periods of time. In this way, the eternal appeal that a painting or an artistically acclaimed motion picture can have can be balanced by the long run and periodicity of a television series.

In contrast to paintings or films, where the impression of a tourist location is created through a unique and evocative emotional connection with the artistic work, television series function differently. In the realm of television, it is the emotional familiarity created by the repeated viewing of the plot's locations that awakens in viewers the desire for experiential contact with certain places.

Of course, it is very difficult to judge whether TV tourism has the power to be a basic motivation for the initial decision to visit a particular destination. Nevertheless, it offers the possibility to produce tourism products, even if the main motivation for visiting a place is different.

Numerous world-famous and financially rewarding tourist excursions revolve around the striking filming locations of popular television series. Notable examples include tours in New York dedicated to series such as *Friends* (1994–2004, by David Crane and Marta Kauffman), *Sex and the City* (1998–2004, by Darren Star) and *Gossip Girl* (2007–2012, by Josh Schwartz and Stephanie Savage). While the main motivation for tourists to visit New York may be the modern architecture and contemporary cultural offerings (art, music, fashion, food, etc.), these special tourist packages, which include visits to the filming locations of the TV series, are also a big draw for visitors.

## 10.4 Summary

In summary, media with dynamic messages, such as cinema, play a central role in promoting tourism by creating compelling narratives that shape perceptions of destinations, influence travel decisions and provide raw material for advertising strategies. In addition, the development of radio and television as mass media that resonate with recipients in their own environment has profoundly influenced the collective imagination by disseminating images of places with a continuous flow and influencing tourism through various programming and promotional initiatives.

## Notes

[1] It is beneficial to compare the power of films to make places attractive for tourism with the tourism industry's commercials, which only run for a certain time and are then no longer picked up by the media (Dore and Crouch, 2003).
[2] The terms used in tourism studies and in the tourism industry when referring to tourism generated by audio-visual media are somewhat blurred. In international literature, the terms 'film tourism', 'film-induced tourism' and 'cinema-induced tourism' are used to refer to tourism generated by all audio-visual media; this is incorrect as cinema cannot be equated with television or other video productions. The terms 'screen tourism' and 'media-related tourism' are also in use, which we believe are correct in terms of the intended grouping. At this point it is important to stress that cinema, television, video productions, etc. are undoubtedly different media, with different names, particular characteristics and different influences on the audience. Most branches of the social sciences (sociology, social psychology, film and mass media studies, cultural studies, feminist studies, etc.) generally use the term media when theorizing mass communication, while they use the specific names of each media when studying a subject in the respective medium (i.e. cinema, television, radio, electronic games, etc.) (Butler, 1990; Evans, 1997; Busby and Klug, 2001; Hudson and Ritchie, 2006; Olsberg/Olsberg/SPI (2007); Connell and Meyer (2009); Connell, 2012).
[3] The *Harry Potter* film series: *Harry Potter and the Philosopher's Stone* (Columbus 2001), *Harry Potter and the Chamber of Secrets* (Columbus 2002), *Harry Potter and the Prisoner of Azkaban* (Cuarón 2004), *Harry Potter and the Goblet of Fire* (Newell 2005), *Harry Potter and the Order of the Phoenix* (Yates 2007), *Harry Potter and the Half-Blood Prince* (Yates 2009), *Harry Potter and the Deathly Hallows – Part 1* (Yates 2010) and *Harry Potter and the Deathly Hallows – Part 2* (Yates 2011). *The Lord of the Rings* film trilogy: *The Lord of the Rings: the Fellowship of the Ring* (Jackson 2001), *The Lord of the Rings: the Two Towers* (Jackson 2002) and *The Lord of the Rings: the Return of the King* (Jackson 2003). *The Hobbit* film series: *The Hobbit: an Unexpected Journey* (Jackson 2012), *The Hobbit: the Desolation of Smaug* (Jackson 2013) and *The Hobbit: the Battle of the Five Armies* (Jackson 2014).
[4] This particular technological application of radio waves, which later became the norm for broadcasters, was developed by Guglielmo Marconi, who in 1884 succeeded in transmitting radio waves from one side of his room to the other, while in 1901 succeeded in transmitting radio messages from one side of the Atlantic to the other. The first major radio station was established in Pittsburgh in the USA in 1920 (Davis, 1996).
[5] According to many researchers, the unidirectional way of using technology to disseminate news chosen for the operation of radio and television, as well as the organizational structure of the companies that developed these media, also reflect the general political perception of the Western world, which makes them the most important mass media for the production of propaganda and public influence (Lasswell, 1927; Ellul, 1965; Bernays, 2005; Herman and Chomsky, 2010).
[6] Be that as it may, the fact is that tourism research has generally overemphasized the sense of sight, leading to a dearth of studies on tourism's relationship to hearing and sound, as well as the other senses (Connell and Gibson, 2004; Waitt and Duffy, 2010).
[7] The first radio broadcast with a quasi-regular programme took place in October 1926, together with the 1st Thessaloniki International Fair (HELEXPO). Since then, HELEXPO has been held every year and serves as a platform for companies, industries and organizations to present their products, services and innovations to a domestic and foreign audience. Participants include domestic and foreign companies from various sectors such as industry, technology, agriculture, tourism, consumer goods and more. As the exhibitions were then under the supervision of the 'Foreigners and Exhibitions Service', which was also responsible for tourism in Greece, this major commercial event acquired tourist significance. A football match was even organized between the Greek and Bulgarian teams, and two foreign countries, the Soviet Union and Hungary, participated in the exhibition with their stands. However, as the cost of owning a radio was extremely high, there were few listeners. Be that as it may, the first public radio station in Greece was established in 1938 in the Attica region under the name Athens Radio Station. Later, on 3 September 1960, on the evening of the opening of the 25th HELEXPO, Greece's first television station began operating at the exhibition centre.
[8] Television is the result of a long process of imagination, research and technological application that led to this medium. In 1884, German scientist Paul Nipkow (1860–1940) invented a system of spinning discs that scanned the moving image and converted it into a series of electrical pulses, creating a poor-quality image on a screen. In 1926, Scottish engineer John Logie Baird (1888–1946) carried out the first

demonstration of television in a Selfridges shop in London, and by 1934 over 10,000 sets had been sold (Davis, 1996). In Greece, the first experimental television operation took place much later, by the State Electricity Company at HELEXPO 1960, while a rudimentary regular programme was broadcast by the National Broadcasting Foundation (EIR) in Athens in 1966.

[9] One of the technologies not covered in this book is video and the video cassette recorder (VCR), which had its heyday in the 1980s and then gradually declined. Video contributed to another important shift in the cultural practices of small-screen entertainment, as it freed viewers from the tyrannical obligation to watch programmes at the time when they aired. On the contrary, they had the choice to watch a film, similar to books or magazines, meaning they could watch it when they wanted and they had control over consumption. There is no doubt that video films, depending on the subject and production location, influence the audience in the creation of tourist images.

## References


Bernays, E.L. (2005) *Propaganda*. Ig publishing, Brooklyn, New York.

Bronfen, E. (2004) *Home in Hollywood: The Imaginary Geography of Cinema*. Columbia University Press, New York. DOI: 10.7312/bron12176.

Busby, G. and Klug, J. (2001) Movie-induced tourism: the challenge of measurement and other issues. *Journal of Vacation Marketing* 7(4), 316–332. DOI: 10.1177/135676670100700403.

Busby, G., Brunt, P. and Lund, J. (2003) In Agatha Christie country: resident perceptions of special interest tourism. *Tourism* 51(3), 287–300.

Butler, R.W. (1990) The influence of the media in shaping international tourist patterns. *Tourism Recreation Research* 15(2), 46–53. DOI: 10.1080/02508281.1990.11014584.

Canudo, R. (1988) Reflections on the seventh art. In: Abel, R. (ed.) *French Film Theory and Criticism*. Princeton University Press, Princeton, New Jersey, pp. 1907–1929.

Cohen, S. (1991) *Rock Culture in Liverpool: Popular Music in the Making*. Oxford University Press, Oxford, UK. DOI: 10.1093/oso/9780198161783.001.0001.

Connell, J. (2012) Film tourism – evolution, progress and prospects. *Tourism Management* 33(5), 1007–1029. DOI: 10.1016/j.tourman.2012.02.008.

Couldry, N. (1998) The view from inside the simulacrum: visitors tales from the set of Coronation Street. *Leisure Studies* 17(2), 94–107. DOI: 10.1080/026143698375178.

Connell, J. and Gibson, C. (2004) World music: deterritorializing place and identity. *Progress in Human Geography* 28(3), 342–361. DOI: 10.1191/0309132504ph493oa.

Connell, J. and Meyer, D. (2009) Balamory revisited: an evaluation of the screen tourism destination-tourist nexus. *Tourism Management* 30(2), 194–207. DOI: 10.1016/j.tourman.2008.06.001.

Croy, W.G. and Walker, R.D. (2003) Rural tourism and film – issues for strategic regional development. In: Hall, D., Roberts, L. and Mitchell, M. (eds) *New Directions in Rural Tourism*. Ashgate Publishing, Aldershot, pp. 115–133.

Davis, E. (1996) *Inventions*. Pataki Publications, Athens. (in Greek).

Dore, L. and Crouch, G.I. (2003) Promoting destinations: an exploratory study of publicity programmes used by national tourism organisations. *Journal of Vacation Marketing* 9(2), 137–151. DOI: 10.1177/135676670300900203.

Edensor, T. (2005) Mediating William Wallace: audio-visual technologies in tourism. In: Crouch, D., Jackson, R. and Thompson, F. (eds) *The Media and the Tourist Imagination: Converging Cultures*. Routledge, London, pp. 119–132. DOI: 10.4324/9780203139295.

Ellul, J. (1965) *Propaganda: the Formation of Men's Attitudes*. Knop, New York.

Evans, M. (1997) *Plugging into TV Tourism*. English Tourist Board, London.

Fiske, J. and Hartley, J. (1978) *Reading Television*. Routledge, London. DOI: 10.4324/9780203356623.

Frost, W. (2006) Braveheart-ed Ned Kelly: historic films, heritage tourism and destination image. *Tourism Management* 27(2), 247–254. DOI: 10.1016/j.tourman.2004.09.006.

Fürsich, E. (2002) Packaging culture: the potential and limitations of travel programs on global television. *Communication Quarterly* 50(2), 204–226. DOI: 10.1080/01463370209385657.

Grihault, N. (2003) Film tourism – the global picture. *Travel & Tourism Analyst* 5, 1–22.

Herman, E.S. and Chomsky, N. (2010) *Manufacturing Consent: the Political Economy of the Mass Media*. Random House, London.
Hudson, S. and Ritchie, J.B. (2006) Film tourism and destination marketing: the case of *Captain Corelli's Mandolin*. *Journal of Vacation Marketing* 12(3), 256–268. DOI: 10.1177/1356766706064619.
Jeanneney, J.-N. (1996) *Une Histoire des Medias: Des origines à nos jours*. Seuil, Paris. (in French).
Kim, D.Y., Hwang, Y.H. and Fesenmaier, D.R. (2005) Modeling tourism advertising effectiveness. *Journal of Travel Research* 44(1), 42–49. DOI: 10.1177/0047287505276590.
Koronaiou, A. (1998) Media and new representations of temporality. In: Panagiotopoulou, R., Rigopoulou, P., Rigou, M. and Notaris, S. (eds) *The Construction of Reality and the Mass Media*. Alexandria Publications, Athens. (in Greek).
Kruger, S. and Trandafoiu, R. (eds) (2013) *The Globalization of Musics in Transit: Musical Migration and Tourism*. Routledge, London. DOI: 10.4324/9780203082911.
Lashua, B., Spracklen, K. and Long, P. (2014) Introduction to the special issue: Music and Tourism. *Tourist Studies* 14(1), 3–9. DOI: 10.1177/1468797613511682.
Lasswell, H.D. (1927) *Propaganda Technique in the World War*. Paul Kegan, London.
Lee, C. (2012) Have magic, will travel: tourism and Harry Potter's United (Magical) Kingdom. *Tourist Studies* 12(1), 52–69. DOI: 10.1177/1468797612438438.
Li, S., Li, H., Song, H., Lundberg, C. and Shen, S. (2017) The economic impact of on-screen tourism: the case of *The Lord of the Rings and the Hobbit*. *Tourism Management* 60, 177–187. DOI: 10.1016/j.tourman.2016.11.023.
Marsh, J.L. (1992) Fitzgerald, Gatsby, and the last tycoon: the 'American dream' and the Hollywood dream factory. *Literature/Film Quarterly* 20(1), 3.
Morgan, N. and Pritchard, A. (1998) *Tourism Promotion and Power – Creating Images, Creating Identities*. Wiley, Chichester, UK.
Olsberg/SPI (2007) *How Film and TV Programmes Promote Tourism in the UK*. Film Council, London.
Oxford Economics (2010) *The Economic Impact of the UK Film Industry*. Oxford Economics, Oxford, UK.
Rhodes, K. (2011) *Booking Hawaii Five-O: an Episode Guide and Critical History of the 1968–1980 Television Detective Series*. McFarland, Jefferson, North Carolina.
Roberts, G. (2007) Dream factory and film factory: the Soviet response to Hollywood 1917–1941. In: Cooke, P. (ed.) *World Cinema's Dialogues with Hollywood*. Palgrave Macmillan, London, pp. 35–51. DOI: 10.1057/9780230223189.
Statista Research Department (2020) Influence of film/TV locations on holiday destination choices in the UK 2017. Available at: https://www.statista.com/statistics/871859/film-tourism-influence-on-holiday-destinationchoice/#statisticContainer (accessed 10 May 2021).
Tooke, N. and Baker, M. (1996) Seeing is believing: the effect of film on visitor numbers to screened locations. *Tourism Management* 17(2), 87–94. DOI: 10.1016/0261-5177(95)00111-5.
Urry, J. (2016) Does mobility have a future? In: Griego, M. and Urry, J. (eds) *Mobilities: New Perspectives on Transport and Society*. Routledge, Abingdon, UK, pp. 3–19. DOI: 10.4324/9781315595733.
Urry, J. and Larsen, J. (2011) *The Tourist Gaze 3.0*. Sage, London. DOI: 10.4135/9781446251904.
Waitt, G. and Duffy, M. (2010) Listening and tourism studies. *Annals of Tourism Research* 37(2), 457–477. DOI: 10.1016/j.annals.2009.10.017.

# 11 New Digital Media and Tourism

**Abstract**

Since the turn of the millennium, communication in tourism has changed dramatically thanks to digital and wireless technologies. This chapter examines the evolution of the industry, focusing on the transition from analogue to digital technology. Digital tools have streamlined operations and changed interactions in the tourism industry. One focus is the role of the Internet in the 'democratization' of destination image building. The influence of user-generated content and the emergence of 'participatory tourism journalists' are key phenomena. Modern means of communication and wireless networks have simplified tourism transactions and improved the travel experience. In addition, the chapter emphasizes the growing importance of new media platforms in tourism research and marketing and highlights the profound impact of technology on the global travel landscape.

As Negropontis (1995) states in his book *Digital World*, humans basically live in an analogue world; our perception and interaction with the physical world are analogue in nature. What the author is essentially saying is that the signals with which humans receive messages from the environment through their senses form a continuum in space and time. Communication (i.e. the exchange of signals), takes place through individuals in a continuous spectrum that has even the smallest fluctuations until the message is completed. The sound of the voices of two people talking to each other is a very typical example of an analogue signal exchange. In this context, it should be noted that some of the typical applications of analogue signal transmission are records, music cassettes, video tapes, older radio and television systems and so on.

In an effort to increase the efficiency of tasks such as information storage, data processing and machine communication, scientists have developed the concept of digitizing signals. This involves simulating the continuous flow of an analogue message and converting it into a binary code. This binary coding is not only simpler, but also more precise and fits seamlessly into the specific activities they wanted to advance.

In simple terms, the digital signal is the binary expression of each element being processed or transmitted (i.e. its encoding with the digits 0 and 1).

Digital technology therefore concerns all applications, systems, devices, data and everything that has to do with digital encoding. In this context, it should be noted that some of the typical applications of digital signal transmission media are DVDs (digital versatile discs), personal computers, digital radio, digital television and social media.

There is no doubt that digital technology now plays a crucial role in most aspects of daily life and forms the basis for computer systems and modern telecommunications. It is widely believed that the technologies of the Internet, new digital media and wireless networking, especially after the turn of the millennium, have completely changed the communications landscape.

Certainly, the term 'new digital media' is dynamic by nature, as the adjective 'new' has a certain mutability and indicates continuous development. Current definitions vary; some emphasize the technical aspects of digital, networked and wireless technologies, while others highlight the cultural forms and contexts in which these technologies are used.

In common parlance, information and communication technology (ICT) is a very general term that refers to all forms of technology used to transmit, process, store, create, analyse, display, communicate or exchange information via electronic communications,



DOI: 10.1079/9781800626034.0011

satellite systems, the Internet and computers, and broadly encompasses a wide range of tools and systems and their applications.[1]

When we refer to 'new digital media' in this book, we generally mean the media that have emerged from today's digital ICTs and their applications that are directly connected to computers, the Internet, wireless networks, smart devices, etc.

## 11.1 ICT and the Tourism Industry

The development of information technology in conjunction with communication in the 20th century has greatly changed all areas of human activity. However, it is noteworthy that the applications of ICT were used very early in the tourism industry.

In the late 1940s, the ENIAC (Electronic Numerical Integrator And Computer), the first large-scale electronic digital computer, was a pioneering machine that occupied a large space and could perform a variety of numerical calculations at high speed; it is believed to have started the adventure of airline reservation systems. At the same time, the UNIVAC-1 (Universal Automatic Computer 1), often considered the first commercially available computer in the USA, helped demonstrate the feasibility of computers for business and scientific applications and marked the introduction of computers to the market in general. Be that as it may, it is a fact that United Airlines in Denver, USA, used the IBM Model 305 Disc Storage as a reservation system as early as 1956 (Pan, 2016a; Ceruzzi, 2003; Sfakianakis, 2008).

One of the first applications of computer information technology that supported the distribution and organization of flight reservation information were computer reservation systems (CRS). These were centralized computer systems used by airlines, and hotels, to manage and book travel-related services.

As early as the early 1960s, American Airlines were using CRS to manage their reservations, and shortly thereafter some major hotel chains did the same. Besides American Airlines, United Airlines, TWA, Eastern Airlines, Hilton Hotels, Hyatt Hotels, Marriott International and many others pioneered the use of CRS technology. Later, car rental companies also used the same systems.

The gradual growth of tourism in the 1970s, but especially in the 1980s, prompted many airlines and hotel chains to link their business systems with travel agencies to make financial transactions faster and easier. Their efforts helped to modernize reservation and booking systems and make travel and accommodation planning more accessible and efficient for professionals. Thus, CRSs have gradually evolved into interactive electronic data systems supported by computer networks, providing access to databases of airlines, hotels, car rental companies, etc. to check availability of products and services and make reservations (Naqvi and Jia, 2016).

As a result, global distribution systems (GDS) have emerged to connect, integrate and disseminate travel information from different travel suppliers internationally, while providing the ability to conduct travel transactions with the companies participating in each system. These systems are intermediary systems that connect all service providers with travel sellers and facilitate the distribution of travel services to a larger international market.

The development of GDS was an important milestone in the evolution of the unified tourism industry. These systems automated, streamlined and unified the booking process globally for various travel services such as flights, hotels and car rentals. They brought together information and booking options from different providers and made it easier for travel agents, airlines and consumers to access and book travel-related services.

There is no doubt that even before the GDS, the concept of a unified tourism industry had emerged, in which travel agencies, hotels, airlines and transport services were active. However, it is important to note that the recognition of tourism as a distinct and integrated industry varies from place to place, depending on the local economy, technological advances, natural attractions and cultural heritage. In some countries, the economic potential of tourism was recognized early due to advanced techno-economic conditions, which facilitated the understanding of the unity of the sector and the linkages between the various relevant private and public sector services. Other countries needed more time to establish it as an independent economic sector.

In any case, the United Nations World Tourism Organization (UNWTO) was founded in 1974, further underlining the importance of tourism as a global industry. Overall, the development of tourism as a recognized industry has been a gradual process, influenced by economic, social, technological and international political factors.[2]

Nevertheless, the systems of GDS played a crucial role in simplifying and connecting the various components of the industry by offering travellers and travel professionals a one-stop solution.

In this respect, tourism, as an information and communication-intensive sector that was an early adopter of ICT, has easily adapted to the development of the Internet, making it one of its main tools for conducting business.

We can justifiably say that at the end of the 20th century, people were living in a rapidly changing new world that had evolved considerably with the advent of sophisticated audio-visual media and the digitization of all forms of communication. In this context, anyone could see a far greater number of images depicting different places (destinations) in just half an hour, than someone living in an agricultural or pre-industrial society could see in a lifetime.

The 1990s saw widespread access to television and the proliferation of cable and satellite channels, while towards the end of the decade the Internet gained popularity. This opened up the world of online travel websites, image galleries and virtual tours that made images of destinations even more accessible.

Anyway, we can argue that in the early years of the so-called digital revolution, the Internet simply provided tourism companies with another way to advertise consumer information on their websites. However, this was a decisive step forward, as it enabled every player in the tourism industry to create and disseminate its image worldwide at low cost by presenting photos and texts on the Internet. Potential tourists could explore destinations with minimal effort, access relevant information and contact businesses directly – whether by phone or e-mail – to facilitate transactions.

In addition, year after year, there was a greater integration of media, with audio, video and text converging on digital platforms. This convergence led to the development of multimedia content and digital entertainment.

It is important to note that transport companies and large hotel chains were not only among the first to adopt and support internal electronic data systems to operate more efficiently, but also among the first to develop 'e-commerce' on a large scale (Mills and Law, 2004).

With the development of GDS, the Internet and its applications, but also with the help of wireless technology, tourism companies have gradually evolved from advanced businesses that only had a global presence to omnipotent platforms where anyone, anywhere, can instantly conduct global electronic transactions.

Indeed, at the beginning of the new millennium and given the growth of the Internet, GDS have evolved into companies specializing in the sale of many different types of tourism services, gradually opening up to consumers in collaboration with online travel agencies. The Sabre, Galileo, Amadeus and Worldspan systems have gained the largest market shares (Buhalis and Licata, 2002; Buhalis, 2003; Pan, 2016b).

The new developments in communication technology thus led to a certain restructuring of the business in the tourism industry. On the one hand, this situation led to the weakening of several traditional intermediaries (i.e. mainly those tourism businesses that did not add value to the tourism product and did not want to adapt to the changing times). On the other hand, it contributed to the creation of new digital interfaces (tourism portals, e-shops, etc.) that drove tourism promotion.

In this context, the new term 'disintermediation' used by many tourism scholars is also worth mentioning. It refers to the fact that tourists increasingly bypass the various intermediaries (travel agencies, etc.) and go directly to the providers of the tourism product via the Internet.

The phenomenon of 'disintermediation' in tourism has been accelerated by online travel websites and apps. These platforms allow consumers to book travel services directly with providers such as airlines and hotels, eliminating the need for intermediaries. This can be convenient and cost-effective for travellers, but it has also led to changes in the business landscape. As we mentioned earlier, traditional travel agencies had to adapt to the new digital landscape to remain relevant, and those that did not were doomed to fail (Graham, 2011).

### 11.1.1 Emergence of digital media

Traditionally, communication media such as newspapers, television and radio served primarily as vehicles for the dissemination of information. However, the emergence of digital media, which includes the Internet and social platforms, heralded a profound change. These channels now serve as dynamic, multifaceted tools that enable two-way communication, interactive engagement and the very important transactional processes. This shift from their historical role as information channels to multifaceted platforms that enable direct business transactions has triggered a revolution in the way travellers conceptualize, plan and book their trips. At the same time, this development has led to considerable changes in the strategies and processes of companies in the tourism sector. Indeed, the exchange of information about tourism offers, the intricacies of travel planning and the buying and selling of tourism products and services via the Internet have quickly become the industry standard and have fundamentally changed the old landscape of tourism transactions from the 'bricks-and-mortar' business of travel agencies.

The transformation of the tourism industry into the digital sphere has essentially given rise to the term 'electronic tourism' (e-tourism), which refers to the digitization of all areas and services of tourism (i.e. the use of digital and network technology by agents, hotels, restaurants, transport, etc.).

E-tourism has eventually evolved into a global system that encompasses many important aspects of travel, from planning and booking to on-site experiences and post-trip feedback. It has become an integral part of the tourism industry, providing convenience to travellers, expanding the reach of tourism businesses and contributing to the development of the global tourism sector (Buhalis, 2003).

According to Katsoni (2017), from the company's point of view, e-tourism moves between three dimensions: (i) the Internet, which allows it to present itself, connect with the whole world and do business with consumers; (ii) the intranet, which supports productivity through intra-company communication; and (iii) the extranet, which it uses to organize its collaboration with tourism business networks.

In particular, through the use of modern applications of digital ICT, the various integrated and globalized tourism systems have expanded unhindered to consumers, facilitating all those who do business in the sector by enabling accurate, secure, instantaneous, transparent and efficient processes for financial transactions, free from the obstacles created by the dimensions of place and time. Therefore, today's digital ICT in the tourism industry encompass a wide range of applications involving the Internet, e-commerce, geographic information systems, location systems, etc., while integrating traditional media (Katsoni, 2017).

From a business point of view, 'e-tourism' is thus understood as 'e-business' in the field of tourism. Therefore, when we examine the use of new ICT and the Internet in the tourism business, we seem to be referring to the broader concept of e-business.

According to Georgopoulos *et al.* (2013), the concept of 'e-business' is much broader than 'e-commerce', as it refers to the provision of other services to customers and collaboration with business partners, in addition to the buying and selling of products or services. 'E-business' also includes the use of technology for decision making and collaboration at the internal company level, as well as for the improvement and automation of all business processes.

Essentially, the term 'e-tourism' has a broader meaning in the field of tourism. It encompasses the use of modern digital ICT in all facets of tourism involving various stakeholders. This extends to the overarching concept of creating a comprehensive system that integrates the entire field of tourism via the Internet. This includes not only the formal tourism businesses, but also the tourists themselves and all those involved in this dynamic phenomenon.

This reality has enabled many companies to efficiently expand their tourism activities. Tourism platforms use state-of-the-art web technologies to inform about travel-related offers and to get in touch with the public in an appropriate way. They can build online communities on tourism, exchange information between users and comprehensively support different aspects of tourism services. This includes geographical and anthropological insights, practical guidance and recommendations, complaint handling, highlighting travel issues and much more. In addition, these platforms play a role in promoting and evaluating tourism products throughout their life cycle, from creation to consumption.

At the same time, potential travellers have access to a much larger amount of trustworthy and detailed information, which facilitates travel planning through direct transactions. It has also enabled the creation of travel blogs where individuals can express their opinions and take on the role of 'participatory tourism journalists'. This term emphasizes the active involvement of citizens in the information-gathering process concerning tourism products. 'E-tourism' has also driven the development of hospitality services online and beyond, whether they involve financial transactions or not.

Indeed, the Internet has enabled several organizations and platforms to offer alternative options far from the practices of the established tourism industry for free or low-cost tourism by connecting with local hosts or like-minded travellers (CouchSurfing, BeWelcome, TrustedHousesitters, Workaway, Warm Showers, etc.).

This makes us very optimistic about the move towards a kind of collaborative tourism (sharing economy tourism, peer-to-peer tourism, etc.). Collaborative tourism platforms allow travellers to connect with hosts or locals to find accommodation, cultural exchange, volunteering or unique travel experiences. This is part of the general trend of the sharing economy and collaborative consumption in tourism, which allows individuals to share resources and experiences while travelling (Dredge and Gyimóthy, 2017).

Nevertheless, it should be mentioned that paradoxically, through the environment of the Internet and with the support of Web 2.0 technologies, the big media companies continue to concentrate and homogenize the cultural environment, but in a somewhat more liberal way in the sense that small companies only get a minimal piece of the pie and are almost invisible in most cases.

In the case of tourism, this can be seen in the identical product advertising strategies of the few companies (booking.com, Expedia, Trivago, Skyscanner, etc.) that have caught on worldwide.

Be that as it may, the whole situation is leading to a kind of new participatory form of tourism marketing, seen by many as more beneficial to the consumer, who has a key role in shaping the tourism programme and greater choice.

## 11.2 The Influence of User-Generated Content

There are some fundamental changes that have altered the construction of tourism images. The progressive development of online technologies has enabled tourists to contribute to the publicity and evaluation of the tourism product through their posts on blogs and later on the broader category of social media, thus significantly influencing the design, consumption and marketing of tourism products.

As recently noted by social media researchers, posts that are broadly related to tourism are the most popular among users, the most shared and the most searched for (Miguéns *et al.*, 2008; Yoo *et al.*, 2016).

The advent of online technologies has led to a more inclusive approach to content creation, where tourists have a significant impact on how we perceive destinations and services. Their insights and first-hand experiences now play a crucial role in helping other travellers make informed decisions about their choice of destination, accommodation, activities and arrangements. Today, travellers increasingly rely on these digital reviews and opinions to make travel plans and consider them part of their information-gathering process, though sometimes not the only one (Gretzel and Yoo, 2008; Cox *et al.*, 2009).

What's more, in our modern digital age, not only tour operators and tourists have the power to shape a destination's image, but virtually everyone. This includes locals, transients, immigrants, foreign students, refugees and many more. The proliferation of social media platforms has given everyone the opportunity to influence perceptions of places and what is considered 'authentic'. Anyone can bring a unique perspective to a place in their social media posts, highlighting hidden treasures, cultural insights and their personal connection to a place.

In addition, posts on social media can receive comments, likes and shares, which serve as indicators of audience admiration. This engagement further enhances the impact of the content. In this way, the image of a destination can be enriched through collective efforts. When many individuals contribute their unique perspectives, a more comprehensive and compelling image of a place emerges.

It is a fact that most people interested in a particular tourist offer or looking for information about hotels, etc. turn to blogs and social media to read the opinions and reviews of those who have already had the experience of travelling there. At the same time, they have the opportunity to ask questions and share their own experiences and photos on online platforms after their trip (Bosangit *et al.*, 2012; Browning *et al.*, 2013; Aluri *et al.*, 2015).

Assuming that freedom of expression is indeed possible on the Internet and that there are no prohibitions on speech, it is easy to conclude that this fact diminishes to some extent the communicative power of traditional large travel agencies and destination management organizations, and their exclusivity in presenting and promoting places. The development of online platforms and social media that allow tourists to speak for themselves thus contributes to an objective, impartial and more global representation of tourism, not to say a 'democratization' of the tourism dialogue.

All these circumstances have led to a change in the way the image of a destination is constructed. Nevertheless, the tourism industries of most countries and all factors related to the phenomenon of tourism have shown great adaptability to the new conditions.

A prime example of the effective creation of an enticing image of a holiday destination through the extensive use of new media is the Indonesian island of Bali. While it initially developed as a tourist hotspot after the Second World War, it owes much of its transformation into a globally recognized and sought-after destination to the influence of new media, particularly social media and online travel platforms. The stunning landscapes and natural beauty of Bali, with its lush green rice terraces, pristine beaches and breathtaking sunsets, are remarkably photogenic. This inherent visual appeal sets this province of Indonesia apart from other destinations and makes it an ideal subject for captivating photos and videos that attract the attention of content creators on platforms such as Instagram and YouTube.

Indeed, Bali has become a focal point for influencers, bloggers and YouTubers looking to raise their profile by showcasing the island's rich culture, lifestyle and attractions to a global audience. In particular, the Instagram hashtags #Bali, #BaliLife and #ExploreBali have generated millions of posts, allowing for easy discovery and interaction between users interested in Bali-related topics, further raising the destination's profile. It is undeniable that Bali's distinctive fusion of culture, spirituality and traditions, as often portrayed in content on platforms such as YouTube, attracts travellers in search of authentic and exotic experiences. In addition, online review platforms such as TripAdvisor and Google Reviews provide convenient access to fantastic deals and offer travellers a platform to share their personal experiences.

Bali is also very popular with digital nomads and many of them have made a habit of describing their experiences online. This subculture has helped to cement this place's reputation as a centre for remote working and adventure.

The island's numerous festivals, including the BaliSpirit Festival and the Ubud Writers and Readers Festival, its traditional Balinese cuisine and its commitment to sustainable tourism, extensively covered by bloggers and Instagrammers, have greatly enhanced Bali's status as a tourist epicentre (Arisandi *et al.*, 2014; Diposumarto *et al.*, 2015).

Looking at Europe, several researchers have also highlighted Lisbon's remarkable success as a destination in the digital age, underlining the central role of social media in shaping the tourist image. The fusion of the city's visual appeal, perfectly tailored to digital screens, with the strategic marketing efforts of tourist boards, actual cultural heritage and, more importantly, the sharing of relevant posts on social media by ordinary users and visitors, has created a captivating image that positions Lisbon as a prime destination for a variety of tourist groups (Miguéns *et al.*, 2008; da Luz Cerdeira and Fernandes, 2020; Miranda and Dias, 2020).

However, it has become difficult to distinguish between pure social tourism media and travel agencies that have developed an e-business, as all online platforms are in a constant process of homogenization in terms of their characteristics and cannot be clearly distinguished from one another. Be that as it may, even though both entities produce similar types of content, such as travel guides, destination reviews and travel tips, social tourism media focuses on creating engaging content and building a community, while travel agencies use digital channels to sell their services. To differentiate between the two, we need to look at the specific features and offerings of both organizations, consider factors such as the main purpose of the platform, the type of content, the presence of booking or transactional features and any links to travel services.

A typical example is the online travel service TripAdvisor, which was founded at the beginning of the new millennium. It was originally a commercial company that wanted to differentiate itself from the growing competition in this area and, after 18 years of successful operation, has transformed itself into a social media model for tourism.

TripAdvisor now relies on 'bloggers' and so-called 'influencers' (i.e. people who are not paid employees of the company), as they have access to a large audience, their 'followers', who they can convince with their authenticity on many topics. On this platform, everyone has the opportunity to freely share their opinions, audio-visual material and personal travel plans, which adds a new dimension to travel planning information (Yoo *et al.*, 2016; Müller, 2018).

It would not be an exaggeration to say that the Internet is possibly the most important arena in which competition in the tourism industry takes place today. This is because it has a major impact on how travellers research, book and experience their trips, making it a crucial battleground for tourism companies.

As we have already mentioned, the application of new technologies has also led to the emergence of a new type of dynamic tourism journalist. The concept of tourism journalism existed long before new media, but the Internet and wireless digital technology have changed it significantly. With the help of new communication technologies, travel journalists have more opportunities to influence the choice of destinations and tourism industry projects and to foster a sense of community among travellers. This change has made travel journalism more dynamic, interactive and diverse than ever before.

While a professional tourism journalist is someone who has formal journalistic training and specializes in reporting on tourism issues on assignment, the term 'participatory tourism journalist' refers to a person who actively participates in tourism events themselves, rather than just reporting on them, and who does not have a work assignment from a media company. The first examples of participatory tourism journalism in the digital age can be traced back to the early 2000s when travel bloggers emerged, but the field has evolved and adapted with technological advances.

Actually, the concept of the participatory tourism journalist is very similar to that of the 'citizen journalist', which is not new as this practice is linked to alternative media theory.

'Citizen journalism' is about ordinary citizens taking an active role in gathering, analysing and disseminating news to provide independent, reliable and accurate information in support of pluralism and democracy (Gillmor, 2006).

It is a fact that with the help of the new technologies of the Internet and smart devices, the institution of the tourism journalist has assumed very large proportions. Anyone who wants to share information about destinations they consider important can do so on their websites, travel blogs, social media (TikTok, Instagram, Facebook, YouTube, etc.) or by publishing audio-visual material on the websites of traditional mass media.

However, today's tourism journalist in social media contrasts with the volunteer journalist in alternative media who is primarily engaged in political activism. The participatory tourism journalist is more reminiscent of the concept of 'prosumer' developed by Toffler (1990), Ritzer and Jurgenson (2010) and other contemporary social researchers. This is because the modern tourism journalist both consumes tourism products and produces free tourism information and advertising that benefits the large commercial online tourism media.[3]

Be that as it may, both amateur and professional tourism journalists contribute to the wealth of information available to travellers and influence their decisions and experiences in the tourism industry.

In addition, the technology of portable communication devices and wireless networks has provided tourists with various applications that are primarily practical tools, such as: (i) location maps that explore the location of markets, attractions, etc.; (ii) various guidance systems; and (iii) software that enables communication with local volunteers who help tourists.

In this way, the benefits of new technologies are not limited to choosing the destination and planning the trip, but also extend to assisting the tourist during their journey. Smart mobile devices are now considered indispensable for the modern tourist and largely determine every tourist activity (Fotis *et al.*, 2011).

Furthermore, the emergence of new digital devices and the proliferation of user-generated content, including comments and photos shared on social media platforms, have significantly changed the landscape of tourism consumer behaviour documentation. These technologies can not only reveal and record tourists' consumption patterns, but also shape and mould their social profiles and identities, which is of great importance for academic and marketing research in tourism (Fodness and Murray, 1999).

## 11.3 Growing Importance of New Media Platforms

### 11.3.1 In tourism research

In academia, the development of digital wireless network technologies has ushered in a new era in tourism research and provided new impetus for academic enquiry. Mobile devices, digital cameras, social media platforms, navigation tools and search engines have intervened in and reshaped the traditional themes of tourism research. For example, the traditional notion of the 'tourist gaze' is now expressed primarily through selfie photos, and the concept of 'authentication' now relies heavily on social media platforms.

The traditional sense of seclusion that characterized tourist escapism has been partially eroded as people are now constantly connected to their social networks through instant messaging and various online applications.

In addition, many specific studies have been conducted to address various issues arising from the profound impact of social media on the tourism sector. For example, many recent studies examined the direct impact of the intensity of online discussions on tourism platforms on travel planning and destination preferences in the context of the rapid spread of the COVID-19 pandemic (Buted *et al.*, 2014; Yu *et al.*, 2021).

Consequently, it is evident that the development of novel technologies plays a crucial role in shaping the current tourism research landscape, which includes the advancement of both theoretical constructs and research methodologies related to tourism products (MacKay and Vogt, 2012; Lo and McKercher, 2015; Dinhopl and Gretzel, 2016; Song and Kim, 2017).

### 11.3.2 In tourism marketing

Digital technologies have significantly advanced marketing research in the corporate sector. It is important to emphasize that the new digital media platforms enable tourism businesses to gain accurate insights into the desires and preferences of today's tourists. They facilitate cost-effective market research by allowing direct data collection from tourists' comments on social media platforms or on the websites of state tourism organizations.

Indeed, metadata from tourism-related social media can serve as invaluable sources of information for tourism research and provide a better understanding of traveller behaviour. Business researchers can use this data to analyse sentiment, track trends and accurately target advertising campaigns based on the interests and travel preferences of specific populations.

Finally, it is important to note that new media platforms also help to increase the influence of traditional media in tourism. They simply complement traditional media by creating an interactive and engaging ecosystem. In particular, they encourage the linking of films, TV shows, festivals and rock concerts that promote related activities on social media sites and offer enticing tourism experiences.

This synergy allows for greater reach, audience engagement and promotion of travel-related content and experiences (Mansson, 2011).

## 11.4 Summary

In summary, the transition to digital technology has revolutionized the tourism industry, initially with the emergence of computer reservation systems (CRS) in the 1960s and global distribution systems (GDS), culminating in the integration of e-commerce and multimedia content on digital platforms, radically changing the landscape of global tourism. The development of the latest communication technologies in tourism has led to a reorganization of the industry, weakening traditional intermediaries as tourists increasingly bypass them and engage directly with service providers online. This has fundamentally reshaped the landscape of tourism transactions and marketing strategies, driving the rise of e-tourism and 'democratizing' the presentation of destinations via social media and online platforms. In the modern digital age the role of tourism journalists has become very important while tourists themselves are increasingly involved in the construction of tourism products. In general, all factors involved in the tourism process interact via digital platforms, which is reshaping traditional academic tourism research and providing insights into tourists' preferences in tourism marketing.

## Notes

[1] The term information and communication technology (ICT) became widely used in the late 20th century as technology began to play a more central role in information processing and communication. It generally covers a wide range of technologies that facilitate the processing and exchange of information. While digital systems are an important part of ICT, the scope of ICT is not limited to digital technologies. ICT encompasses both digital and non-digital technologies used for communication, data management, information processing and various applications. The landscape of ICT is dynamic and is constantly adapting to new developments and innovations.

[2] Before the establishment of the UNWTO, the International Union of Official Travel Organizations (IUOTO) functioned as a global coordination and cooperation body for tourism. The IUOTO was founded in 1925 and was active until 1974, when it merged with the International Travel and Tourism Federation (ITTF) and founded the World Tourism Organization (WTO). Later, in 2003, the WTO was transformed into the UNWTO after becoming a specialized agency of the United Nations. The UNWTO is the main global authority on tourism and recognizes and supports the central role of ICT in tourism. This recognition extends to technologies such as CRS and GDS. These technological facets not only characterize the international tourism sector, but also drive it forward. Their integration contributes significantly to the efficiency, accessibility and sustainability of tourism and fits seamlessly into the overarching goals of the UNWTO, which promotes responsible and inclusive tourism practices on a global scale.

[3] 'Prosumer' is a hybrid word of 'producer' and 'consumer'. It refers to a person or entity that both produces and consumes goods or services. In the context of business and marketing, the term often refers to people or organizations that not only consume products but also play an active role in their production or adaptation. This concept is particularly relevant in the context of the modern information age and the advent of user-generated content, where consumers often create and share content, blurring the line between consumption and production. Alvin Toffler argued that as society progresses, there is a shift from consumer to prosumer, where individuals are not only passive consumers but also active producers of information and goods (Toffler, 1980).

## References

Aluri, A., Slevitch, L. and Larzelere, R. (2015) The effectiveness of embedded social media on hotel websites and the importance of social interactions and return on engagement. *International Journal of Contemporary Hospitality Management* 27(4), 670–689. DOI: 10.1108/IJCHM-09-2013-0415.

Arisandi, N.G.A.D.P., Wirawan, I.G.P. and Linawati, L. (2014) The image of Bali tourism in social networking media. *Image* 1(2), 128–138.

Bosangit, C., Dulnuan, J. and Mena, M. (2012) Using travel blogs to examine the postconsumption behavior of tourists. *Journal of Vacation Marketing* 18(3), 207–219. DOI: 10.1177/1356766712449367.

Browning, V., Fung So, K.K. and Sparks, B. (2013) The influence of online reviews on consumers' attributions of service quality and control for service standards in hotels. *Journal of Travel & Tourism Marketing* 30(1/2), 23–40. DOI: 10.1080/10548408.2013.750971.

Buhalis, D. (2003) *eTourism: Information Technology for Strategic Tourism Management*. Pearson Education, Harlow, UK.

Buhalis, D. and Licata, M.C. (2002) The future etourism intermediaries. *Tourism Management* 23(3), 207–220. DOI: 10.1016/S0261-5177(01)00085-1.

Buted, D.R., Gillespie, N.S., Conti, J.B., Delgado, B.A., Marasigan, R.M.P., *et al.* (2014) Effects of social media in the tourism industry of Batangas Province. Asia Pacific Journal of Multidisciplinary Research 2(3), 123–131.

Ceruzzi, P. (2003) *A History of Modern Computing*. MIT press, Cambridge, Massachusetts.

Cox, C., Burgess, S., Sellitto, C. and Buultjens, J. (2009) The role of user-generated content in tourists' travel planning behavior. *Journal of Hospitality Marketing & Management* 18(8), 743–764. DOI: 10.1080/19368620903235753.

da Luz Cerdeira, J.H. and Fernandes, A. (2020) Estudo da imagem da região turística de Lisboa. *Revista Iberica de Sistemas e Tecnologias de Informacao* 40(40), 89–112. DOI: 10.17013/risti.40.89-111.

Dinhopl, A. and Gretzel, U. (2016) Selfie-taking as touristic looking. *Annals of Tourism Research* 57, 126–139. DOI: 10.1016/j.annals.2015.12.015.

Diposumarto, N.S., Purwanto, W. and Ramdan, I. (2015) Analysis of tourism destination image and promotion through social media towards purchasing decision for Bali tourism product by foreign tourist. *Mediterranean Journal of Social Sciences* 6(5 S5), 175. DOI: 10.5901/mjss.2015.v6n5s5p175.

Dredge, D. and Gyimóthy, S. (2017) *Collaborative Economy and Tourism*. Springer, Cham, Switzerland.

Fodness, D. and Murray, B. (1999) A model of tourist information search behavior. *Journal of Travel Research* 37(3), 220–230. DOI: 10.1177/004728759903700302.

Fotis, J., Buhalis, D. and Rossides, N. (2011) Social media impact on holiday travel planning: the case of the Russian and the FSU markets. *International Journal of Online Marketing* 1(4), 1–19. DOI: 10.4018/ijom.2011100101.

Georgopoulos, N.V., Kopanaki, E.D., Pantazi, M.-A.A., Nikolarakos, X.T. and Vagelatos, I.X. (2013) *E-business, Programming and Design*. Benu S.E. Publications, Athens. (in Greek).

Gillmor, D. (2006) *We the Media: Grassroots Journalism By the People, For the People*. O'Reilly Media, Sebastopol, Russia.

Graham, M. (2011) Disintermediation, altered chains and altered geographies: the internet in the Thai silk industry. *The Electronic Journal of Information Systems in Developing Countries* 45(1), 1–25. DOI: 10.1002/j.1681-4835.2011.tb00321.x.

Gretzel, U. and Yoo, K.H. (2008) Use and impact of online travel reviews. In: O'Connor, P., Höpken, W. and Gretzel, U. (eds) *Information and Communication Technologies in Tourism 2008*. Springer, Vienna, pp. 35–46. DOI: 10.1007/978-3-211-77280-5.

Katsoni, B. (2017) *e-Tourism: Online Transactions in Tourism*. Greenbooks, Athens. (in Greek).

Lo, I.S. and McKercher, B. (2015) Ideal image in process: online tourist photography and impression management. *Annals of Tourism Research* 52, 104–116. DOI: 10.1016/j.annals.2015.02.019.

MacKay, K. and Vogt, C. (2012) Information technology in everyday and vacation contexts. *Annals of Tourism Research* 39(3), 1380–1401. DOI: 10.1016/j.annals.2012.02.001.

Mansson, M. (2011) Mediatized tourism. *Annals of Tourism Research* 38(4), 1634–1652. DOI: 10.1016/j.annals.2011.02.008.

Miguéns, J., Baggio, R. and Costa, C. (2008) Social media and tourism destinations: Tripadvisor case study. *Advances in Tourism Research* 26(28), 1–6.

Mills, J. and Law, R. (2004) *Handbook of Consumer Behaviour, Tourism and the Internet*. Harworth Hospitality Press, New York.

Miranda, S. and Dias, D. (2020) Lisbon brand image perception: the perspective of tourists. In: *Conference Proceedings of the International Conference on Innovation and Entrepreneurship in Marketing and Consumer Behavior (ICIEMC), 12–13 November, Universidade Aveiro, Portugal*, ICIEMC, Aveiro, Portugal, pp. 136–144.

Müller, F. (2018) TripAdvisor relaunches as a social network. *Deutsche Welle*, 13 November 2018. Available at: https://www.dw.com/en/tripadvisor-relaunches-as-a-social-network/a-46273147 (accessed 12 March 2022).

Naqvi, M.A. and Jia, H. (2016) Computer reservation system. In: Jafari, J. and Xiao, H. (eds) *Encyclopedia of Tourism*. Springer, Cham, Switzerland, pp. 177–178. DOI: 10.1007/978-3-319-01384-8.

Negropontis, N. (1995) *Digital World*. Kastaniotis Publications, Athens. (in Greek).

Pan, B. (2016a) eTourism. In: Jafari, J. and Xiao, H. (eds) *Encyclopedia of Tourism*. Springer, Cham, Switzerland, pp. 324–327. DOI: 10.1007/978-3-319-01384-8.

Pan, B. (2016b) Global distribution system. In: Jafari, J. and Xiao, H. (eds) *Encyclopedia of Tourism*. Springer, Cham, Switzerland, pp. 396–397. DOI: 10.1007/978-3-319-01384-8.

Ritzer, G. and Jurgenson, N. (2010) Production, consumption, prosumption: the nature of capitalism in the age of the digital prosumer. *Journal of Consumer Culture* 10(1), 13–36. DOI: 10.1177/1469540509354673.

Sfakianakis, M. (2008) *Introduction to Computational Thinking*. Keydarithmos Publications, Athens. (in Greek).

Song, S.-G. and Kim, D.-Y. (2017) How was your trip experience while you were obsessed with social media? The influence of compulsive social media usage on trip experience. In: Schegg, R. and Stangl, B. (eds) *Information and Communication Technologies in Tourism*. Springer, Cham, Switzerland, pp. 669–681. DOI: 10.1007/978-3-319-51168-9.

Toffler, A. (1980) *The Third Wave*. Morrow, New York.

Toffler, A. (1990) *Powershift: Knowledge, Wealth, and Violence at the Edge of the 21st Century*. Bantam, New York.

Yoo, K.H., Sigala, M. and Gretzel, U. (2016) Exploring TripAdvisor. In: Egger, R., Gula, I. and D. and Walcher, D. (eds) *Open Tourism*. Springer, Berlin, pp. 239–255. DOI: 10.1007/978-3-642-54089-9.

Yu, M., Li, Z., Yu, Z., He, J. and Zhou, J. (2021) Communication related health crisis on social media: a case of COVID-19 outbreak. *Current Issues in Tourism* 24(19), 2699–2705. DOI: 10.1080/13683500.2020.1752632.

# Part Three – Media and Tourism in Greece

As mentioned in Parts One and Two, states today try to improve their image on the international stage through diplomatic relations, effective communication, tourism promotion, cultural events, cultural and educational exchanges, and international cultural diplomacy in general. The Greek state also attaches particular importance to all these practices as part of its foreign policy. In this international atmosphere, the Greek Ministry of Foreign Affairs emphasizes that Greek culture is a global link in the search for common origins and historical ties between countries.

Within this framework, Greek cultural diplomacy in general aims, among other things, to promote the country's culture and cultivate a Greek image. Some of the main institutions and factors that contribute individually, or in collaboration, to the shaping of Greek cultural diplomacy are: (i) the Hellenic Ministry of Foreign Affairs; (ii) the Hellenic Ministry of Culture and Sports; (iii) the Hellenic Ministry of Education and Religious Affairs; (iv) the Ministry of Tourism and the Greek National Tourism Organisation; (v) Greeks who have emigrated abroad; (vi) the Ecumenical Patriarchate; and (vii) various institutions interested in Greek culture abroad (e.g. university departments of classical and modern Greek studies). Greek cultural diplomacy is also carried out by institutions that are subordinate to, or cooperate with, the Ministry of Culture, or that in one way or another play an important role in the promotion and preservation of Greek culture. This includes both national and municipal regional theatres, the Greek National Opera, the Athens State Orchestra, the National Gallery, the Greek Film Centre, the National Book Centre of Greece and the Chamber of Fine Arts of Greece. It also includes institutes such as the Foundation of the Hellenic World, the Hellenic Institute for Cultural Diplomacy, the Hellenic Foundation for Culture, the Benaki Museum, the Onassis Foundation and the Hellenic Foundation for European and Foreign Policy (ELIAMEP). However, many scholars in the field of political and public relations believe that there was no significant and systematic cultural diplomacy in modern Greece until the 1980s to build a flirtatious image of the country (Tzumaka, 2005).

Be that as it may, the image of Greece, especially the tourist image of Greece in which we are interested, is not only a creation of the cultural diplomacy of the modern Greek state.

The process of building the image of Greece began a long time ago, without the mediation of modern cultural diplomacy or tourism promotion. Rather it began with elements from the depths of the centuries, based on the culture and cultural products of the various Greek and other peoples who lived in this region and whose ideas, techniques and works were spread throughout the world. The ancient monuments, ancient philosophy and political beliefs, ancient Greek literature, the theatre, ceramics, statues, clothing, murals, traditional folk culture, and modern Greek literature and art (e.g. modern painting, music, cinema) are only some of the cultural elements that contribute to the composition of the tourist image of the modern Greek state.

On the other hand, it is not only the Greeks who contribute to the creation of the image of Greece. The peculiarity of Greece is that all other peoples who admired and assimilated elements of Greek culture have contributed significantly to this image since the time of the Roman Empire.

In Part Three of this book, we will examine some of the factors that contributed to this continuous formation of the Greek image. As we have discussed in previous parts, the media has the power to shape the image of a place by


DOI: 10.1079/9781800626034.PART3

providing information, images and narratives that influence people's perceptions, cultural understanding and travel decisions. Now we will see how the various actors and media from ancient times to the present have contributed to shaping the image of Greece over time. At this point, however, it must be emphasized that we will not go into detail about the intensity of the effect and the influence of each individual media product on this image, because there are so many that it would probably fill volumes to do so.

We will not attempt to clearly define the contemporary 'Greek image', or the composition of the Greek image at particular stages, as these were shaped at different times by cultural contributions from antiquity to the present. Nor will we examine the extent to which the image is more the result of one influence over others; we believe all factors (i.e. ancient Greek civilization, work of literary scholars, artists and film directors, etc.) played a role. However, we will identify key historical points that contributed to the present-day Greek image, as it is important to understand the broader context and interconnections of these points.

For our analysis in Part Three, we look at the relationship between communication and travel in Greece in several phases. More specifically, we will examine this relationship:

- in antiquity (Chapter 12, this volume);
- in the Roman Empire and Byzantium (Chapter 13);
- in the Renaissance and Enlightenment (Chapter 14);
- at the founding of the Greek state (Chapter 15);
- in the period when tourism enters the Greek world as an economic activity (Chapter 16);
- in the years of mass tourism in Greece (Chapter 17); and
- in the new digital age (Chapter 18).

The word tourism does not appear in Greek dictionaries until several decades after the beginning of the 20th century (Koumanoudis, 1900; Giannaris, 1905). This does not necessarily mean that the concept of tourism was completely absent from Greek culture. We should not confuse the absence of a word in a language with the absence of the idea that word expresses in the people who use that language. As Castoriadis (2004) would say, this is almost like saying that the ancient Greeks had no subconscious because they had no word to name it.

This is not the case, however, because Greek also had the words *periigisi* for visiting a place for leisure, research or study, and *periigitis* for someone visiting a place for recreational purposes or to learn about and study the sights and way of life of its inhabitants. These words come from the ancient Greek verb *periigoumai*, which means to travel to a place for recreation or to get to know and study the sights and way of life of its inhabitants (Babiniotis, 1998).

So, we can say that *periigitis* has the positive meaning that many scholars give to the traveller in relation to the tourist. In other words, there was some kind of meaning of tourism as a concept in the Greek language since ancient times, but the adoption of the European words 'tourist' and 'tourism' was perhaps due to the xenophilia of the Greeks at that time and their tendency to use foreign words. Alternatively it was for other reasons, namely the pejorative character and the inclusion of commerciality in this phenomenon.

Katsigiannis (2017) considers that the Greek word *periigitismos* was eventually replaced by the word 'tourism' around 1926–1927. As a starting point for this change, he suggests that it was probably the book *General Principles of Tourism Science* (1929) by Nikolaos Aeginitis, a lawyer and profound expert on the tourism industry.

Dritsa (2019) also argues that the term 'tourism' seems to have gradually gained acceptance among the Greek public at around the same time, coinciding with the establishment of the Greek National Tourism Organisation in 1929, and that Greeks at that time were more or less aware of the negative connotation of mass tourism, as opposed to the term *periigitismos*, which was overlaid with a cultural undertone.

According to brief research we have carried out, the word tourism seems to have officially appeared in the Greek language (i.e. in Greek dictionaries) in the interwar period, imported from abroad, already with the double meaning of the term (i.e. both as a form of human mobility for pleasure and as a business activity).[1] Thus, from the time when the word tourism started to be used in the Greek language, the word was linked to trade and the development of capitalism.

## Note

[1] More precisely, the term tourism was introduced in Greek dictionaries from the beginning as the movement of travellers from one place to another for recreation, sightseeing, admiring works of art or natural beauty, but also as the organization that attracts visitors and excursionists and the services offered by this organization (Eleftheroudakis, 1931).

## References

Aeginitis, N. (1929) *General Principles of Tourism Science*. Private Edition, Athens. (in Greek)

Babiniotis, G. (1998) *Dictionary of the Modern Greek Language*. Lexicology Publications, Athens. (in Greek)

Castoriadis, C. (2004) *Ce qui fait la Grece, tome 1: D'Homere à Heraclite*. Editions de Seuil, Paris. (in French)

Dritsa, M. (2019) *Mosaics of Greek Tourism.* Economia Publishing, Athens. (in Greek)

Eleftheroudakis (1931) *Encyclopaedic Dictionary*. Eleftheroudakis Publishing, Athens. (in Greek)

Giannaris, A.N. (1905) *A Brief Greek Dictionary: After an Introduction to the Greek Language and Writing.* Anestis Konstantinidis, Athens. (in Greek)

Katsigiannis, K.X. (2017) *Hellenic Tourism Organization: Travel Through Time*. Private Edition, Athens. (in Greek)

Koumanoudis, S. (1900) *Collection of New Words Molded by Scholars: From the Fall (of Constantinopole) to Our Own Years*. P.D. Sakellarios Press, Athens. (in Greek)

Tzumaka, E. (2005) *Cultural Diplomacy: International Facts and Greek Perspectives*. I. Sideris, Athens. (in Greek)

# 12 Travelling in the Ancient World

**Abstract**

This chapter explores the profound influence of ancient Greece on the concept of travelling through visual art, oral traditions and the works of epic poets such as Homer and Hesiod. The narrative explores the cultural elements that inspired wanderlust, such as thalassophilia, *philoxenia* and divine protection for travellers, embedded in ancient media representations. The depictions in literature and philosophical discussions of the Greeks not only described foreign lands, but also shaped perceptions and fostered the belief that travelling expanded human consciousness. This cultural influence also extended to the public sphere, facilitating open conversations, and theatre performances. The legacy of these ancient depictions has endured through the centuries and continues to fuel the desire for discovery and curiosity about faraway destinations.

## 12.1 Influence of Visual Art and Oral Traditions on Travelling

Journeys to distant lands were very often the subject of visual art in ancient Greece. Paintings on vases, plates, amphorae and other ceramic objects, or engravings on metal, often contain depictions of heroic deeds associated with distant lands, or other depictions directly related to travel for one reason or another. It can thus be argued that all these visual representations of wanderers, horses, carriages, ships, as well as mythical figures and real-life personalities associated with faraway places, helped to convey and internalize the core ideas of the concept of travel to the people who lived where Greek civilization flourished.

In a geographically complex area with rugged mountains, forests and lakes, surrounded by seas and thousands of islands, such messages encouraged people to travel. All these representations were thus very early constructed images through which essential features associated with the movement of people from one place to another were put into a general order that structured the concept of travel, as it was perceived in ancient Greece.

It is generally recognized that the spoken word precedes the written word, and no one today questions the importance of oral speech for the reproduction of cultural memory and the continuity of human societies. In non-literate societies, oral language, together with monuments, works of art, customs and rituals, linked yesterday with today and tomorrow.

The Greek language developed from the common Indo-European language after a long process of change that took place over a protracted period of time. According to archaeologists, before the emergence of the Greek language, the Balkan Peninsula was inhabited by various peoples who spoke different languages, which are now called ‘pre-Greek’. All these languages influenced the Greek language to some extent (Andreou, 2014; Clackson, 2014; Duhoux, 2014a; Duhoux, 2014b; Sacks, 2014).

As far as written records from this period are concerned, numerous pre-alphabetic texts have been unearthed on Crete and in other regions, dating back to the late 19th century BC. These finds are mostly tablets, plaques, medals, etc. with graphic systems, found mainly near the palaces and whose content related to certain administrative tasks. The numerous texts in these archaeological finds contain various pre-Greek scripts (i.e. Cretan hieroglyphs, Linear A, etc.)

After the arrival of the Proto-Greeks in the region, a series of migrations of Greek-speaking peoples occurred. Soon the Mycenaeans on the




DOI: 10.1079/9781800626034.0012

Greek mainland began to use Linear B script, adapting the Linear A script of Minoan Crete to the Greek language. Linear A, which refers to an extinct non-Greek language, is still largely unexplained (Andreou, 2014; Clackson, 2014; Duhoux, 2014a; Duhoux, 2014b).

The ancient Cretans and Mycenaeans most likely used language and stories to describe their journeys, just as we do today. Storytelling was crucial to the sharing of knowledge and experiences in their societies. These stories probably included descriptions of their travels, encounters with people and adventures, and spread information about different regions. Although written travel accounts are scarce, they probably relied on language and oral traditions to pass on their travel experiences to those back home.

The development of shipbuilding in the second half of the 16th century and the advancement of knowledge in navigation enabled Cretan and Cycladic ships to be present throughout the Mediterranean. Such a voyage is described and depicted in the fresco known as the 'fleet mural' in Akrotiri on Santorini (Fig. 12.1). The numerous ships and the state of the sea depicted provide a unique seascape from the end of the 16th century BC. These depictions show in a cinematic way the story of a sea voyage with very clear geographical features of the areas visited by the ships.

On their decks, these ships had a covered area for the passengers, who sat on benches under a kind of awning. Murals like these were certainly not created only with the intention of serving as historical documents about the course of the voyages of the time, but they were representations of daily life and were intended to remind the people of the time that travelling to the Mediterranean was also a pleasure that helped to open the mind and expand knowledge.

In this way, the artistic representations contributed to the development of social ideas about sea travel as early as the late Bronze Age, and motivated people to undertake these journeys.[1]

The Mycenaean era produced collective memories of the journeys and the various faraway places of that time, which were later brought to life through writing. The resulting

**Fig. 12.1.** Detail of the 'fleet mural' in Akrotiri on Santorini. The particular series of wall paintings at Akrotiri depicts a fleet travelling to distant lands, with the entire flotilla, consisting of small and large ships and the people in them, playing the leading role. The fresco ends in a city, probably Akrotiri, and raises questions about triumph, search and surveillance, territorial supremacy, or trade. Ultimately, it refers to the central role of Akrotiri in the Minoan Empire (Marinatos, 2015). (Ship fresco from Room 5 on the upper floor of the West House of the archaeological site in Akrotiri on Santorini, Cyclades, Greece. By unknown author, between 1650 and 1500 BC. Public Domain.)

myths and stories, which always had to do with journeys and wanderings, were read, copied, embellished and commented on again and again (Beaton, 2021).

However, the decline of the Mycenaean civilization was followed on the Greek mainland by a period from about 1100 to 800 BC, which is referred to as 'dark'. This is because it is unclear what exactly happened during these years, while the finds from this period paint a picture of widespread introversion, devastation and impoverishment. The literary and artistic culture of the Mycenaeans seems to have all but disappeared, and the population shifted from cosmopolitan palaces to small, isolated villages (Sacks, 2014).

Be that as it may, the collective memory of the Mycenaean period was strong and recitations to audiences were one of the main means of communication to spread the idea of travel. Many epic poems about the past that referred to travel were recited by the *aoidoi* (i.e. singers and poets with an excellent memory and a beautiful voice), in front of their audience, accompanied by a musical instrument. The *aoidoi* sang before the social events of the aristocracy, at religious festivals, musical competitions, but also in the evenings at gatherings of small communities, in front of houses or in public squares. These skilled entertainers were of great importance to early Greek societies, for in the 'absence of writing' their memory was one of the few ways of preserving the past and describing distant places that the general public had not seen.[2]

We can say, then, that the *aoidoi* were among the first institutionalized means of shaping the image of distant and nearby places and the culture of their inhabitants, and that from a structuralist perspective they were more generally an agency for the construction of the world.[3]

## 12.2 The Written Word

After the Greek Dark Ages, during which archaeologists have found virtually no written evidence, the creators of Greek civilization began to use an alphabet adopted from the Phoenicians, which they reconstructed by adapting it to their language, but this probably did not happen until the ninth century. In particular, they imitated the general forms, names and order of the Phoenician letters. As the Phoenician alphabet contained no vowels, they converted some of the letters into vowels and created the Greek alphabet.

Interestingly, the Greek alphabet was also reconstructed and adopted by the Etruscans and Romans. Again, we should open a parenthesis to note that the influence of Greek, either directly or through Latin, on all European languages was enormous. As the Spanish linguist Francisco Adrados (2005) argues, the ancient Greek language lives on in Greece, but it also has a second life throughout the Western world. Its vocabulary, syntax and the literary genres it created are present in all languages, so the ancient Greek language lives on in them. Suffice it to point out how many words in English, French and most other languages of our time, especially in scientific terminology, derive from Greek, such as democracy, aristocracy, politics, rhetoric, philosophy, history, geography, anthropology, drama, music, Europe, planet, galaxy, melancholy, panic and phobia.[4]

It would not be an exaggeration to say that no one can think of the word 'philosophy' without thinking of ancient Greece, because philosophy is a Greek creation. So, it would be very naïve to assume that this does not have a positive effect on the image people have of modern Greece, at least those who understand this fact.

At this point it should be noted that in 2002 the Greek National Tourism Organisation launched an original advertising campaign presenting some basic Greek terms that appear in almost all European languages. In this way, the promotion of the country was combined with the indirect highlighting of its contribution to world culture. The words used were mathematics, horizon, odyssey, athletics, cosmopolitan, energy, magnetism, theatre, logic, architecture, symbol, phenomenon, harmony, gastronomy, photography and eros (Katsigiannis, 2017).

Be that as it may, it is not correct to speak of a single Greek language, for there were many forms of different dialects (Aeolic, Doric, Ionian, etc.), and it is not certain that there was a single universal language that could be called Greek (Andreou, 2014; Clackson, 2014; Duhoux, 2014a; Duhoux, 2014b; Sacks, 2014).

However, some scholars believe that later, in the Alexandrian period, a kind of common Greek, based on the highly regarded dialect of classical Athens, had established itself as the universal language for administration, trade and culture (Beaton, 2021).

To make a clarification at this point, ancient Greek literature included poetic works, which can be divided into epic, lyric and dramatic poetry, and prose works dealing with history, philosophy and rhetoric. Most of the written works of the ancient Greeks were rich in narratives, which were able to create images of distant worlds and motivate the inhabitants of ancient Greece to go on adventures to see the world. The importance of Homer, Hesiod, Herodotus, Thucydides, Xenophon, Strabo, Diodorus Siculus, Callimachus of Cyrene, Antipater of Sidon and Pausanias in creating images of distant places and ideas about the culture of other peoples has already been highlighted earlier in this book (e.g. see Chapter 8, this volume).

### 12.2.1 Influence of epic poets on travelling

Let us take another look at the great epic poet Homer, whose work from the eighth century BC describes the past periods of Greek world. Of particular importance to us is that Homer talks a lot about the Greek sun and the sea, two of the elements that make up the tourist myth of modern Greece. While the sun or 'Helios' as a god seems to play a very important role in Greek mythology with myths such as the myth of 'Phaethon' or in tales from the *Iliad* and the *Odyssey*, the sea was often described as 'salty water', as 'beautiful' and 'deep sea', but also as 'rough' and 'noisy' and 'wine-like', among many other characterizations. Indeed, Homer invented a plethora of adjectives to convey a vivid impression of the sun and the sea that long characterized the image of Greece in the world (Linardaki and Aslanides, 2020).

We can argue that Homer's timeless imagery can serve as a source of inspiration for contemporary travel experiences and can influence the way destinations are marketed, explored and appreciated, including by modern travellers seeking a connection to the natural beauty and cultural richness depicted in Greek epic poetry.

Another theme related to communication and travel that appears frequently in the Homeric poems concerns *philoxenia* and the reception of guests/strangers. And let us not forget that Homer's two epics, the *Iliad* and the *Odyssey*, were not only poetry, but also the main teaching tools for the education of generations of Greeks after the seventh century.

According to the poet, the Greek word *xenos* denotes both the guest and the stranger. Depending on the circumstances, the *xenos* in Homer's works is received with friendliness or hostility. In his poem the *Odyssey*, the first thing Odysseus, the king of Ithaca, asked himself when he arrived at a place during his wanderings was whether he was among wild and fearless people who knew no law, or whether he was among those who respected the gods and were hospitable according to the rules of *philoxenia*. It is also interesting to note that Telemachus, the son of Odysseus, who had travelled in search of him, once saw the goddess Athena disguised as a man at the entrance to the palace in Ithaca, ran to her full of hospitality, greeted her and told her that his house belonged to her. As can be seen from the literature of the time, then, hospitality was a characteristic of the ancient Greek nobility as well as the common people (Coulet, 1996).

Elsewhere in the same poem, Homer again shapes the concept of hospitality in the minds of the Greeks as a fundamental element of the civilized world. When Odysseus reaches an island next to the island of the Cyclopes with his soldiers, he calls a meeting in which he informs them that the main army will stay put, while he himself will go over to the other island with a few men to see what kind of people live there. The purpose is to establish whether these people are abusive, disregard every rule, are wild and unjust, or whether they are distinguished by *philoxenia* and fear of God.

Among the ancient Greeks, the term *philoxenia* referred to the friendly reception and especially the provision of free accommodation, food and care for one or more persons during their temporary stay in a foreign region. Hospitality was an important institution in ancient Greek cities, as described in the Homeric epics. It was a great necessity for Greek societies, where people travelled for various reasons and needed to secure accommodation in the places where they stayed. At this time, travellers were very unsafe on their way through various regions, as many thieves were on the move and wars between cities were commonplace. The *philoxenia* was therefore of great importance as an institution to protect travellers in these conditions. In any case, it was believed that this institution was protected by Xenius Zeus, Athena Xenia and the Dioscuri.

As a result, something like a hospitality law was created. However, this law was an absolutely general and popular law and had nothing to do with the subject of today's official legislation. In other words, the *philoxenia* was rooted in customary law and tradition. On the other hand, it was considered disrespectful to the gods to treat a stranger indecently, and it was believed that this act would incur their wrath and general contempt. The sense of hospitality was thus deeply rooted in the consciousness of the Greeks also as a spontaneous spiritual impulse and religious belief.

Homeric hospitality typically involved codified behaviour on the part of the 'host' (i.e. the one receiving the guest). First, the stranger was welcomed into the best room, offered the host's best chair, bathed, given food and drink, and often given a feast in his/her honour. After being asked about his/her background and his/her wishes, his/her requests were granted and he was given accommodation for as many days as he needed. Finally, the whole thing was sealed with the exchange of gifts (Tzermias, 1997; Petrounias, 1998; Maronite *et al.*, 2012).

It is also worth mentioning at this point the connection of *philoxenia* with the term 'symbol', which later acquired a completely different meaning. The 'symbol' was then an object, the actual connecting element of *philoxenia*. If there was a host and a guest, the two were connected by the sacred bond of *philoxenia*. Before they said goodbye, they divided an object (cube, ring, etc.) in half and kept one piece each. When the former host or his/her children travelled to the former guest's country, they took half of the object with them. If the two pieces matched, *philoxenia* was shown again and the new guests were honoured in a special way. Thus, these two parts were called 'symbols' because they contributed to the formation of the original object of *philoxenia* (the Greek verb for contribute is *simválo*). Over time, the process with the symbol disappeared, but *philoxenia* remained an essential feature of the temperament of the people who lived in the places where ancient Greek civilization developed, until its reification by the tourism industry (Tzermias, 1997).

Hesiod, an influential figure in early Greek literature of the eighth century BC, made an important contribution to the understanding of ancient Greek culture with his central works *Theogony* and *Works and Days*. In the *Theogony* he deals with the creation of the world and the emergence of the Olympian gods, while *Works and Days* serves as a pragmatic guide that resembles a farmer's almanack. In this latter work, Hesiod not only reflects the social mentality of his time, but also gives practical advice on how to lead a virtuous life through diligent endeavour.

Hesiod provides an insight into the prevailing social ethos and daily life of the people of his time, thus conveying a vivid picture of ancient Greek society. Although *Works and Days* briefly touches on the theme of travelling, particularly in relation to the different ages of mankind, and the journeys of the heroic generation, these references do not form the main focus of the work through his central works.

All these myths told by Hesiod and Homer found their way into Western civilization. As we shall see, they were later represented in paintings, sculptures, mosaics, plays and literature throughout Europe, helping to put Greece on the European agenda of cultural discussions and creating a cultural memory that connected Europeans to Greece (Sacks, 2014).

From the above, it is clear that Greeks were not only inspired to travel by various narratives, but that Greek culture and the various regions and cities inhabited by Greeks were constantly presented by contemporary media. Although this presentation differed from modern tactics of tourism promotion, it gave Greece a competitive advantage over other destinations early on.

### 12.2.2 Other Greek writers who contributed to images of foreign places

Many other ancient Greek writers contributed to the formation of images of the foreign and the distant with their works, such as the lyric poets Callinus (seventh century BC), Archilochus (c.680–645 BC), Alcman (seventh century BC), Pindar (c.518–438 BC) and many others with the imaginative descriptions and dazzling images of their poems.

There were also a number of scholars who showed great interest in the geography and structures of distant empires. Among them, Anaximander of Miletus (c.611–547 BC) stood out as a pioneer in the creation of a world map based on theoretical principles. Hecataeus of Miletus (c.550–476 BC), a student of Anaximander who is often referred to as the 'father of geography', made a decisive contribution to geographical research. In particular, he created an influential map covering the continents of Africa, Asia and Europe. Hellanicus of Lesbos (c.490–405 BC) deserves mention for his groundbreaking work on the legendary foundation of Rome and for his numerous writings on Greek territories and foreign cultures. Thales of Miletus (c.626–545 BC), one of the Seven Sages of ancient Greece, contributed to this intellectual tapestry by advocating the bold idea that the Earth rests on water. Together, these and other luminaries enriched the landscape of early geographical research.[5]

In addition to all the names mentioned above, who as historians and geographers described foreign places and cultures in their works, it would be remiss not to mention that the sophists and philosophers also liked to travel and often stayed for some time in different cities of the Greek world, which of course influenced their work and ideas.

Sophists such as Protagoras of Abdera, Hippias of Elis and Gorgias of Leontinoi developed the art of rhetoric (i.e. the art of persuasion), and travelled from city to city to teach and spread their knowledge for a fee. They often spoke to large audiences about places and cultures. Therefore, we are dealing with a form of dynamic mass communication reminiscent of the modern media, with the difference that there was more direct interaction with the audience. With the development of philosophical schools, visits and stays abroad became a fundamental element of the education of the sages. Many philosophers with strong Eastern influences contributed to the development of travel to the East and thus to the formation of images of these places (Coulet, 1996).

## 12.3 Spread of the Cultural Influence of Greek Philosophy and Art

At this time, the entire archipelago, the cities of Asia Minor and the inland coasts of the Black Sea, the cultural centres of Sicily and Greater Greece, North Africa and the western coast of the Mediterranean formed the great platform on which the creative force of Greek philosophy and art took root and developed. Alongside philosophy, the artistic creativity of the Greek mind reflected in the most diverse fields of expression what in later times will be called the measure of all art in Europe (Uscatescu, 1986).

The gods, demigods and heroes who protected the ancient Greeks on their journeys were also widespread in the temporal media. Depictions of Poseidon, Hermes, Triton, Proteus, Apollo, the Dioscuri and many others protecting travellers in relative narratives in ceramics, frescoes, sculptures, etc., helped to convey a sense of security that was a central element of the social representation of travel in ancient Greece.

For the ancient Greeks, even the gods travelled and had adventures, and this was often portrayed in the media of the time. It would be remiss not to mention the 'Cup of Dionysus', a ceramic cup estimated to have been made by Exekias around 540–530 BC and now on display in the Staatliche Antikensammlungen in Munich. Inside the cup is a depiction of a ship in which Dionysus is sailing, while vines with seven grapes grow around him and next to the mast. In the same scene, seven dolphins swim around the ship. The depiction goes back to a legend about Dionysus, according to which the god, when he fell into the hands of pirates, transformed them into dolphins, while the mast of the ship was extended into a vine. The journey through the sea was not easy, as the legend tells us, because it held surprises even for the gods, but they had their own way of dealing with the unexpected (Hofmann, 1962).

The Dioscuri, the mythical twin heroes Castor and Polydeukes, who were worshipped like gods and were considered protectors of guests, sailors and travellers, were a popular motif on many amphorae and vases, but sometimes they were also engraved on architectural complexes. They often appear as horsemen, dressed in a cloak or chiton with egg-shaped hats or helmets, crowned with stars and holding spears in their hands. Examples include the *Dioskouri Fighting Giants*, a red-and-black-figure Athenian amphora from the fourth century BC now in the Louvre, or a sculptural representation of the ship Argo with the Dioskouroi on the metopes of a demolished building of the Sicyonian treasury at Delphi from the early fifth century. The Dioscuri are also frequently seen on Etruscan mirrors and coins and on Roman sarcophagi. Another famous representation of the Dioscuri is the 4 m-high statue of the Twin Gods of the Seafarers in front of the Quirinal in Rome (Elderkin, 1934; Walker, 2015).

All this shows that the Greek deities and heroes depicted in various works, which influenced the Greeks' ideas about the safety of seafaring, were also adopted by other peoples, which contributed to the further spread of this belief.

The places where people in ancient Greek cities conversed, informed themselves, told their problems, held political discussions and generally communicated were the agora, the sanctuaries, the theatre and the gymnasium. It is worth noting, however, that usually only the citizens of the city had the right to participate in these places (i.e. women, children, slaves and immigrants were often excluded from any functional engagement).

The agora was the place for political discussions, for conversations about public problems and belief in the gods, and for commercial transactions; in short, it was the central point where most of the community's activities converged. It can be concluded that conversations about faraway places and foreign cultures often took place in the space of the agora, while the surrounding architectural buildings sometimes represented deeds and adventures in foreign lands that gave people images of other places.

The sanctuaries were places of communication with the gods, but also between ordinary mortals. People came from far and wide to ask questions of all kinds and to receive the oracles with their answers. After victories in battles, sporting achievements or successes of any kind in the social life of the Greeks, offerings were often made to the sanctuaries, so that they became a kind of museum with their wealth and collected objects. Thus, sanctuaries eventually became places of worship, asylums, propaganda agencies, museums and, in general, places where people received information, and this information very often concerned distant worlds (Coulet, 1996).

The Parthenon is an ancient temple on the Acropolis of Athens, built in the fifth century BC. It was dedicated to the goddess Athena and is often considered a fundamental symbol of Greece's tourist image. Even in ancient times, people from different parts of the Greek world and beyond visited Athens to see the Parthenon and other remarkable buildings on the Acropolis. Since then, the Acropolis has not only been a religious site, but also a symbol of Athens' cultural and artistic achievement. Many people came to admire the architectural beauty and elaborate sculptures and to pay homage to the goddess Athena. Today, the iconic silhouette of the Parthenon against the Athens skyline is a distinctive tourist landmark worldwide. The image of the Parthenon has been reproduced in various media, from photographs to postcards and paintings, and has become synonymous with Greece. The combination of historical, architectural, artistic, cultural and educational significance makes the Parthenon a natural and enduring symbol of Greece's tourist image (Neils, 2005).

In the classical period, thousands of people flocked to ancient Greek theatres at great festivals, theatrical competitions and other events to see plays depicting adventures. Many of these were set in distant lands, such as the oldest surviving tragedy, Aeschylus' (*c.*525–456 BC) *Persians*, which is set in Susa, the great centre of the Persian king Xerxes (519–465 BC).[6]

In ancient Athens, the Dionysia was a great festival in honour of the god Dionysus, at which first tragedies and later comedies and satirical dramas were performed, and to which all the citizens of Athens and many visitors came from all over the Greek world.

Theatre performances were a place of communication par excellence; they were an essential part of social life that brought all citizens together. The participation of women and children is a topic that has been much debated in research, and many scholars now believe that they were probably allowed; however, the question is likely to remain open. Metics could attend the performances as long as they paid for their seat, while slaves were only allowed to enter the theatre when accompanied by their master. The performances were attended by many visitors from other cities who were attracted by the famous festivities (Coulet, 1996).

Many ancient Greek playwrights such as Aeschylus, Sophocles (*c.*497–406 BC), Euripides (*c.*480–406 BC) and Aristophanes (*c.*446–386 BC) developed themes of travel, wandering and hospitality, and their works still inspire performances around the world.

Journeys play a central role in the narrative of many scenarios, especially in stories rooted in the traditions of Greek mythology and embedded in the scripts of great tragedies. These journeys are often depicted as captivating adventures, such as the arduous journey from the sacred temple of Apollo in Delphi to Athens. In Aeschylus' tragedy *Eumenides* in particular, this expedition becomes a central point where the Erinyes, driven by their quest for justice, attempt to capture Orestes in order to tear him to pieces. This emphasizes the thematic importance of the journey as a dynamic force in the unfolding drama, weaving a tapestry that connects both the mythical and the human worlds.

Moreover, *philoxenia* was a term frequently mentioned in plays, as in Sophocles' tragedy *Oedipus at Colonus*, where the blind and exiled former king of Thebes arrives in Athens after a long wandering through the Greek lands, accompanied by his daughter Ismene, and avails himself of Athenian hospitality by asking the city's king, Theseus, to let him stay in Athens.

In the tragedy *Bellerophon* by Euripides, the hero travels to Olympus on Pegasus, the winged horse, to become a god, but the horse crashes and Bellerophon, regretting his infidelity, dies.

In the first part of *Ornithes* [*The Birds*], the well-known comedy by Aristophanes, the two friends Peisthetairos and Evelpides fly through the woods to find a hoopoe (*tsalapeteinos*) and ask in which city one can live quietly, richly and peacefully.

From all these and many other tales describing various journeys of gods and mortals, it is clear that travelling was an important part of the lives of the ancients and was always held in high esteem. Moreover, all these tales, with their lyrical breath, the splendour of the language, the acting skills of the actors and the expression of the chorus, aroused the audience's interest in travelling, as they transport them with their imagination to all the faraway places described in the play.

Later, Lycurgus (390–324 BC), the Athenian politician and orator, used his position in the state to arrange for the theatre of Dionysus on the edge of the Acropolis to be rebuilt in limestone from a wooden structure and for the statues of the three most important tragedy writers, Aeschylus, Sophocles and Euripides, to be erected there. Many scholars see this as part of an effort to establish the theatre repertoire of the past century as Athens' 'trademark', its legacy for the entire Greek-speaking world. Athens, which has added this feature to its image since antiquity, is a unique destination for theatre lovers, even today with performances in the original ancient Greek theatres and modern theatres.

In addition to theatre, music was also an important part of the cultural landscape of ancient Greek society, attracting people to certain events and festivals. Of particular importance is that it had a great influence on later music in Europe, but also in the Middle East and especially on Arabic music (Düring, 1956; Mathiesen, 1999).

Physical training in ancient Greece was a true educational system in the modern sense. In this context, the gymnasium was of great importance because it was not only a place of sport, but also an intellectual centre, comparable to today's university, which the Greeks attended daily. It should be noted, however, that it was a place for educational discussions and information that was only visited by men. From the fifth century onwards, the gymnasium became a meeting place for new ideas, as the sophists, Socrates and other philosophers often taught there (Coulet, 1996).

In this sense, the conversations that took place in the gymnasiums helped the Greeks to gather information about distant lands from well-travelled citizens, to understand the importance of travel, to learn about other cultures and to establish social ideas about travel.

Explorers, sailors, merchants, ambassadors, heralds and messengers were, of course, also important intermediaries in the dissemination of information about faraway places and the culture of their inhabitants, and helped to consolidate the image of foreign lands. Herodotus described that in the seventh century BC, the Samian merchant Colaeus opened trade in the Far West and was blown by a great storm across the Mediterranean and through the Strait of Gibraltar into the Atlantic, becoming the first Greek to pass the 'Pillars of Hercules'.[7]

### 12.3.1 Early Greek exploration across oceans

Much of the early Greek exploration of the Atlantic goes back to the time of the Greek colonies, and mainly from Marseilles. The *Natural History* of Pliny the Elder (23–79 AD) mentions an early Greek Atlantic traveller, a certain Midakritos, who was the first to import *plumbum album* (tin) from the island of Cassiteris; although no date is given, he must have been not much later than Colaeus.[8]

Scylax of Caryanda, another Greek explorer and writer from the late sixth and early fifth centuries BC, also travelled east across the Mediterranean during the reign of Darius and crossed the Indian Ocean for the first time to reach India (Lagadas, 1990).

Euthymenes of Massalia, who explored the coast of West Africa, was probably also one of the first Greeks to sail through the Strait of Gibraltar and navigate the ocean around 500 BC. He is credited with being the first to use the term 'Atlantic', but his writings have not survived.

Around 310–306 BC, another Massaliot, merchant, explorer and geographer named Pytheas travelled to the British Isles, reached Iceland, touched the frozen sea, set foot on the German coasts, the coasts of Denmark and the Baltic Sea. Pytheas wrote the work *On the Ocean*, in which he recorded important information on physical geography, oceanography, climatology and anthropology. The book has not survived, but it aroused great interest in its time. We learn about the content of the work mainly from scattered secondary sources.[9]

Therefore, it seems that despite the limited resources for seafaring and the fear of the unknown, people in antiquity were driven by an innate urge to explore. Driven by an inherent urge to move and a basic instinct of 'seeking' rooted in the quest for an expansion of consciousness and vitality, these bold souls wanted to surpass the 'Pillars of Hercules' and set out in search of new realms (Edwards *et al.*, 1982; Coulet, 1996; Roller, 2015; Roller, 2016).[10]

### 12.3.2 Alexander the Great: spread of Greek culture and knowledge of distant lands

With his conquests, Alexander the Great built a bridge between East and West and created, so to speak, one of the first examples of the 'globalization' (or cultural imperialism) of the then-known world, the archetype of the first world empire, whose universal cultural influences are undeniable.

The Greek language became not only the main means of communication between so many and so diverse peoples, but also a means of rapprochement and an attempt to resolve their common differences and problems. Greek culture, with its rational character, spread in East and West, in Asia, Africa and Europe; for centuries the Greek heritage lived on mysteriously in the background, in the collective memory of peoples so different and scattered over great distances.

It should also be mentioned here that Alexander the Great's admiral, Nearchos (356–300 BC), was the first European to sail around the south-west coast of India around 326 BC. The navigation of the Indus by Alexander's fleet with Admiral Nearchos and the long voyage from the mouth of the Indus to the Persian Gulf are considered one of the greatest epics of the ancient Greek fleet. Both the voyage of Nearchos and the conquests of Alexander the Great brought a wealth of knowledge about the geographical data of the time (Uscatescu, 1986; Lagadas, 1990).

In the Alexandrian and Hellenistic periods, the ancient Greeks' knowledge and ideas about distant lands were expanded, as was natural after the conquests of Alexander the Great. The great conqueror made distant lands known to the Greeks and the West, which today include Iran, Iraq, Afghanistan, Tajikistan, India and many other countries that were largely unexplored by the Greeks until then. Philosophers, scholars, geographers and engineers travelled with him, recorded everything and then passed it on to Greece.

In the early third century BC, during the reign of the Ptolemies, the Macedonian Greek royal dynasty, the great Museum of Alexandria was founded. It was part of a larger complex that included the Library of Alexandria, which was one of the most famous libraries in the ancient world. The main purpose of the Alexandrian Museum was to serve as a centre of learning, research and intellectual exchange. It was dedicated to the Muses, the nine goddesses of the arts and sciences in Greek mythology. This is where its name came from, which was later adopted by all institutions that exhibit and preserve significant objects and play an important role in the tourism industry.

The museum attracted scholars, philosophers, scientists and intellectuals from different regions and became a vibrant centre of knowledge and cultural exchange. Together with the Library of Alexandria, it played a crucial role in the preservation and production of knowledge. They housed an extensive collection of texts, including literature, scientific works, historical records, philosophical treatises and various works on geography, discoveries and travel. While we do not have a complete list of all the books and guidebooks that were kept in the library, it is generally believed that there were indeed numerous texts that can be considered travel literature or guidebooks (Lagadas, 1990; Encyclopædia Britannica, 2006, p. 44).

During this period, Eratosthenes of Alexandria (*c.*276–194 BC), who is often referred to as the founder of scientific geography, constructed the geographical map with parallels and meridians, produced a catalogue of the stars, worked out a calendar with leap years, and developed his views on the size and shape of the Earth. His work inspired other scholars and researchers to further explore the Earth's surface and expand knowledge about different regions, cultures and civilizations.[11]

In fact, a form of traveller developed in the Hellenistic period that was very similar to the modern tourist who wanted to visit wonderful places, guided by people or books. It is significant that where there were famous monuments, visits were organized systematically and under the guidance of someone who guided visitors through the sites and explained their history (Coulet, 1996).

In the third century BC, Heraclides the Critic, a Greek critic, traveller and geographer who probably came from Crete, made remarkable contributions with his works such as *On the Cities of Greece*, *On the Mountains of Greece* and *Register of Greece*. His writings not only helped his contemporaries to understand the various regions of Greece, but also laid the foundations for modern research into the historical designation of Greek territories in antiquity.

In one of his texts, Heraclides vividly describes Athens as a city of unparalleled beauty in its time, highlighting iconic landmarks such as the Acropolis and the Theatre of Dionysus. But he also provides insights into the practicalities of daily life and describes the city's narrow, winding streets, which have been preserved since ancient times. Heraclides points out the challenges of the city, such as the limited water supply and food shortages, and emphasizes that access to these essential goods often depended on one's financial means, with exorbitant prices prevailing. He even warns that at first glance, it might be difficult for a foreign visitor to believe that this is the famous city of Athens (Beaton, 2021).

Since the Greek media were spread throughout the known world at that time, Greek culture also influenced other peoples and made Greek cities very attractive.

The evidence suggests that people throughout the known world gained knowledge of regions outside their immediate neighbourhood through various forms of media. This not only stimulated their curiosity and imagination, but also created a desire to explore these distant lands.

## 12.4 Summary

To summarize, the artistic and narrative traditions of ancient Greece, ranging from myths and epics to historical accounts, not only served as sources of inspiration but also actively promoted travelling. These narratives depicted distant lands, heroic deeds on journeys and the divine protection of travellers, presenting travel as a means of adventure, cultural enlightenment and safety. At the same time, in ancient Greek society, communal spaces such as the agora, sanctuaries, theatres and gymnasiums were lively centres of dialogue, cultural exchange, political discourse and intellectual pursuit. These places often provided a dynamic environment for discussions about distant places, influencing collective perceptions and fuelling the population's shared ambition to travel. In essence, the artistic and communal spheres of ancient Greece worked synergistically to promote and shape the ethos of travelling, enriching both individual experience and collective understanding.

## Notes

[1] Although there are no written records, we know from archaeology that as early as the 14th century BC the Mycenaeans, and before them the Cretans, traded with their ships in a very large radius around their homeland, which encompassed the entire Mediterranean and the Black Sea, some believe even further. Thus, the inhabitants of mainland Greece, who once roamed as nomadic shepherds, after cultivating the land out of necessity and founding cities, rediscovered the power of life through mobility on the sea and became travellers, traders and colonists reaching distant places. As they sailed their ships across the west coast of the European Mediterranean, North Africa and the Middle East to the Black Sea, they encountered other peoples and countries and transmitted information about them to their compatriots, which formed the basis for the construction of social representations about sea voyages and the important centres of the then-known world (Beaton, 2021).

[2] When we speak of the 'absence of writing' we are referring to the Greek Dark Ages, for as we have already noted, archaeologists have discovered pre-Greek written texts from much earlier years. Linear A and Linear B are thought to have been the first syllabic and iconographic writing systems used in the region between the 19th and 12th centuries BC. However, these systems were probably only used for recording goods, keeping records and other administrative purposes, but not for writing historical literary texts or poetry, and then somehow fell into oblivion during the Dark Ages.

[3] Later, after the seventh century, it was the rhapsodists who travelled through Greece and recited or performed the epics accompanied by a stringed instrument on certain holidays or other occasions.

[4] Greek, along with Chinese and Hebrew, is one of the three languages in the world that are still spoken and written and have a continuous written tradition of more than 3000 years (Beaton, 2021).

[5] The Greek term 'geography' was coined relatively late, probably by Eratosthenes, the famous polymath and chief librarian of Alexandria, in the late third century BC. However, some claim that Democritus (*c.*460–370 BC), the ancient Greek pre-Socratic philosopher from Abdera, was the inventor of the word 'geography' (Lagadas, 1990; Geus, 2018).

[6] The *Persians* is an ancient Greek tragedy, Aeschylus' greatest anti-war work, is considered the oldest surviving tragedy, and is in some ways a hymn to the Athenian fleet and the seafaring abilities of the Athenians, depicting the despair of the Persians when they learn of the defeat of Xerxes' fleet at Salamis (Lagadas, 1990).

[7] In ancient times, the 'Pillars of Hercules' were the two headlands at the eastern end of the Strait of Gibraltar. More precisely, they were the Rock of Gibraltar in Europe and Jebel Musa in North Africa. According to legend, they were placed there by Hercules to commemorate the capture of the cattle of the three-headed giant Geryon (Encyclopædia Britannica, 2006).

[8] The exact location of Cassiteris is still uncertain, and scholars argue about its identity. Some theorize that Cassiteris could refer to islands in the British Isles or the Isles of Scilly, where tin was actually mined in ancient times. Others suggest locations in the Mediterranean or even on the Iberian Peninsula.

[9] Although Pytheas' works have not survived, later ancient authors such as Polybius (*c.*200–118 BC), Strabo and Pliny the Elder have mentioned and quoted his accounts. Be that as it may, it is believed that Pytheas provided the first written mention of the British Isles and introduced the idea of far-off Thule in the far north into the geographical imagination; subsequently, the name Thule has symbolized the northernmost point of the world since antiquity. There are various hypotheses as to where Thule was located; it probably refers to Iceland, Norway, the Faroe Islands or the Shetland Islands (Chevallier, 1984; Cunliffe, 2002; Roller, 2015).

[10] Similar attempts to navigate beyond the then-known sea routes were made by Necho II of Egypt, king of the 26th Dynasty (610–595 BC), who hoped to circumnavigate Africa by sending Phoenician sailors from the Red Sea to circumnavigate the continent. However, there is no evidence that this expedition was ever

successful. The same is true of Himilco, a Carthaginian navigator and explorer who lived in the late sixth or early fifth century BC, or of Hanno the Navigator, another Carthaginian explorer from the fifth century BC, who also crossed the Strait of Gibraltar (Lagadas, 1990).

[11]Around the same time, Aristarchus of Samos (*c*.310–230 BC) taught that the sun is at the centre of the world and the Earth revolves around it. His contribution is also considered enormous for the much later development of knowledge about the world and the progress of astronomy, which is undoubtedly intertwined with navigation, since for a long time it was impossible to travel without knowing the world of the stars (Lagadas, 1990).

## References

Adrados, F.R. (2005) *A History of the Greek Language: From its Origins to the Present*. Brill, Leiden, Netherlands.

Andreou, S. (2014) Prehistory: the context. In: Christidis, A.F. (ed.) *History of the Greek Language*. Institute of Modern Greek Studies, Thessaloniki, Greece (in Greek), pp. 193–201.

Beaton, R. (2021) *The Greeks: a Global History*. Basic Books, New York.

Chevallier, R. (1984) The Greco-Roman conception of the north from Pytheas to Tacitus. *Arctic* 37(4), 341–346. DOI: 10.14430/arctic2217.

Clackson, J. (2014) The formation of the Greek language. In: Christidis, A.F. (ed.) *History of the Greek Language*. Institute of Modern Greek Studies, Thessaloniki, Greece. (in Greek), pp. 169–174.

Coulet, C. (1996) *Communiquer en Grèce Ancienne: Ecrits, Discours, Information, Voyages*. Les Belles Lettres, Paris. (in French).

Cunliffe, B. (2002) *The Extraordinary Voyage of Pytheas the Greek: the Man Who Discovered Britain*. Penguin, Harmondsworth, UK.

Duhoux, Y. (2014a) Greek and pre-Greek languages: introduction. In: Christidis, A.F. (ed.) *History of the Greek Language*. Institute of Modern Greek Studies, Thessaloniki, Greece (in Greek), pp. 202–204.

Duhoux, Y. (2014b) Linear A. In: Christidis, A.F. (ed.) *History of the Greek Language*. Institute of Modern Greek Studies, Thessaloniki, Greece (in Greek), pp. 210–214.

Düring, I. (1956) Greek music. Its fundamental features and its significance. *Cahiers d'Histoire Mondiale [Journal of World History]* 3(2): 302.

Edwards, I.E.S., Crook, J.A. and Boardman, J. (eds) (1982) *The Cambridge Ancient History (Vol. 4)*, Vol. 4. Cambridge University Press, Cambridge, UK.

Elderkin, G.W. (1934) The Dioscuri on an early Protocorinthian aryballos. *American Journal of Archaeology* 38(4), 543–546. DOI: 10.2307/498189.

Encyclopædia Britannica (2006) *Britannica Concise Encyclopedia*. Encyclopædia Britannica, Inc, London.

Geus, K. (2018) Greek and Greco-Roman geography. In: Taub, L.C. and Jones, A. (eds) The Cambridge History of Science *(Vol 1, Part III)*. Cambridge University Press, Cambridge, UK, pp. 402–412.

Hofmann, U. (1962) The chemical basis of ancient Greek vase painting. *Angewandte Chemie (International Edition in English)* 1(7), 341–350. DOI: 10.1002/anie.196203411.

Katsigiannis, K.X. (2017) *Hellenic Tourism Organization: Travel Through Time*. Private Edition, Athens. (in Greek).

Lagadas, C. (1990) *The Ancient Greeks and the Sea*. Chrysafis I. Panezis Publications, Athens. (in Greek).

Linardaki, C. and Aslanides, A. (2020) From poem and song to cultural diplomacy: challenges and opportunities for place branding and tourism promotion. *Place Branding and Public Diplomacy* 16(4), 304–315. DOI: 10.1057/s41254-020-00176-1.

Marinatos, N. (2015) *Akrotiri, Thera and the East Mediterranean*. Militos, Athens.

Maronite, D.N., Polka, L. and Touloumi, K. (2012). Greek Language Center, Thessaloniki, Greece. (in Greek). Available at: https://www.greek-language.gr/digitalResources/ancient_greek/encyclopedia/epic/page_111.html (accessed 16 October 2019).

Mathiesen, T.J. (1999) *Apollo's Lyre: Greek Music and Music Theory in Antiquity and the Middle Ages*. University of Nebraska Press, Lincoln, Nebraska.

Neils, J. (ed.) (2005) *The Parthenon: From Antiquity to the Present*. Cambridge University Press, Cambridge, UK.

Petrounias, E. (1998) *Dictionary of Standard Modern Greek*. Centre for the Greek Language. Available at: http://www.greek-language.gr/greekLang/modern_greek/tools/lexica/triantafyllides/search.html?lq=%CF%86%CE%B9%CE%BB%CE%BF%CE%BE%CE%B5%CE%BD%CE%AF%CE%B1&dq= (accessed 15 October 2019).

Roller, D. (2016) Atlantic Ocean, Oxford Classical Dictionary, 7 March. Available at: https://oxfordre.com/classics/view/10.1093/acrefore/9780199381135.001.0001/acrefore-9780199381135-e-8005 (accessed 23 August 2022).

Roller, D.W. (2015) *Ancient Geography: the Discovery of the World in Classical Greece and Rome*. I.B. Tauris, New York. DOI: 10.5040/9780755624300.

Sacks, D. (2014) *Encyclopedia of the Ancient Greek World*. Infobase Publishing, New York.

Tzermias, P.N. (1997) *The Image of Greece in the Foreign World*. Sideris Publications, Athens. (in Greek).

Uscatescu, G. (1986) *Brief Theory and History of Culture*. Ekdosis Grigori, Athens. (in Greek).

Walker, H.J. (2015) *The Twin Horse Gods: the Dioskouroi in Mythologies of the Ancient World*. I.B. Tauris, New York.

# 13 Travel in the Roman Empire and Byzantium

**Abstract**

From the third century BC, Rome's expansion beyond Italy had a decisive influence on Western civilization. The Romans, who admired Greek culture, not only spread Greek elements in the conquered territories, but also created an extensive transport network that actively promoted the culture of travelling by land and sea. After the advent of Christianity, European travellers passed through the southern Balkans on their way to the Holy Land, gaining valuable insights into geography, ethnography, history and culture. The Byzantine Empire, which emerged in 330, represented a fascinating fusion of Greek, Roman, Christian and Eastern influences and piqued the interest of Europeans in travelling to its opulent capital. Western travel literature reporting on journeys to the East inspired more visitors and also helped to shape perceptions of the southern part of the Balkan Peninsula.

By the end of the third century BC, the Romans dominated all of Italy, and after their victory over the Carthaginian general Hannibal (*c.*247–181 BC), they were able to expand unhindered into Europe and the Mediterranean in general, as well as into the East and Asia.

According to Uscatescu (1986), the conquest of the Greek world by the Romans and the subsequent spread of Greek culture under the rule of Rome to all parts of Western Europe and the Mediterranean countries is one of the most significant events in cultural history.

In Roman times, travel was made possible above all by a highly developed and extensive transport network consisting of roads built by the Romans and sea and river routes.

Later in the Middle Ages, longer journeys were not uncommon, even if most journeys involved very short distances.

Thus, during this long period, countless travellers crossed the south of the Balkan Peninsula for various reasons (trade, diplomacy, religion, education, etc.), and it is worth noting that at the time of the Byzantine Empire, the influx of foreigners to Constantinople was even so great that it required higher officials, such as the *quaestor*, who supervised new arrivals in the imperial capital, or the *logothetes tou dromou*, who was responsible for the road network and movements within the Byzantine territory, and so on (Lekkas, 1924; Verdon and Holoch, 2003).

## 13.1 Early Written Accounts by Travellers to the Greek Area

Before we begin discussing this period, we should highlight the work of Kyriakos Simopoulos (1921–2001), a Greek journalist and historian, who identified and recorded a large number of works of travel literature from 333 AD to the Greek Independence Revolution of 1821, written by thousands of foreign visitors who travelled the area where ancient Greek civilization developed. There are indeed too many who left their impressions of the region and its inhabitants, and some of them were widely read by the educated Europeans of the time, while others remained unknown to the public. The Greek scholar points out in particular that those who visited the Greek lands and seas during the Middle Ages and Ottoman rule, were men and women of all ages, many of whom were pilgrims on their way to the Holy Land. There were also smugglers in search of sacred relics and religious artefacts, robbers of ancient works of art, merchants travelling to



DOI: 10.1079/9781800626034.0013

the East, crusaders, diplomats, spies, archaeologists, geographers, doctors, botanists, artists, missionaries, pirates, adventurers, gourmets, refugees, bishops and monks, but also ordinary people thirsting for adventure and knowledge.

Of course, there were many other scholars who dealt with the subject of the bibliography of texts by visitors to the Greek area, either in general or only in specific areas that interested them. These include the Greek physician Nicholas G. Mavris (1899–1978), who compiled the literature on travellers and geographers in the Dodecanese between 1102 and 1967 (Mavris, 1973), Vourazeli (1939) and Paschalis (1964).

The writings of all these travellers are usually referred to as 'travels', 'narratives', 'memoirs', 'descriptions', 'journeys' or 'diaries'. Most travel texts contain chronicles of geographical, ethnographical, topographical, historical, archaeological, folkloristic or physiognostic information.

More specifically, they contain details on a variety of topics about the country and its people, including: (i) the composition of the population; (ii) the health of the inhabitants; (iii) economic life; (iv) ecclesiastical affairs, temples and monasteries; (v) administration; (vi) trade; (vii) handicrafts; (viii) agricultural products; (ix) taxes; (x) pirate raids; (xi) the sufferings of enslaved Greeks; and (xii) folk customs, dress, dances, songs, traditions, legends and superstitions. Many of them are enriched with pictures, maps, geographical, nautical or hydrographical plans, views of cities and ports, sketches of costumes and jewellery or artefacts, drawings of archaeological monuments, castles and various representations showing images of daily life.

Mavris (1973) notes that the authors of these texts differ not only in terms of their ethnicity, profession, the purpose of their journey and their particular interests, but above all and primarily in terms of their education, mentality, writing ability, reliability and powers of observation. Be that as it may, these texts are valuable and extremely important, and many of them feed into the periods examined below, a large part of which concern a period for which there are few or no sources due to Ottoman rule.

It should be noted, however, that until the last decades of the 17th century there was a lack of travelogues that dealt exclusively with the place where modern Greece is today. As a rule, the information is scattered in accounts of journeys to the Holy Land, Asia, Egypt, Constantinople, Asia Minor, etc., while the information about mainland Greece comprises only a few pages and sometimes only paragraphs (Simopoulos, 2003a).

As this remarkable work by Simopoulos (2003a, b, c, d, e, f) shows, with the increasing frequency of travel for various reasons, more and more educated travellers began to record their impressions of the country and its people, thus playing a crucial role in shaping the image of the Hellenic peninsula in people's minds. In this sense, there was a prolonged period in which an educated European elite perceived the views described in these early travel texts about the place and the people who lived where ancient Greek civilization flourished, cultivated the opinions and then influenced their attitudes and behaviour. These elites then passed on their own interpretation of the information alongside the actual content of the literature to the general public in Europe. In this way, an image of modern Greece gradually developed.

## 13.2 External Influences Began to Shape Perceptions of the Greek World

If we begin our investigation at the beginning of the period in question, it is noticeable that during the Roman Empire external influences began to play a greater role in shaping perceptions of the Greek world. This is in contrast to antiquity, where the construction of this image was primarily the prerogative of those who called themselves Hellenes and who exercised a certain control over the media of their time.[1]

Nevertheless, the Romans managed to imitate the culture of the Greeks in an ostentatious manner. It is not difficult to see the influence of Greek art and architecture on Roman style. And although the language of imperial administration was Latin, Greek was either spoken as the first language or was considered the language of art and commerce in all the eastern provinces. The prevailing opinion at this time was that *humanitas* had been invented by the Greeks and passed on by them to the Romans.[2]

Horace (65–8 BC), one of the most famous Roman poets, formulated the well-known saying that the conquered Greeks had captured their savage conquerors and brought the arts to rural Latium (Woolf, 1994; Beaton, 2021).

Be that as it may, the media of the Romans established their own views on the image of Greek culture. Very important sources for gathering information about Greek space and travel to Greece in the early Roman centuries are the works of the Greeks, including Strabo (*c.*63 BC–24 AC), Plutarch (*c.*46–119 AC), Dion Chrysostom (*c.*40–115 AC) and Philostratos (*c.*170–250 AC). However, it also includes works of the Romans, such as Horace (65–8 BC), Ovid (*c.*43 BC–18 AC), Seneca (four BC–65 AC), Pliny the Elder (*c.*24–79 AC), Juvenal (*c.*55–127 AC), Tacitus (*c.*56–120 AC), Pliny the Younger (*c.*61–113 AC), Suetonius (*c.*69–122 AC) and the Jewish philosopher Philo Judaeus (*c.*20 BC–50 AC) (Simopoulos, 2003a).

We have already mentioned Strabo (Chapter 12, this volume), the great admirer of Homer, who lived in the transition period from the Roman Republic to the Roman Empire, but at this point it is worth saying two or three more things about him. Strabo was a Greek geographer, philosopher and historian who lived in Asia Minor, travelled extensively and was later, in the Middle Ages, revered as the 'geographer of the world'. Strabo himself regarded his work as a form of philosophy that gives a comprehensive overview of the world and leads the reader to virtue and wisdom. One of his works, the '*Geography*', consisting of 17 books, described the entire then-known world, including the regions where Greece, Turkey, Cyprus, India, Iraq, Iran, Syria, Palestine, Israel, Saudi Arabia, Egypt, Libya, Italy, Switzerland, Germany, France, Spain, Great Britain, Ireland and many other countries are located today (Theodoridis, 1994).

Strabo's work was translated and read by many scholars in Europe from the middle of the 15th century. His ideas are mentioned in numerous texts on cosmography, descriptive geography, history, pedagogy and political thought of the late Middle Ages. Thus, like the works of other Greek and Roman geographers, his ideas contributed significantly to the reception of the European geographical area. An indication of Strabo's widespread use are the various Strabonian toponyms and terms that can be found in hundreds of literary works. In general, he provided a paradigm for geographical writing that inspired many writers and navigators in Renaissance Europe who broke new ground. Therefore, his contribution to the creation of an image of Greece and the Greek 'view' of the world is very great (Dalché, 2017b).

Dionysius Periigetes, a poet of the second century AC from Alexandria who lived in Rome, wrote the work *Ecoumenis Periigesis* (Orbis description) in hexameter verse, which soon enjoyed great popularity in antiquity and occupied an important position as a textbook. Dionysius in his poem, describes various countries, cities and tribes in a way that gives the reader a clear, systematic and understandable picture of the then-known world. Since Dionysius' poem was regularly used in teaching throughout the Middle Ages, numerous annotated manuscript copies and translations have survived to preserve the text (Ilyushechkina, 2016).

As already mentioned, since the rise of the Roman Empire, there was a more intense involvement of foreign peoples who used their own means of communication to present Greek culture and land. This continued in the following periods, in the Middle Ages with pilgrims, writers and artists, as we will see below, in the Renaissance with painters and so on. In other words, with the foundation of the Roman Empire and afterwards, the construction of the image of the 'Greek world' was strongly influenced by non-Greek factors.

## 13.3 Byzantine Culture Attracted Travellers

The period of the Eastern Roman Empire (i.e. Byzantium) extends from 324, when the Roman Emperor Constantine I the Great (*c.*280–337) defeated his rival and ally, the other Roman Emperor Valerius Licinianus Licinius (*c.*265–325) on the Asian shores of the Bosphorus, and founded Constantinople in 330 AD, until 1453.[3] This was when the Byzantine Emperor Constantine XI Palaiologos (1405–1453) was gloriously killed defending Constantinople during the attack of Mehmed II the Conqueror (1432–1481).

The elements that make up Byzantine culture are diverse, but their fusion into a homogeneous culture is undeniable. Ancient Greece, Hellenism, the Latin tradition, the ecumenical Christianity of the synods, the rebirth of Persian culture by the Sassanids, Eastern religious and state ideas, the objective role of the state in preserving the human heritage and passing it on to Europe are some of the dynamic elements of Byzantine culture (Uscatescu, 1986).

Even in antiquity, although there were images of destinations that could motivate people to travel to the region where Greek civilization was born (or other images of distant places that encouraged Greeks to travel far afield), the idea of leisure travel was not widespread among the common people in the Middle Ages and Byzantine period. This is the case if we exclude pilgrimage tours to the Holy Land or journeys to religious sites as well as educational and cultural trips to cultural centres such as Antioch, Beirut, Alexandria and Athens and later to Constantinople from the concept of leisure travel. With regards to the latter, we can speak of mass travel and a medieval Christian 'industry' that supported it (European Center for Byzantine and Post-Byzantine Monuments (ECBP-BM), 2021).

The Asia Minor and Greek provinces of the empire, the coasts, as well as the deserts of Africa, Palestine and Syria, were full of life at this time. Martyrs, priests, monks, monasteries and believers marvelled at the pillars of the Stylites and countless faithful travelled along prescribed pilgrimage routes from monastery to monastery to the place of Christ's martyrdom. It can therefore be assumed that the Christian religion also brought life and movement to the wider Mediterranean region, especially after the adoption of Christianity by the empire in the fourth century (Ahrweiler, 2003a).

During this period, Byzantine churches with their exotic domes and rich mosaics as well as monasteries were important means of transmitting knowledge and ideas. The various churches scattered throughout the Byzantine Empire were decorated with frescoes, portable paintings in the form of icons and mosaic floors with scenes from the Old and New Testaments (Guillou, 1993).

It can be assumed that all of these depictions, together with the sermons, led the faithful spiritually to the holy places; but also to other places where the saints had lived, or even to the centre of the empire, Constantinople, and motivated people to travel there. These depictions followed the typical tradition of Christian iconography but were in the Byzantine style, such as the mosaic of the monastery of St Luke in Boeotia, which was regarded by pilgrims as one of the most important Byzantine monuments (Fig. 13.1).[4]

Moreover, the music of the Orthodox Church became famous, and its melodies were loved by people all over Europe and the Middle East. And we should not forget that many contemporary tourism scholars believe that music is an important attraction for a tourist destination.[5]

Thus, Byzantine architecture, icons and music, were central expressions of Byzantine culture and undoubtedly shaped the image of the empire that spread throughout the region where the ancient Greek civilization had developed, while the modern Greek state, founded centuries later, occupied some of Byzantium's most important territories. Consequently, we can claim that the Byzantine Empire and its culture contributed as another layer to the modern image of Greece, or rather, were included as another primary cell or dimension for its development.

By this time, a very large pilgrimage movement to the holy places had already developed. It is said to have assumed these enormous proportions after Flavia Julia Helena Augusta or Saint Helena (*c.*246–330), Empress of the Roman Empire and mother of Constantine I the Great, undertook a journey to find the mortal remains of Jesus Christ and to venerate the places where he had spent his earthly life.

Of course, numerous believers from all parts of Christendom had already flocked to these sites. They believed that by visiting these sites they would atone for their sins; others saw it as a form of religious exercise or punishment; and still others believed that this journey would give them prestige in medieval society. As a result, stable transport and communication networks were created throughout Europe and the Mediterranean: (i) standard routes were established; (ii) transport structures with accommodation for pilgrims were developed; and (iii) an important link was created between

**Fig. 13.1.** Greek National Tourism Organisation (GNTO) poster, *Detail from the mosaic of St Luke's Monastery, 11th century* (1976). It is fair to say that the modern tourist representation of Greece is unimaginable and inaccurate if it does not include elements of Byzantine art, architecture and culture in general. The Byzantine Empire, which formed an important bridge between classical Greek and Roman civilization and the subsequent periods of Greek history, has left its indelible mark. Of course, the significance of Byzantine influence goes beyond mere aesthetics, as it encompasses a broader cultural and historical continuity. The persistence of specific traditions, linguistic nuances and cultural practices from the Byzantine era contributes significantly to cultivating a tangible sense of historical continuity in Greek history. A comprehensive understanding of Greece's tourism image therefore requires the inclusion and appreciation of the enduring legacy of the Byzantine Empire. Image used with permission from the GNTO.

all the numerous ecclesiastical centres, not only in the holy places but also in other places that were considered sacred and required a pilgrimage. With a little imagination, one could say that this network represents a primitive form of the tourism industry, or at least an embryonic state in that direction.

From the beginning, there were also Greek pilgrims who travelled to the Holy Land, and some of them recorded their memories and motivated other people to travel there (Irmscher, 1993).

However, the particular circumstances of their journeys, the routes and sights, the legends and traditions associated with certain buildings and places, and other information were recorded in detail by Greek monks in the *Proskynetaria*. These manuscripts resemble travel guides and were written mainly at the famous monastery of St Sabbas near Jerusalem. The *Proskynetaria* can be considered travel media of the time and were actually geographical commentaries on the Old and New Testaments, a kind of travel literature intended to guide and direct pilgrims through the Holy Land (Kislinger, 1993; Kuelzer, 2017).

Pilgrims from Europe were usually given boat connections or used refuelling stops to get to Palestine from places like Sicily, Venice or the Peloponnese. Others walked across the Greek mainland, and often some were given permission to visit Constantinople just to see the great capital of the East, with the help of local guides who had some basic knowledge of a foreign language sufficient to show them the sights (Ciggaar, 2017).

What is of great interest to our study is the fact that many of these Western pilgrims wrote down their experiences as they travelled through the area where the ancient Greek civilization had developed, resulting in a series of books that in turn shaped the image of the place.[6]

Moreover, the Byzantine scholars of this period searched with great devotion for ancient Greek texts, copying them by the thousands and thus creating part of the image we have today of Greek antiquity. In the bibliographic workshops, the scriptoria of Byzantium, all kinds of ancient texts were copied and sent to palace and university libraries, but also to the libraries of the patriarchate and monasteries as well as to private individuals (Ahrweiler, 2003b; Lemerle, 2003).

Between 547 and 550, Cosmas Indicopleustes, a merchant from Alexandria who later probably became a monk, wrote the *Christian Topography*, which was not only a kind of travelogue about places that might be of interest to a Christian, but included both cosmology and geography. His main aim was to propagate a truly biblical worldview, to prove that the Earth is ultimately flat and that every inhabited part of the Earth conforms to a divinely ordained pattern derived from the Bible. After all, he had crossed the borders of the Byzantine Empire and seen with his own eyes how much larger the world is. Later readers judged that almost everything about his personality can be doubted, but his concept is remarkable for its geography, which is credible and offers enormous interest in the trade data of a vast area, as well as flora, fauna and cultural things of various kinds that probably impressed many of his contemporaries on their travels and shaped the core images for these faraway places (Faller, 2011; Beaton, 2021).

While Athens at this time had degenerated into a forgotten and insignificant settlement whose image was only alive in stories and myths, cosmopolitan Constantinople had become the largest and richest city in Europe and probably in the whole world. A visitor of Jewish origin from Muslim Spain, Benjamin of Tudela (1130–1173), in his work documenting Jewish communities, their lives and their problems in the then-known world, praised the city's wealth, the gold and jewels of its kings, its imposing churches, adding that its Greek inhabitants were richer and more elegant than anywhere else in the world. Descriptions like these clarify the image that Constantinople, the centre of Hellenism at the time, left on travellers: everyone wanted to go there, and many Western and Eastern armies wanted to conquer its riches.[7]

It is clear from this evidence that the Byzantine capital, the coveted city, represented a different vantage point, a different core image in the collective memory of Hellenism than when it was later occupied by the Ottomans (Adler, 2018).

During a relatively long period spanning the 11th, 12th and 13th centuries, crusader armies were organized in Europe to move eastwards and liberate the Holy Land, which had previously been conquered by the Muslims in the seventh century.

According to Macrides (2017), however, of the thousands of crusaders who travelled through Byzantine territory, only a few saw Constantinople itself within its walls, and of those in turn, only four produced significant and credible writings about their impressions. These were the French priest Fulcher of Chartres (*c.*1059–1128), the French historian Odo of Deuil (*c.*1110–1162), the French historian Geoffrey of Villehardouin (*c.*1150–1213) and the French cleric Robert of Clari (*c.*1170–1216). It should be noted here that Hellenism at this time was considered to be closely linked to the Eastern Roman Empire and Constantinople, and Constantinople was thus considered part of the picture of the Greek world.[8]

Be that as it may, towards the end of the Middle Ages, travelling to the region returned to the centre of Western literature and a separate literary branch gradually emerged, which was linked to chivalry, but also to journeys to the East. This new literary genre could take various forms: (i) sometimes they were simple and quite accurate descriptions, following the personal view of the knight as narrator; (ii) others were fantastic and adventurous, in the form of short stories in the third person; and (iii) still others were fictions that tried to create intermediate states. And, of course, literary narratives of this kind stimulated the reader to creative compositions and enriched his/her imagination with new images about the territory where the ancient Greek civilization had developed, as well as about its inhabitants.

Another medium that shaped the image of the Byzantine world in the minds of travellers were the *portulans*, a kind of nautical chart first published in the 13th century. The *portulans* of the Mediterranean were widely used at the time and were soon used not only by those interested in day-to-day navigation, but became aids to travel preparation and companion books for travellers with different interests: royal officers preparing crusading expeditions, merchants, pilgrims and so on. They were also used for geographical descriptions of landscapes, bays, sandy beaches, rocky coasts, mountains and so on. The *portulans* of the Mediterranean provided a relatively accurate knowledge of the land and islands and painted a more complete picture of the places. Their cultural, but also their practical, role should not be underestimated (Dalché, 2017a).

The 14th century was a time of crisis and upheaval in Europe: there was a great famine, the Black Death claimed millions of lives, temperatures dropped worldwide making travel difficult, the Knights Templar who protected pilgrims on their way to holy sites in Muslim lands disbanded, and so on. But it was also a time of new beginnings, technological advances and growing empires. Towards the end of the century, Italian scholars sought to recover ancient Greek texts, both to improve their language skills and to better understand the philosophical, scientific and historical annotations of the classical authors.

The Byzantines rarely used the older term 'Hellines' to refer to themselves, but Georgios Gemistos or Plethon (*c.*1360–1452), one of the most important Neoplatonic philosophers of the late Byzantine period, used the term in his work that anticipated the concept of national identity for the inhabitants of the southern part of the Balkan Peninsula. In 1439, he accompanied the penultimate Byzantine emperor, John VIII Palaiologos, to the Council of Florence, where his discussions and lectures succeeded in sensitizing the Italian humanists and the wider European intellectual community to the value of Plato and in curbing the omnipotence of Aristotle. The educated of Florence were fascinated by Gemisto's Platonic lectures and called him the great teacher. His importance for the progress of Platonic studies throughout Europe is considered enormous (Antonakopoulos, 1980; Margaris, 1982; Beaton, 2021).[9]

As we have seen, although in the Byzantine era there were various influences and motivations for travel to the vast area where the ancient Greek civilization had developed, the concept of modern mass media as we understand it today did not yet exist. Therefore, word of mouth, written texts, Byzantine art, iconography, architecture and, in general, personal narratives and collective memory should be mentioned as basic means. But another factor also played an important role in attracting visitors – the magnificent festivals, fairs and cultural events of the Byzantines.

Entertainment was one of the most important aspects of daily life in the Byzantine era. The most famous Byzantine festivals and celebrations include Easter, Christmas, the Feast of the Exaltation of the Cross, the Day of the Accession of the Byzantine Emperor and the Byzantine New Year. Also the chariot races and games held at the Byzantine Hippodromes in Constantinople were great public spectacles and celebrations that attracted large crowds. However, there were many other local and regional festivals and celebrations throughout the empire that reflected the rich religious and cultural diversity of Byzantine society.

Some Byzantine festivals and cultural events were known for their opulence and splendour, often with lavish decorations, music, dance and theatrical performances that attracted elite guests. Cultural events provided a platform for the exchange of ideas, art forms and traditions, and visitors from other regions of the empire and beyond were eager to experience and learn about this particular mix of cultures. Visitors often saw these events as opportunities to interact with influential people such as government officials, scholars and artists. In addition, many Byzantine celebrations were associated with religious festivals and holidays, so devout Christians travelled to Greece to participate. In this respect, the Byzantine authorities recognized the potential of cultural events and festivals to attract visitors. They often promoted these events as part of a broader diplomatic strategy to encourage travel to the region and boost the local economy.

In summary, the great festivals, banquets and cultural events of the Byzantine era acted as visitor magnets for the entire southern part of the Balkan Peninsula due to their entertainment value, cultural significance, social relations and the opportunity to experience the splendour and complexity of Byzantine society (Graf, 2015; Alysandratou, 2019).

All these elements led Europeans to have a complex image of 'cosmopolitan' Constantinople, characterized by a mixture of admiration and envy for its wealth and culture, its religious significance, its political power and its trading opportunities. The city occupied a unique place in the European imagination as the 'Rome of the East', and its influence on the art, culture and geopolitics of Western Europe cannot be underestimated, which is why the city exerted a great attraction on travellers.

The fall of Constantinople in 1453 brought the production and circulation of ancient Greek manuscripts to a halt, making Italy the most important centre for these texts. At this time, the surviving manuscripts began to be copied with fervour, which had cultural consequences in Europe, as these texts were a significant source of ideas and knowledge that inspired many scholars during the Renaissance and led to new creations (Dalché, 2017b).

So, we see that during this period the writings of ancient Greece were filtered and reproduced through the Renaissance framework and the image of the Hellenic world was reshaped.

However, it is believed that after the fall of Byzantium, Western travel literature again reported more objectively on the continent where Greek civilization had developed, while many Western travellers found the Ottomans and their culture a parallel and very interesting subject.

On the other hand, some travellers focused their attention mainly on the Greek past. Among other things, many of them felt that Byzantium had failed in the task of preserving the material remains of its classical past. And finally, the opinions of foreign media about the Greek region and culture gradually intensified. All these views soon led Western humanists to define themselves as preservers of classical antiquity, which resulted in more and more Western visitors coming to Greece to admire the ancient monuments (Angold, 2017).

## 13.4 Summary

In short, during this period the Romans adopted many elements of ancient Greek culture, reproducing and disseminating their own view of the Greek world. In the Byzantine era, new elements (Christianity, Latin tradition, oriental culture, etc.) were added to the image of the place where the ancient Greek civilisation had developed. Literature, architecture and art were the means that gave the region a new image. Basically, the glory of Byzantium was added as another layer to the image of the region. Be that as it may, during these years there was a lot of traffic from Western travellers and travel reports in the Western media about the region, mainly because it represented the transition to the Holy Lands.

Undoubtedly, travelling to the Greek and eastern countries of the time was a life experience for travellers. Many chose to write down their memories of the adventures, describing the places and people they encountered, the different cultures and strange things they saw, the living conditions of the inhabitants, or the difficult times and problems they experienced on their journey. Other works are short and contain only lists of towns and distances; others report on markets and local products, as they were addressed to merchants; and still others are more narrative and convey the particular impressions of their authors. Travel works are valuable sources not only for the personality and narrative qualities of their authors, but above all for cultural history (ECBP-BM, 2021).

## Notes

[1] As already mentioned, the image of the 'Greek world' was for a long time mainly represented and preserved by the Greeks themselves, although in the past other peoples from countries near or far also contributed to this process (Darbandi and Zournatzi, 2008). This included: (i) Persian sources such as the Behistun inscription, written during the reign of Darius I the Great (*c.*522–486 BC), refer to the Greeks as *Yauna* and praise their military prowess; (ii) the ancient Egyptian priest Manetho of Sebennytos, who lived in the early third century BC, wrote about them in his work *Aegyptiaca*, describing them as a foreign people who brought about both positive and negative changes in Egypt; and (iii) the Chinese historian Sima Qian (*c.*145–86 BC) from the early Han dynasty mentioned the *Yuezhi* people living in the area, while other Chinese scholars such as Zhang Qian, an official and diplomat who lived as an imperial envoy to the world outside China in the late second century BC, regarded the Greeks as a distant civilization with advanced culture and military technology.

[2] The concept of *humanitas* has its roots in classical Greek education (i.e. in a series of general education courses that prepared young people for citizenship). Cicero (106–43 BC) used the word to organize a training programme for orators, which began in 55 BC and which he considered necessary for the possession of the virtues and character for an active political life in public, but also for a dignified private life. Today we use the term to refer to branches of knowledge that deal with people, their culture and the way they express themselves and include the study of languages, literature, art, history and philosophy. Some modern definitions of the humanities also include law and religion (Encyclopædia Britannica, 2006).

[3] Constantine I the Great, officially Flavius Valerius Constantinus or Constantine I, was Roman emperor from 306 to 337 and the first to profess Christianity. In 330, he moved the capital from Rome to Byzantium, an ancient Greek city on the Bosphorus, which he renamed Constantinople.

[4] Depictions of Christ, the Virgin Mary, saints and other religious subjects were venerated as sacred objects and were believed to possess spiritual powers. During the years of iconoclasm, iconographers limited the number of icons with religious themes in and around Byzantine churches scattered throughout the empire. They partially replaced the icons of the saints with other representations such as horse races, hunting scenes, depictions of birds and animals in beautiful gardens, if not in the central area of the sanctuary, then at the peripheral points and in the outer surrounding buildings. In the context of our study, we can assume that this served as a kind of propaganda or advertisement for the empire to provide churchgoers with images of daily life in the Byzantine capital and to attract them to Constantinople (Guillou, 1993).

[5] Great cantors, composers and teachers of Byzantine music who later appeared and spread it in Europe include Ioannis Koukouzelis (1300–1350) and some centuries later Petros Peloponnesios (1735–1778) and Petros Byzantios (1770–1808).

[6] In the period 900–1300, as Berger (2017) describes, there were also texts by Arab travel writers about the Byzantine Empire and Constantinople. They reported on what they had seen and experienced themselves, but also other texts in Arabic based on chronicles or geographical treatises from older sources. It is also worth noting that in the 14th and 15th centuries there was a large flow of Russian pilgrims to the Middle East who visited Constantinople and other places in Greece and wrote down their impressions of these places, which helped to shape the Russians' social representations about the place (Majeska, 1984).

[7] To understand the image that travellers had of Constantinople, then the capital of Byzantium, and its connection with Hellenism, a quotation from the text of Benjamin of Tudela is reproduced:

> All kinds of merchants come here, from the land of Babylon, from the land of Shinar, from Persia, Media and the whole dominion of the land of Egypt, from the land of Canaan and the kingdom of Russia, from Hungary, Patzinakia, Khazaria and the land of Lombardy and Sepharad. It is a busy city to which merchants come from all lands and seas, and there is no comparable city in the world except Baghdad, the great city of Islam. In Constantinople is the church of Santa Sophia and the seat of the Pope of the Greeks, since the Greeks do not obey the Pope of Rome. There are also churches arranged according to the number of days in the year. Immense wealth is brought here year after year as tribute from the two islands and the castles and villages there. Such wealth is found in no other church in the world. And in this church, there are pillars of gold and silver and lamps of silver and gold, more than a man can count. Near the palace walls there is also a pleasure place belonging to the king called the Hippodrome. Every year on the anniversary of the birth of Jesus, the king holds a great spectacle there, where men from all the nations of the world, with and without jugglery, come before the king and queen and make lions, leopards, bears and wild asses fight each other; the same is done with birds. There is no such entertainment in any other country. This King Emmanuel built a great palace on the coast as the seat of his government, in addition to the palaces his fathers had built, and called it Blachernae. He covered its columns and walls with gold and silver and had depictions of battles before his time and of his own battles engraved on them. He also had a throne made of gold and precious stones and a golden crown hung on a golden chain above the throne so that he could sit under it. It was set with jewels of inestimable value, and no lights were needed at night, for everyone could see through the light emitted by the stones. There are countless other buildings in the city. Every year tributes are brought here from all parts of the Greek Empire and they fill the fortresses with robes of silk, purple and gold. Nothing in the whole world can match these warehouses and this wealth. It is said that the tribute of the city amounts to 20,000 pieces of gold a year, derived both from the rents of the shops and markets and from the tribute of the merchants who come by sea or by land. The Greek inhabitants are very rich in gold and precious stones and dress in robes of silk with gold embroidery, ride horses and look like princes. Indeed, the country is very rich in all fabrics and in bread, meat and wine. Such wealth as that of Constantinople is not to be found in the whole world. Here, too, there are men learned in all the books of the Greeks, and everyone eats and drinks under his vine and his fig tree.
>
> (Adler, 2018, pp. 21–23)

[8] 'Hellenism' in this book is understood in its simplest and broadest sense as Greek culture and the totality of Greeks as the bearers of this culture (Babiniotis, 1998). The shift of the centre of the empire from Rome in the west to Constantinople in the Greek-speaking east shows the close relationship and privileged ties that the Byzantines had with ancient Greece. Constantinople was a cosmopolitan capital, Greco-Roman, Byzantine and Christian, a centre of political and religious power with universal appeal, which at the same time included many other ethnicities (Armenians, Slavs, etc.). Towards the end of the 14th century, the Byzantines began to use the term Greek as an integral part of their identity without abandoning their Roman identity; the terms Greek (Éllin) and Roman were not seen as opposites or incompatible in the Byzantine world at that time. Be that as it may, the Greek tradition and Greek historical memory were always present in the Byzantine world, and therefore Constantinople was considered the cradle of Hellenism, and Byzantium an experience of Hellenism (Glykatzis-Arveler, 1978; Minakakis and Tsevrenis, 2011).

[9] There were certainly many other notable Byzantine figures who played a significant role in shaping the image and history of the Byzantine Empire in the centuries before the fall of Constantinople. Here are some of them: (i) the scholar and statesman Theodore Metochites (1270–1332); (ii) the icon painter and artist Theophanes the Greek (*c*.1340–1410); (iii) the scholar, philosopher, professor and translator Manuel Chrysoloras (*c*.1350–1415); (iv) the Emperor Manuel II Palaiologos (1350–1425); (v) the Emperor John VIII Palaiologos (1392–1448); (vi) noblewoman Anna Notaras (*c*.1400–1500); and (vii) noblewoman and scholar Theophano Martinakia (*c*.1410–1480).

## References

Adler, M.N. (2018) *The Itinerary of Benjamin of Tudela*. Philipp Feldheim Inc., New York.

Ahrweiler, H. (2003a) From paganism to Christianity. In: Ahrweiler, H. and Aymard, M. (eds) *The Europeans, Volume I*. Savallas, Athens. (in Greek), pp. 123–133.

Ahrweiler, H. (2003b) A new European world. In: Ahrweiler, H. and Aymard, M. (eds) *The Europeans, Volume I*. Savallas, Athens. (in Greek), pp. 162–180.

Alysandratou, M. (2019) Entertainment in Byzantium during the period of the Komnenians and Angelos (1081–1204). Doctoral dissertation, National and Kapodistrian University of Athens, Athens. (in Greek).

Angold, M. (2017) The decline of Byzantium seen through the eyes of Western travellers. In: Macrides, R. (ed.) *Travel in the Byzantine World*. Routledge, London, pp. 213–232. DOI: 10.4324/9781315235646.

Antonakopoulos, A. (1980) *Introduction to the History of Modern Political Ideas*, Vol. 1. Sakkoulas Publications, Athens. (in Greek).

Babiniotis, G. (1998) *Dictionary of the Modern Greek Language*. Lexicology Publications, Athens. (in Greek).

Beaton, R. (2021) *The Greeks: a Global History*. Basic Books, New York.

Berger, A. (2017) Sightseeing in Constantinople: Arab travellers, c.900–1300. In: Macrides, R. (ed.) *Travel in the Byzantine World*. Routledge, London, pp. 179–192. DOI: 10.4324/9781315235646.

Ciggaar, K. (2017) Bilingual word lists and phrase lists: for teaching or for travelling. In: Macrides, R. (ed.) *Travel in the Byzantine World*. Routledge, London, pp. 165–178. DOI: 10.4324/9781315235646.

Dalché, P.G. (2017a) Portulans and the Byzantine world. In: Macrides, R. (ed.) *Travel in the Byzantine World*. Routledge, London, pp. 59–71. DOI: 10.4324/9781315235646.

Dalché, P.G. (2017b) Strabo's reception in the West (fifteenth–sixteenth centuries). In: Dueck, D. (ed.) *The Routledge Companion to Strabo*. Routledge, London, pp. 367–383. DOI: 10.4324/9781315696416.

Darbandi, S.M.R. and Zournatzi, A. (eds) (2008) Ancient Greece and Ancient Iran: Cross-Cultural Encounters. In: *1st International Conference, Athens, 11-13 November 2006*. Available at: http://www.eie.gr/greece-iran-conference/Proceedings_AncientGreeceAncientIran_2008.pdf (accessed 24 May 2020).

Encyclopædia Britannica (2006) *Britannica Concise Encyclopedia*. Encyclopædia Britannica, Inc, London.

European Center for Byzantine and Post-Byzantine Monuments (ECBP-BM) (2021) Exploring the Byzantine World. Forms of Communication. Available at: http://exploringbyzantium.gr/EKBMM/Page?name=ypomeleti&id=5&sub=431&lang=gr&level=3 (accessed 21 June 2022).

Faller, S.A. (2011) The world according to Cosmas Indicopleustes – concepts and illustrations of an Alexandrian merchant and monk. *The Journal of Transcultural Studies* 2(1), 193–232.

Glykatzis-Arveler, E. (1978) *General Introduction – Hellenism and Byzantium I.E.E,* vol. G. Ekdotiki Athinon, Athens. (in Greek).

Graf, F. (2015) *Roman Festivals in the Greek East: From the Early Empire to the Middle Byzantine Era*. Cambridge University Press, Cambridge, UK. DOI: 10.1017/CBO9781316135778.

Guillou, A. (1993) Communication in the Byzantine Empire. In: Moschonas, N.G. (ed.) *Communication in Byzantium*. Center for Byzantine Research, Athens. (in Greek), pp. 31–49.

Ilyushechkina, E. (2016) *Dionysius Periegetes. Description of the Known World with Introduction, Text, Translation, and Commentary*. Oxford University Press, Oxford, UK.

Irmscher, J. (1993) Byzantine pilgrimages. In: Moschonas, N.G. (ed.) *Communication in Byzantium*. Center for Byzantine Research, Athens. (in Greek), pp. 347–350.

Kislinger, E. (1993) Sightseeing in the Byzantine Empire. In: Moschonas, N.G. (ed.) *Communication in Byzantium*. Center for Byzantine Research, Athens. (in Greek), pp. 457–468.

Kuelzer, A. (2017) Byzantine and early post-Byzantine pilgrimage to the Holy Land and to Mount Sinai. In: Macrides, R. (ed.) *Travel in the Byzantine World*. Routledge, London, pp. 149–161. DOI: 10.4324/9781315235646.

Lekkas, N. (1924) *Hotels in Greece, from Ancient Times to the Present*. Ministry of National Economy, Athens. (in Greek).

Lemerle, P. (2003) The continuity of the European consciousness. In: Ahrweiler, H. and Aymard, M. (eds) *The Europeans, Volume I*. Savallas, Athens. (in Greek), pp. 181–183.

Macrides, R. (2017) Constantinople: the crusaders' gaze. In: Macrides, R. (ed.) *Travel in the Byzantine World*. Routledge, London, pp. 193–212. DOI: 10.4324/9781315235646.

Majeska, G.P. (1984) *Russian Travelers to Constantinople in the Fourteenth and Fifteenth Centuries*, Vol. 19. Dumbarton Oaks, Georgetown, Washington, DC.

Margaris, V. (1982) *Great Neo-Greeks since 1400*. Epiktitos, Athens. (in Greek).

Mavris, N.G. (1973) *Travelers and Geographers in the Dodecanese (1102-1967)*. Dodecanese Bibliography, Athens. (in Greek).

Minakakis, V. and Tsevrenis, G. (2011) *Constantinople – the Cradles of Hellenism*. Kathimerini Ekdosis – Alpha Bank, Athens. (in Greek).

Paschalis, D. (1964) Tours in Greece from the 12th to the 1st century. *Yearbook of Cycladic Studies*. (in Greek) 4, 231–324.

Simopoulos, K. (2003a) *Foreign Travelers in Greece: Public and Private Life, Popular Culture, Church and Economic Life, from Travel Chronicles*, Vol. 1. Stachy, Athens. (in Greek).

Simopoulos, K. (2003b) *Foreign Travelers in Greece: Public and Private Life, Popular Culture*, Vol. 2. Stachy, Athens. (in Greek).

Simopoulos, K. (2003c) *Foreign Travelers in Greece: Public and Private Life, Popular Culture, Church and Economic Life, from Travel Chronicles*, Vol. 3. Stachy, Athens. (in Greek).

Simopoulos, K. (2003d) *Foreign Travelers in Greece: Public and Private Life, Popular Culture, Church and Economic Life, from Travel Chronicles*, Vol. 4. Stachy, Athens. (in Greek).

Simopoulos, K. (2003e) *Foreign Travelers in Greece: Public and Private Life, Popular Culture, Church and Economic Life, from Travel Chronicles*, Vol. 5. Stachy, Athens. (in Greek).

Simopoulos, K. (2003f) *Foreign Travelers in Greece: Public and Private Life, Popular Culture, Church and Economic Life, from Travel Chronicles*, Vol. 6. Stachy, Athens. (in Greek).

Theodoridis, P. (1994) Strabo's geography or a Greek world. In: Strabo. *Geographical 3, Spain*. Cactus, Athens. (in Greek), pp. 11–28.

Uscatescu, G. (1986) *Brief Theory and History of Culture*. Ekdosis Grigori, Athens. (in Greek).

Verdon, J. and Holoch, G. (2003) *Travel in the Middle Ages*. University of Notre Dame Press, Notre Dame, Indiana.

Vourazeli, H.D. (1939) *The Life of the Greek People During the Turkish Rule on the Basis of Foreign Travelers*. Self-published, Athens. (in Greek).

Woolf, G. (1994) Becoming Roman, staying Greek: culture, identity and the civilizing process in the Roman East. *The Cambridge Classical Journal* 40, 116–143. DOI: 10.1017/S0068673500001875.

# 14 The Renaissance, Enlightenment and the Tours in Greece

**Abstract**

After the Middle Ages, Europeans developed a strong fascination for ancient Greek civilization. This keen interest was triggered by various factors, including the study of ancient language and philosophy at European universities, Byzantine literature and the rise of intellectual movements such as humanism and classicism. All of these factors had a decisive influence on Europeans' perception of the southern Balkans. Despite the dominance of the Ottoman Empire, notable Greek figures played an important role in preserving the Greek heritage and creating a lasting image of the place where ancient Greek civilization flourished. A wide range of artistic forms and visual representations, as well as travellers' accounts of their journeys, had a great influence on the global perception of this place. This historical period can be seen as highly significant in the elevation of the southern Balkan peninsula and its islands to a captivating tourist destination in the European consciousness.

The image of the region where Greek civilization flourished in antiquity was shaped by various influences in European perception after the Middle Ages. Some influences were the study of the ancient Greek language, the classical texts of ancient Greece, ancient Greek political philosophy, the founding of more and more universities and educational institutions during this period that included the study of Greek literature in their programmes, Byzantine literature and the scholars who escaped the fall of Constantinople in 1453. These elements, along with the decaying ancient monuments rooted in the earth, travellers and 'Grand Tourists' who wrote down their experiences, and the intellectual currents such as humanism, classicism and romanticism, all conveyed a certain idea of the southern Balkan region and its inhabitants.

The European travellers and explorers who visited the region and the travel literature they wrote, with its vivid descriptions and images of Greek landscapes, cities and people, strongly influenced the way the rest of the world perceived this land. The classical architecture and art of ancient Greece, which served as a model for the artists and architects of the Renaissance and Neoclassical periods, intensified Europeans' fascination with Greek culture and enriched the image of Greece at the time.

This revived admiration for Greek culture and ancient Greek thought had a lasting impact on the intellectual and artistic development of Europe and contributed to a broader movement to revive classical studies and the study of ancient civilizations, bringing this place back into the consciousness of the world public.

The media that played a crucial role in the dissemination of knowledge, ideas and artistic representations about this region were therefore: (i) manuscripts, books, travel diaries, the press – and here it is worth remembering that the invention of new printing processes in the 15th century revolutionized the dissemination of knowledge; (ii) the various art forms and visual representations, paintings, sculptures and illustrations that recreated ancient Greek scenes and architecture; and, of course (iii) the spoken word.

Undoubtedly, the convergence of all these media contributed to the creation and dissemination by Europeans at this time of a complex image of the southern end of the Balkan peninsula, which primarily saw older elements through a new lens, but also contained entirely new features of the current culture.

 *Communication and Tourism: Reflecting on the construction of the tourist image of Greece* (Michael Tsangaris)
DOI: 10.1079/9781800626034.0014

Be that as it may, after the Middle Ages, all these cultural references to ancient Greece, with many references to the natural landscape and fewer to modern culture, were the fundamental factors that helped to establish the region at the end of the Balkans, the islands and the coasts of Asia Minor, in the minds of Europeans as attractive destinations.

## 14.1 Shaping Perceptions of the Southern Balkans in This Period

The Renaissance marked the beginning of a transformative epoch that triggered far-reaching changes in all areas of European society. The revolution of the 'Quattrocento' had a profound impact on the arts as well as on social and political dynamics and heralded a very special era. This transformative cultural change was characterized by a renewed emphasis on schematic representation, which led to a departure from previous conventions.

In the visual sphere, there was a gradual transition from the representation of surface to the representation of depth with the help of anatomy and proportion, with light and shadow creating dramatic effects, and with the *sfumato*. Portraiture became an important genre, and there was a preference for realistic and naturalistic representations of the human figure and the environment, inspired by classical Greek and Roman art, and themes that glorified the value of the human being, with landscape painting gaining more and more importance and attention. The same spirit of creative freedom, critical attitude and mastery of sensory experience can be observed in literature, historiography, criticism and philological research, cosmology, geographical exploration and philosophy (Uscatescu, 1986).

It is important to emphasize that this post-medieval revival of 'art and writing' was largely linked to the foundations of classical antiquity. Thus, ancient Greece once again became the ideal model, the stable point of reference and the standard of comparison for the cultural reflections of the new age. In other words, after a long period of marginalization of Greek culture through religious dogmatism, it returned to the agenda of cultural discussions in Europe.

This period is characterized by a particular aspect in the perception of the regions in which ancient Greek civilization flourished. The Greeks lacked the institutional authority of a state to formally shape the image of the regions in which they lived. Under the rule of the Ottoman Empire, which set other priorities, the shaping of the image of the region fell primarily into the hands of the European media, which influenced both the perception of the region and the evaluation of the culture of its inhabitants.[1]

### 14.1.1 Greek personalities who contributed to portraying Greek heritage

There were nevertheless notable Greek personalities who, on their own initiative, made important contributions in various fields during this period and helped to create an image of the undefined territory inhabited by Greeks. Some of them were recognized by the elite and scholars of the West, while others may not have been well known outside the Greek community, but they played an important role in preserving Greek culture and heritage.

Thus, we can mention a large number of personalities from the time of the fall of Constantinople, most of whom desperately turned to the West. Among others this included: (i) the humanist, theologian and Catholic Cardinal Bessarion (1403–1472); (ii) the scholar Demetrios Chalkokondyles (1423–1511); (iii) the scholar and teacher Constantine Lascaris (1434–1501); (iv) the scholar Janus Lascaris, also called John Rhyndacenus (*c.*1445–1535); (v) the philosopher Marcus Musurus (*c.*1470–1517); (vi) Dionysius the Philosopher (*c.*1471–1546); (vii) the monk, publicist, writer, scholar and translator Maximus the Greek (*c.*1475–1556); (viii) the icon and fresco painter Theophanes the Cretan-Strelitzas (1490–1559); (ix) the mathematician and astronomer Francesco Maurolico (1494–1575); (x) the navigator Juan de Fuca (1536–1602); (xi) the painter, sculptor and architect Domenikos Theotokopoulos, known as El Greco (1541–1614); (xii) the scholar and cleric Meletius Pegas (1549–1601); (xiii) the Patriarch of Alexandria and later Patriarch of Constantinople Cyril Lucaris

(1572–1638); (xiv) the philosopher, physician and archaeologist Leo Allatius (*c*.1586–1669); (xv) the poet and scholar Parthenios of Nicaea (1602–1690); (xvi) the Greek adventurer who became prime minister to King Narai in the Kingdom of Ayutthaya, Constantine Phaulkon, also known as Costantin Gerachi (1647–1688); (xvii) the physician and consul in Smyrna Jacobus Pylarinus (1659–1718); and (xviii) the Indologist Dimitrios Galanos (1760–1833) (Margaris, 1982; Holton, 1991).

It is a fact that in these centuries the role and opportunities of women in European society were limited, which was even more evident in the area that included the southern part of the Balkan Peninsula and the coast of Asia Minor, which was cut off from Europe during the Ottoman occupation. As a result, there are relatively few well-documented Greek women from this period who became known beyond their local communities. There were, however, some of them who contributed to Greek culture and society, even if they did not achieve great fame in their lifetime. These included the religious poet and hymnographer Agathe Gratsia (*c*.1430–1500), poet Martha Argyriadi (16th century), the intellectual Irene Choumnaina Palaiologina (*c*.1523–1569), poet Eleni Boukoura (16th century), noblewoman and poet Isabella Kottounia (*c*.1520–1570) and intellectual Theodora Prodromou (16th century). These were followed in the 17th century by noblewoman and scholar Katerina Notaras, painter Parthenia Zografou, and poet and scholar Theodora Kotsigianni. Women who made notable contributions in the 18th century were the intellectual Evgenia Voulgaris; poets Zoe Karelli, Doudou Ypsilanti and Domnas Katinko; painter Elena Bouzmani (*c*.1720–1800); playwright, poet and feminist Evanthia Kairi (1799–1866); painter and iconographer Eleni Papadaki (*c*.1756–1826); and writer Elizabeth Moutzan-Martinegou (1801–1832).

It is important to point out once again that Greek women had very limited opportunities to gain social recognition compared with men during this period, and that information about their contributions is often missing from the historical record. Even when there are exceptions, their achievements are usually insufficiently documented compared with those of their male counterparts and are difficult to find in the literature. Their lives and work reflect the resilience, perseverance and resourcefulness of Greek women during a relatively dark and complicated period in Greek history. Undoubtedly, however, their contribution to the construction of the Greek image was crucial, even if they have been neglected by historians (Holton, 1991; Commire, 2002; Smith, 2008; Denisi, 2014).

### 14.1.2 Advent of printing and the spread of knowledge

Throughout the period when the Greeks were under the rule of the Ottoman Empire, some of them practised the art of typography with great skill and realized that printing could play an important role in preserving a certain Greek identity. Thus, sometimes within the framework of Italian printing houses and sometimes by attempting to establish their own printing house, they succeeded in making the typographic art a national diplomacy and an informal tool for the formation of the image of Hellenism.

An example of this is the printing house of the Italian Aldus Manutius (*c*.1449–1515), which was surrounded by great Cretan scholars such as Marcus Musurus (*c*.1470–1517), Demetrius Ducas (*c*.1480–1527) and Ioannis Grigoropoulos. There were also the purely Greek printing houses in Venice of the Cretans Alexandros Georgiou and Laonikos Protopapas, Zacharias Kalliergis (1473–1524), Nikolaos Sofianos, Andreas and Petros Kounadis, and Nikolaos Glykis (1619–1693).

From all this it is clear that even in the years of the Ottoman conquest there was an urge, a rudimentary cultural and artistic production that brought the Greek spirit to life before the eyes of the world. At that time, for example, Crete was under Venetian rule for many centuries and therefore received significant influences of Western culture, especially from Italy, in all areas of life, including art and literature.

The Cretan theatre that flourished during this period was thus the work of an educated ruling class that was in direct contact with the Italian Renaissance. At the end of the 16th and beginning of the 17th centuries, an excellent Greek literature emerged in Crete, printed mainly in Venice, which can rightly be called neo-Greek, since it broke away from the mentality of the Middle Ages. Georgios Chortatzis (*c.*1545–1610) with *Erofili*, Ioannis Andreas Troilos with *King Rodolinus*, Vitsentzos Kornaros (*c.*1553–1614) with *Erotokritos*, and Marinos Tzane Bougialis (*c.*1613–1686) with *The Cretan War* are some representatives of this direction (Vitti, 2003).

The above-mentioned works are of particular importance if we consider that a large part of Greek cultural production came from the inhabitants of an island that was one of the most important transport hubs in the world at that time. We can also say that other maritime centres inhabited by Greeks, such as the Ionian Islands and Cyprus, which had a similar cultural development, also contributed to the spread of a certain image of modern Greek culture.

The adventure of Greek printing continued later in other European cities such as Vienna and Paris, with protagonists such as Georgios Vendotis (1757–1795) and the Markides Pouliou brothers, who published the first edition of the *Efimeris* in Vienna in 1790, the oldest Greek-language newspaper of which copies have survived. Be that as it may, certainly there were also many others who can be considered fundamental components in the creation of the idea of Greek unity (Skiadas, 1976; Staikos and Sklavenitis, 1999).[2]

Other well-known journals-newspapers in Greek that appeared before the revolution for Greek independence were: (i) *Hermes o Logios*, founded by Anthimos Gazis (1758–1828), Theoklitos Farmakidis (1784–1860) and Konstantinos Kokkinakis (1781–1831) in Vienna in 1811; (ii) *Eidiseis dia ta anatolika meri*, founded by Joseph Franz Hall and Effronios Popovits in Vienna in 1811; (iii) *Ellinikos Tilegrafos*, founded by Joseph Franz Hall and Dimitrios Alexandridis (1784–1851) in Vienna in 1812; and (iv) *Athina* and *Mouseion*, both founded by Panagiotis Ioannidis in Paris in 1819 (Balta, 2005; Koumarianou, 2005).

The Greek political newspapers of the late years of this period were motivated mainly by their political ideas and their support for the emancipation of a Greek state, rather than by the financial gain of the editor. Despite their political differences, they shared common goals and priorities, namely the defence and promotion of national issues and the education of the public in matters of law and institutional support, while the cultural journals were focused on the promotion of a Greek cultural identity. The newspapers were purely opinion sheets, less concerned with current affairs and the latest news, and initially almost entirely free of advertising. The countless articles, most of which were of essay character, reflect the course of Hellenism throughout its duration, but also the personality of the intellectuals with their reflections on questions of education, art, writing, etc., on national issues, on political or social life in relation to Western political thought and philosophy (Balta, 2005; Kariatoglou, 2005; Koumarianou, 2005).

While these printing houses, as well as newspapers and cultural journals on the European continent, did not directly influence modern tourism as we understand it today, they did help to lay the intellectual and cultural foundations for the emergence of Greece as an independent nation in the 19th century. It can also be assumed that the Greek press laid a kind of modern cultural cloak over the image of Greece, which had always been shaped by ancient Greek culture. This in turn played a role in the later development of tourism in Greece, when travellers began to explore the country's ancient heritage, modern Greek culture and picturesque landscapes.

### 14.1.3 Role of visual arts in constructing the image of modern Greece

As far as the European visual arts are concerned, a brief look at the various artistic representations in the spectrum of our subject shows that the Greek element was also present in the preferences of the European artistic circles of that time.

While the themes of European painting in the early Middle Ages were always limited to the church, religion and the Holy Land within the framework of patronage, this began to change towards the end of the era and artists gradually drew inspiration from influences of Greco-Roman mythology as well. With the transition from the Middle Ages to the Renaissance and then to early modernity, European artistic representation increasingly drew on the technique and ideas of classical antiquity. As examples of how the themes of Greco-Roman mythology gradually developed in painting, consider:

- Sandro Botticelli (1445–1510) with his works *Primavera* (1482), *The Birth of Venus* (1485) and *Venus and Mars* (1485);
- Raphael (1483–1520) with *Parnassus* (1511) and *The School of Athens* (1509);
- Giorgione (1477–1510) with *Sleeping Venus* (1510);
- Tiziano Vecelli (1488–1576) with *Bacchus and Ariadne* (1520), *Venus and Adonis* (1554), *Perseus and Andromeda* (1556) and *The Rape of Europa* (1562);
- Caravaggio (1571–1610) with *Bacchus* (1596), *Medusa* (1597) and *Narcissus at the Spring* (1597);
- Rubens (1577–1640) with *The Fall of Phaeton* (1604), *The Feast of Venus* (1636) and *The Rape of the Daughters of Leucippus* (1617);
- Anthony van Dyck (1599–1641) with *Cupid and Psyche* (1638);
- Diego Velázquez (1599–1660) with *The Triumph of Bacchus/The Drunkards* (1629) and *The Forge of Vulcan* (1630); and
- Nicolas Poussin (1594–1665) with *Venus and Adonis* (1628), *Et in Arcadia ego – Les Bergers d'Arcadia* (around 1637–1638), *Landscape with Orpheus and Eurydice* (1650) and *Helios and Phaeton with Saturn and the Four Seasons* (1635).

To help understand the importance of these paintings in constructing images of the places that are now part of modern Greece, let us take as an example the work of the French Baroque painter Nicolas Poussin and his work *Et in Arcadia ego – Les Bergers d'Arcadia* [The Arcadian Shepherds] (around 1637–1638), which depicts a scene from Arcadia.

Arcadia is a region in the centre of the Peloponnese, dominated by mountains and plateaus, with running water, streams and rivers coming from the forested mountain slopes, but it is also partly wetted by the Mediterranean. Its remote valleys, mountain slopes and pastoral character made it a kind of myth, a paradise for Greek and Roman bucolic poetry.

Poussin depicted Arcadia as he had dreamed it, as did many artists before and after him. In general, the term 'Arcadia' came to symbolize a pure, bucolic, idyllic and natural life (Fig. 14.1). Many intellectuals also used the term, such as the German Johann Wolfgang von Goethe in his book *Italian Journey: 1786–1788*, when he wrote '*Et in Arcadia ego*' and described his experiences in Italy. It seems that the meaning of Arcadia has been extended to any region that evokes feelings of happiness and bliss due to ideal living conditions in nature. In this case, then, all these mythical and historical references to Arcadia evoke associations that equate the landscape of mainland Greece with an earthly tourist paradise.

Paintings such as these mentioned above carry complicated meanings that had to be deciphered by the artists for the viewers. Effective communication therefore depended on a shared understanding of cultural conventions between the painters and their audience. Deciphering most of these scenes required a profound knowledge of Greek mythology, a knowledge originally reserved for the elite and the privileged ranks of the aristocracy. The British art critic Berger's assertion that art serves the interests of the ruling class is echoed here (Berger, 1972). It is clear from the above-mentioned works that the Greek element occupied a prominent position of particular prestige among the powerful elite of European nations at the time. Originally a privilege of the nobility, it became increasingly accessible with the advent of reproduction through printing and the founding of museums. Nevertheless, certain paintings remain as central images in the social representation of Greco-Roman culture in the subconscious of today's tourists.

It can be argued that in all these works of art the world of Greek and Roman mythology is viewed in a highly poetic way through Western spectacles. However, it can also be argued that the spectacular epic representations of gods, demigods, ancient Greek philosophers and heroes in the artworks of the Renaissance and the following centuries, compared with the everyday reality in the Greek region of the time, were a harbinger of how we arrived at the image of Greece projected by some foreign mass media. This image is so complex and contested, as it is torn between an idealized antiquity and a deglamourized contemporary Greek reality.

### 14.1.4 The 'Grand Tour' and growing interest in Greece

In the atmosphere of the Renaissance, which glorified ancient Greek culture, the 'Grand

**Fig. 14.1.** Nicolas Poussin *Et in Arcadia ego – Les Bergers d'Arcadia* (around 1637–1638). Although this may not have been the only factor that prompted people to visit the southern part of the Balkan Peninsula, the artistic and literary representations of an Arcadian ideal may have played a role in shaping the perception of Greece as a destination with a rich cultural and natural heritage. The desire to experience the beauty and history associated with Arcadia in Greece may have influenced the interest of travellers during and after the Renaissance. Source: https://commons.wikimedia.org/wiki/File:Nicolas_Poussin_-_Et_in_Arcadia_ego_(deuxième_version).jpg

Tour' became more and more extended so that it would include Greece in its itinerary. As a rule, the participants travelled first to France and then in a south-easterly direction through Switzerland to Italy. After stops in various Italian cities, they continued north through Germany and then west to the Netherlands. However, since ancient Greece, along with Rome, was now considered by the nobles to be the origin of Western civilization, the journey to Italy took a different route for many of them and continued in Greece (Brodsky-Porges, 1981).

In fact, it was during this period that the word 'tourist' was invented and used with a completely different meaning (one could even say the opposite) than in the 19th and 20th centuries. The inhabitants of the various regions of Central Europe began to refer to the visitors who took part in the 'Grand Tour' simply as tourists (Leiper, 1983).

In other words, the tourists were simply the travellers or rather the visitors who took part in the 'Grand Tour', but the term was later to take on a negative connotation as a feature of the developing mass society.

Interest in Greece continued to grow as several prominent scholars and artists undertook their own Grand Tours, sponsored by aristocratic art societies. The 'Society of Dilettanti' and many other organizations funded numerous trips to Greece to record archaeological ruins of ancient Greek civilization (Eisner, 1991; Mayfield, n.d.).

Moreover, the southern end of the Balkan Peninsula was of particular interest, not only as an area of study, but also as a destination for the search for a cultural identity. Europeans began to see in their own new secularized identity a continuation of the ancient Greek spirit; they saw themselves as heirs of Greek culture and began to regard ancient Greece as the starting point of European civilization. It should be noted that this turning of Europe towards ancient Greece took place in conditions where there were no competitors, as Constantinople, which already in the 13th century saw itself as the descendant of ancient Greece alongside its Roman origins, had been conquered by the Ottomans (Giakovaki, 2014).

## 14.2 Travellers to the Region in the Renaissance and Enlightenment

It would be useful to mention some of the many European travellers, diplomats, military men, artists and scientists who visited the area during the Renaissance and the Enlightenment. Among many others, these included: (i) Italian Franciscan priest Cristoforo Buondelmonti (1385–1430); (ii) Italian humanist and antiquarian Ciriaco Pizzeccolli Anconitano (1391–1452); (iii) German nobleman and politician Bernhard von Breydenbach (1440–1497); (iv) French naturalist, topographer and translator Pierre Gilles (1490–1555); (v) French geographer Nicolas de Nicolay (1517–1583); (vi) Prussian pharmacist Reinhold Lubenau (1556–1631); (vii) British poet and translator George Sandys (1578–1644); (viii) Scottish writer and alleged spy William Lithgow (1582–1645); (ix) German mathematician, cartographer and writer Johann Laurenberg (1590–1658); (x) French writer Jean de la Vega (16th century); (xi) French gem merchant Jean-Baptiste Tavernier (1605–1689); (xii) French writer and actor Georges Guillet de Saint-George (1624–1705); (xiii) British diplomat and historian Sir Paul Rycaut (1629–1700); (xix) Dutch sailor Jan Janszoon Struys (1630–1694); (xx) French painter Guillaume-Joseph Grelot (1630–1680); (xxi) French linguist, naturalist and botanist Jean de Thévenot (1633–1667); (xxii) Italian writer Cornelio Magni (1638–1692); (xxiii) British clergyman and scientist John Covel (1638–1722); (xxiv) British merchant Bernard Randolph (1643–1690); (xxv) French physician Jacob Spon (1647–1685); (xxvi) French painter and draftsman Jacques Carrey (1649–1726); (xxvii) Italian Franciscan friar, cosmographer, cartographer, publisher and encyclopedist Vincenzo Maria Coronelli (1650–1718); (xxviii) Dutch artist Cornelis de Bruijn (1652–1726); (xxix) French antiquarian and priest Michel Fourmont (1690–1746); (xxx) French diplomat Sieur du Loir (17th century); and (xxxi) French diplomat Julien Bordier (17th century). Each of them, in his field of interest, described and analysed the nature, geography and people of the larger area of ancient Greece, contributing to create a new image of this place (Simopoulos, 2003a, b, c, d, e, f; Vingopoulou, 2005).

Of particular interest is the work of the French physician Jacob Spon, who travelled through the region and wrote a travel book entitled *Voyage d'Italie, de Dalmatie, de Grèce et du Levant* (1678), which, in addition to a systematic description of the places, also contains a glossary of modern Greek words of the time that is useful for travellers. As Giakovaki (2014) notes, this glossary is important evidence of the tendency of Europeans to familiarize themselves with 'vulgar Greek', as opposed to a revered dead language, Ancient Greek, which was taught in universities.[3]

Thus, in addition to the ancient Greek monuments and the Greek landscape, the modern Greek language was also depicted in the abovementioned work and in many others, although it had changed from its ancient form. The depiction of a living language as a descendant of ancient Greek in these writings was an important element in the recognition of 'Greekness' in the culture of the region's inhabitants and contributed to the formation of the image of modern Greece as part of the modern world.

### 14.2.1 The neoclassical movement

At the beginning of the 18th century, many European artists were obsessed with classical Greek and Roman antiquity and considered Greek and Roman art to be the highest form of art. They therefore tried to imitate in their works the representation of the human figure and its movements as well as the architecture as they had been conceived in Greek and Roman antiquity. This approach crystallized in the neoclassical movement. At the same time, philosophers believed that the works of art of antiquity could serve as a basis for the theoretical thought of the Enlightenment.

Since the neoclassicists believed that ancient Greek and Roman art reflected nature and reality most perfectly and represented the highest art, they based their works exclusively on it, creating an important medium for spreading the idea of Hellenism.

Furthermore, they believed that humanity could be led to ideal morality, a perfect political state, and social prosperity if it followed the rules of ancient Greece in these areas.

According to Johann Joachim Winckelmann (1717–1768), one of the most important theorists of neoclassicism, the greatness of the soul, the beauty of the body, moral values, political virtue, the euphony of language and the climate are the main elements on which the perfection of Greek art is based (Winckelmann, 2006).

Thus, the artists of neoclassicism took all these elements from antiquity and, using all the means of their time, transferred them to the Western European population, which reflected the Greek ideal. Consequently, the aesthetic and philosophical values of neoclassicism brought ancient Greece more to the fore, and in this way the collective consciousness of Europeans preserved the Greek spirit, which later helped to form a kind of positive image for the modern Greek state, despite the many contradictions it contained.

The notorious neoclassicist French painter Jacques-Louis David (1748–1825) showed a particular predilection for themes of Greco-Roman antiquity, such as *The Duel of Athena* (1771), *Apollo and Artemis at the Arch of the Children of Niobe* (1772), *The Burial of Patroclus* (1778) and *The Death of Socrates* (1787). Other artists famous for the Greek themes of their works were the painter Angelika Kauffmann (1741–1807) with *The Sorrow of Telemachus* (1783) and *Penelope Awakened by Eurykleia* (1772), and the sculptor Antonio Canova (1757–1822) with *Orpheus* (1777) and *Theseus and the Minotaur* (1781).

Similarly, subjects relating to specific Greek places and Mediterranean landscapes in general were very often used as backgrounds for paintings by various other art movements and painters. This included Pierre Henri de Valenciennes' (1750–1819) *Landscape of Ancient Greece* (1786) and Antoine Watteau's (1684–1721) *Embarkation for Cythera* (1717) and *The Pilgrimage to Cythera* (1717) (Mayfield, nd). The way these painters perceived the elements of the environment as landscape, and that they expressed their imaginary position to the southern end of the Balkan Peninsula, ultimately represents the way they defined the image of Greek space. We can argue that the modern tourist/viewer can still look for some elements of this relationship in the landscape.

Much of the European architecture of this period can also be described as influenced by Greco-Roman antiquity. The neoclassical buildings are characterized by geometric purity, dynamism and grandeur, clear surfaces with few Greco-Roman ornaments and repoussé elements, Greek pediments and colonnades and many other details used in the Greek temples. Most of these works were based on original or hybrid models, such as the 'Tower of the Winds' or the 'Kyros Clock' in the Roman Agora of Athens, which are among the most famous models of neoclassicism. They were used in many public buildings such as universities, churches and private houses like the Radcliffe Observatory at Oxford University (1773) by James Wyatt (1746–1813) (Petridou and Ziro, 2015).

The neoclassical movement not only influenced art and architecture, but also played a role in the early development of a trend towards travel to Greece, as people wanted to explore the cradle of Western civilization and experience the fascination of the ancient past.

## 14.3 Accounts of Travellers

Important European figures of the time continued to visit the area. Among others, this included: (i) French merchant, naturalist, physician and antiquarian Paul Lucas (1664–1737); (ii) French diplomat Aubry de La Motraye (1674–1743); (iii) Scottish archaeologist, architect and artist James 'Athenian' Stuart (1713–1788); (iv) French scholar, writer and numismatist Jean-Jacques Barthélemy (1716–1795); (v) British statesman John Montagu 4th Earl of

Sandwich (1718–1792); (vi) British architect Nicholas Revett (1720–1804); (vii) British antiquary Richard Chandler (1737–1810); (viii) German diplomat and writer Johann Hermann von Riedesel (1740–1785); (ix) British painter William Pars (1742–1782); (x) French astronomer and archaeologist Jean Baptiste LeChevalier (1752–1836); and (xi) British botanist John Sibthorp (1758–1796) (Tzermias, 1997; Vingopoulou, 2005).

As we can see from the previous sections, most of the people who were traditionally documented as visitors to Greece and who provided testimonies and perspectives were men. This pattern not only reflects the patriarchal framework that prevailed in European societies of the time, but also emphasizes that those who preserved travel memoirs, which were predominantly written by male authors, were also shaped by gender stereotypes. Undoubtedly, many women also travelled to the area where ancient Greek civilization had previously developed, but their names and works have not remained as vivid in 'historical memory' (Mitsi, 2002; Kamberidou, 2013).

With this in mind, it is worth highlighting the names of some important European women who travelled to Greece during the Renaissance and the Age of Enlightenment, often out of a combination of cultural exploration, intellectual curiosity and a desire to engage with Greece's classical heritage. A few of these women are:

- the British aristocrat and writer Lady Mary Wortley Montagu (1689–1762);
- the German physician Dorothea Christiane Erxleben (1715–1762);
- the British writer and playwright Lady Elizabeth Craven (1750–1828);
- the French aristocrat and writer Sophie de Marbois-Lebrun, Duchess of Plaisance (1785–1854);
- the British aristocrat and adventurer Lady Hester Stanhope (1776–1839); and
- the German archaeologist and art historian Sibylle Mertens-Schaaffhausen (1797–1857).

The writings and observations of all these women, who helped to shape the image of this place for Europeans, also contributed to a general understanding of the cultural aspects and gender relations in Greece at their respective times.

Be that as it may, the works of all these people were full of useful information about the region and the people of that time, but they also exuded nostalgia for the land of ancient Greek culture in a stimulating way. Some wrote accurate descriptions of the various provinces and cities and drew maps; others composed poetry, music or prose about the area. A few authors noted their thoughts about society and described the prevailing social, political and cultural codes, or discussed how they saw the Christian population as oppressed. Other writers described the area levelled by the Ottomans and the antiquities reduced to rubble; many others looked at the ruins with a romantic eye or immortalized the antiquities with their art, while others stole them. But most of them perceived and passed on the magic of the Mediterranean landscape and its people, combined with the veneration of the ancient Greek culture.

As we have seen above, in this period the southern end of the Balkan Peninsula, with its rich fauna and flora, was a destination of great interest for naturalists and nature lovers who wanted to study and enjoy the nature of this place. One could even say that at that time there was already a longing for the unmediated contact with nature, a first idea of the need to preserve the natural resources, especially the forests, which were largely destroyed by deforestation. Over the years, and especially in the later stages of this period, biologists, zoologists and botanists who studied and classified the natural environment confirmed the continuing extinction of species (Dalton, 1998; Watson and Watson, 2007).

In addition to the naturalists and nature lovers already mentioned who visited the site, there were some particularly prominent ones, whose names are symbolic for ecologists, who travelled to Greece and played a key role in documenting and researching natural history. Among others, these included: (i) the French explorer, nature writer and diplomat Pierre Belon (1517–1564); (ii) the French botanist and physician Joseph Pitton de Tournefort (1656–1708); (iii) the British clergyman, traveller and travel writer Richard Pococke (1704–1765); and (iv) the French naturalist Jean-Baptiste Bory de Saint-Vincent (1778–1846) (Gudger, 1934; Ferrière, 2009; Klaver, 2020).

It can be argued that many philosophers and naturalists of the time supported the responsible management of forests based on principles that reflected some of the ideals of the ancient Greeks, such as the concept of balance and respect for nature. In addition, the picturesque landscapes of Greece, as depicted in Romantic paintings and writings, also contributed to the European public's fascination with the need to preserve the beauty of nature. The Romantic depiction of Greek landscapes, which particularly moved the public, undoubtedly helped to strengthen appreciation for natural landscapes and develop an ecological consciousness.

From all of this, it is clear that the Greek image also contained symbols of the beauty of nature, while the broader cultural appreciation for the beauty of natural landscapes as depicted by scholars and travellers and projected by the various media of the time, inspired in part by the appeal of Greece, played a key role in shaping early thinking about the preservation and protection of scenic and historic sites in later years.

According to Simopoulos (2003c), there is no aspect of Greek life in the 18th century that is not illuminated by the multitude and variety of information offered by travel books of the period. As mentioned in Chapter 8, this volume, the 18th century is considered by many to be the golden age of travel literature. It would not be an exaggeration to say that all fields of scientific enquiry can draw from these works – history, geography, archaeology, architecture, agronomy, oceanography, zoology, botany, biology, medicine, mineralogy, sociology, anthropology, jurisprudence, pedagogy, as well as philosophy. However, above all these books provided details about culture and folklore, thanks to the wealth of information on customs, songs, dances, dialects, costumes, jewellery, furniture and all manifestations of popular culture of this place.

Also worth mentioning is a study by Tolias (1997), who examined the portrayal of Greece in French periodicals between 1794 and 1815, such as the *Magasin Encyclopédique*, the *Mercure de France* and the *Mercure Etranger*. He tracked down relevant articles and examined the image that the French press conveyed of Hellenism.

Augustinos (1994) took a similar approach but examined French travel literature about Greece from the arrival of the first French travellers in the mid-16th century to the outbreak of the Greek Independence Revolution in 1821. She identified and examined commonplaces about Greece in Enlightenment-era travel literature, and highlighted and interpreted the changes in content that emerged from earlier accounts of medieval religious pilgrimages under the influence of the secularization of European thought. She argued that a positive image of Greece can also be found in the writings of some important French literary figures of the last decades of the 18th century. These include: (i) the 18th-century philosopher Voltaire (1694–1778), who had never visited Greece; (ii) the merchant and classicist Pierre Augustin Guys (1721–1799), a member of the Academy of Letters and Sciences of Marseilles; and (iii) the young Count Auguste de Choiseul-Gouffier (1752–1817), a member of the Académie française. Their common interest in modern Greece was based primarily on their admiration for antiquity. Olga Augustinos assumed that after the middle of the 18th century, in line with the development of European literature, travel texts were enriched by sentimental reflections, scenic descriptions of nature and monuments, and lyrical expressions that gave a subjective coloration to the previously factual narratives (Augustinos, 1994).

In this context, we must mention that in 1776 Count de Choiseul-Gouffier travelled through Ottoman-occupied Greece and the coasts of Asia Minor, sketching archaeological sites, castles and temples, making maps and plans and recording the customs of the inhabitants and natural and astronomical phenomena. The contribution of Choiseul-Gouffier with his three-volume monumental work *Voyage Pittoresque de la Grece* (1782–1802) was decisive in the formation of the image of Greece at the time, for through his depictions the European public became visually familiar not only with the 'picturesque' landscape but also with the modern inhabitants of Greece. This monumental work contains a considerable number of copper engravings as well as valuable archaeological, historical and geographical material, offering the reader rare information about the period of Greek servitude under the Ottomans.

## 14.4 Summary

In this chapter the inclusion of various historical events, personalities, works of art, studies, narratives and memories associated with the contemporary location of the new Greek state is not intended to provide a definitive account of the image of the Hellenic peninsula as it was perceived by Westerners during the period under consideration.

Undoubtedly, the perception of the place where ancient Greek civilization flourished and its inhabitants varied considerably from person to person, from country to country and between different cultures. Many people may have felt that while the Greek world existed within the Ottoman Empire and not in the West, it was also distinct from the East. This distinction stemmed from differences in language, religion and culture, coupled with a distinct classical heritage that Europeans considered part of their own cultural heritage. It would perhaps be more accurate to say that in the European mentality of the time, Greece was sometimes seen as East and sometimes as West.

In short, many of the intrepid travellers who ventured into the territories of the southern Balkan Peninsula against all odds were confronted with an unavoidable duality. On the one hand, they encountered an economically depressed country with inhabitants trying to escape Ottoman oppression; on the other, they dreamily entered the realm of the ancient Greeks, the cradle of civilization, a heritage that the West treasured as part of its culture (Skarpelos, 2005).

## Notes

[1] Certainly, the Ottoman Empire as a political entity was not primarily interested in acquiring an image of the ancient Greek territories in Europe. Its main objectives were the administration and preservation of its vast empire. However, it can be said that for much of this period, the Ottoman Empire also unconsciously promoted a certain image of Greece in Europe through, for example: (i) the activities of Greek scholars who held high positions in the administration of the empire; (ii) interaction with European travellers and diplomats; (iii) the preservation and dissemination of Greek manuscripts from monasteries; and (iv) the integration of Greek elements into Ottoman culture.

[2] As Koumarianou (2005) points out, there was also an earlier but ultimately unsuccessful attempt by Georgios Vendotis (1757–1795) to publish a Greek newspaper. However, he faced the hostile attitude of the Austrian authorities, who bowed to Ottoman pressure and very soon prohibited the circulation of his newspaper. We only have information from the Austrian archives about Vendotis's short-lived newspaper, which does not seem to have appeared for more than 2 months in the summer of 1784, and we do not know its title.

[3] Here we can also mention Johannes Meursius (1579–1639) with his *Glossarium graeco-barbarum* (1614) and Charles du Cange (1610–1688) with his *Glossarium ad scriptores mediae et infimae Graecitatis* (1688), who also contributed to this perspective with their works.

## References

Augustinos, O. (1994) *French Odysseys: Greece in French Travel Literature from the Renaissance to the Romantic Era*. John Hopkins University Press, Baltimore, Maryland.

Balta, N. (2005) The evolution of news in Greek newspapers in the 19th century (1830 onwards). In: Droulia, L. (ed.) *The Hellenic Press, 1784 to Today*. Athens Institute for Neohellenic Research/ National Hellenic Research Foundation, Athens. (in Greek), pp. 63–70.

Berger, J. (1972) *Ways of Seeing*. BBC and Penguin Books, London.

Brodsky-Porges, E. (1981) The grand tour travel as an educational device 1600–1800. *Annals of Tourism Research* 8(2), 171–186. DOI: 10.1016/0160-7383(81)90081-5.

Commire, A. (ed.) (2002) *Women in World History: a Biographical Encyclopedia*. Yorkin Publications, Waterford, Connecticut.

Dalton, R.J. (1998) The environmental movement in Western Europe. In: Stavrakakis, G. (ed.) *Nature, Society and Politics*. Nisos, Athens. (in Greek), pp. 344–377.

Denisi, S. (2014) *Detecting 'Invisible' Writing: Women and Scripture in the Years of the Greek Enlightenment-Romanticism*. Nefeli, Athens. (in Greek).

Eisner, R. (1991) *Travelers to an Antique Land*. The University of Michigan Press, Ann Arbor, Michigan. DOI: 10.3998/mpub.7867.

Ferrière, H. (2009) *Bory de Saint-Vincent: L'évolution d'un Voyageur Naturaliste*. Syllepse, Paris. (in French).

Giakovaki, N. (2014) The European encounter with the vernacular. In: Christidis, A.F. (ed.) *History of the Greek Language*. Institute of Modern Greek Studies, Thessaloniki, Greece, pp. 1002–1006.

Gudger, E.W. (1934) The five great naturalists of the sixteenth century: Belon, Rondelet, Salviani, Gesner and Aldrovandi: a chapter in the history of ichthyology. *Isis* 22(1), 21–40. DOI: 10.1086/346870.

Holton, D. (ed.) (1991) *Literature and Society in Renaissance Crete*. Cambridge University Press, Cambridge, UK. DOI: 10.1017/CBO9780511519666.

Kamberidou, I. (2013) The East in the eyes of Western women travellers of the 18th and 19th centuries: solidarity and understanding the East. In: *'The East in The Eyes of The West', International Conference of the Faculty of Arts*, Kuwait University, Kuwait, 26–28 November 2013. Available at: https://www.researchgate.net/publication/261143461_The_East_in_the_Eyes_of_Western_Women_travellers_of_the_18th_and_19th_centuries_solidarity_and_understanding_the_East (accessed 25 January 2024).

Kariatoglou, E.I. (2005) Advertisements. In: Droulia, L. (ed.) *The Hellenic Press, 1784 to Today*. Athens Institute for Neohellenic Research/National Hellenic Research Foundation, Athens. (in Greek), pp. 196–200.

Klaver, J.M.I. (2020) Richard Pococke and the natural curiosities of the East. In: Loop, J. and Kraye, J. (eds) *Scholarship between Europe and the Levant: Essays in Honour of Alastair Hamilton*. Brill, Leiden, the Netherlands, pp. 260–275. DOI: 10.1163/9789004429321.

Koumarianou, A. (2005) Historical path of the Greek press, 1784–1922. In: Droulia, L. (ed.) *The Hellenic Press, 1784 to Today*. Athens Institute for Neohellenic Research/National Hellenic Research Foundation, Athens. (in Greek), pp. 55–61.

Leiper, N. (1983) An etymology of tourism *Annals of Tourism Research* 10(2), 277–280. DOI: 10.1016/0160-7383(83)90033-6.

Margaris, V. (1982) *Great Neo-Greeks Since 1400*. Epiktitos, Athens. (in Greek).

Mayfield, J.M. (n.d.) The Grand Tour Beyond: Greece and the Holy Land. *Piranesi in Rome*. Available at: http://omeka.wellesley.edu/piranesi-rome/exhibits/show/grand-tour-beyond (accessed 2 August 2022).

Mitsi, E. (2002) "Roving Englishwomen': Greece in women's travel writing. *Mosaic* 35(2), 129–144.

Petridou, B. and Ziro, O. (2015) *Arts and Architecture from the Renaissance to the 21st Century*. Kallipos, Open Academic Editions, Athens. (in Greek).

Simopoulos, K. (2003a) *Foreign Travelers in Greece: Public and Private Life, Popular Culture,Church and Economic Life, from Travel Chronicles*, Vol. 1. Stachy, Athens. (in Greek).

Simopoulos, K. (2003b) *Foreign Travelers in Greece: Public and Private Life, Popular Culture, Church and Economic Life, from Travel Chronicles*, Vol. 2. Stachy, Athens. (in Greek).

Simopoulos, K. (2003c) *Foreign Travelers in Greece: Public and Private Life, Popular Culture, Church and Economic Life, from Travel Chronicles*, Vol. 3. Stachy, Athens. (in Greek).

Simopoulos, K. (2003d) *Foreign Travelers in Greece: Public and Private Life, Popular Culture, Church and Economic Life, from Travel Chronicles*, Vol. 4. Stachy, Athens. (in Greek).

Simopoulos, K. (2003e) *Foreign Travelers in Greece: Public and Private Life, Popular Culture, Church and Economic Life, from Travel Chronicles*, Vol. 5. Stachy, Athens. (in Greek).

Simopoulos, K. (2003f) *Foreign Travelers in Greece: Public and Private Life, Popular Culture, Church and Economic Life, from Travel Chronicles*, Vol. 6. Stachy, Athens. (in Greek).

Skarpelos, G. (2005) *Image and Society*. Nefeli, Athens. (in Greek).

Skiadas, E.N. (1976) *Chronicle of Greek Typography*. Gutenberg, Athens. (in Greek).

Smith, B.G. (2008) *The Oxford Encyclopedia Women in World History*. Oxford University Press, New York.

Staikos, K.S. and Sklavenitis, T. (1999) The Greek printing houses of the Diaspora and the Near East (16th–19th centuries). In: Sklavenitis, T., Staikos, K.S. and Kremmidas, V. (eds) *Treasures of the National Library: Exhibition Catalogue*. National Library of Greece, Athens. (in Greek), pp. 95–149. Available at: https://hdl.handle.net/10442/7676 (accessed 12 August 2022).

Tolias, G. (1997) *La Médaille et la Rouille: L'image de la Grèce Moderne dans la Presse Littéraire Parisienne (1794–1815)*. Hatier, Paris. (in French).

Tzermias, P.N. (1997) *The Image of Greece in the Foreign World*. Sideris Publications, Athens. (in Greek).

Uscatescu, G. (1986) *Brief Theory and History of Culture*. Ekdosis Grigori, Athens. (in Greek).
Vingopoulou, I. (2005) *The Greek World Through the Eyes of Travelers (15th–20th Century): an Anthology from the Dimitrios Kontominas Collection: Exhibition Catalogue, Benaki Museum, February 8, 2005–March 6, 2005*. Benaki Museum Exhibition, Kotinos, Athens. (in Greek). Available at: https://hdl.handle.net/10442/14283 (accessed 12 August 2022).
Vitti, M. (2003) *History of Modern Greek Literature*. Odysseus, Athens. (in Greek).
von Goethe, J.W. (1816 [1982]) *Italian Journey: 1786–1788*. Penguin Books, London.
Watson, R.N. and Watson, R. (2007) *Back to Nature: the Green and the Real in the Late Renaissance*. University of Pennsylvania Press, Philadelphia, Pennsylvania.
Winckelmann, J.J. (2006) *Essays on the Philosophy and History of Art*. Thoemmes Continuum, New York.

# 15 The Foundation of the Greek State

**Abstract**

At the beginning of the 19th century, the term 'tourism' found its way into the dictionaries of Western countries and marked the beginning of an era in which the concepts of mass tourism and the growing tourism industry began to capture the imagination as promising economic prospects. Although the establishment of the Greek state changed the sociopolitical landscape at the southern end of the Balkan Peninsula and offered the chance to create a conscious tourism image for Greece, this opportunity remained largely unrealized. The tensions of the 1821 War of Independence and the lack of systematic administrative organization in this area by the public sector, as well as the absence of methodical private tourism initiatives, prevented the development of a targeted tourism image for Greece. This chapter examines the factors that influenced the development of Greece's image during this period. It focuses on the role of intellectuals and Western travel literature, the artwork of foreign and Greek artists, and the influence of newspapers, photographs, advertising, postcards and international events.

The 19th century was marked by the 'national awakening' of European nations. This period was politically and institutionally turbulent for the vast area where the ancient Greek civilization had developed. The various ethnic groups of the Balkan Peninsula were struggling against the Ottoman Empire, and the Greeks were eager to unite in a modern nation state to give political substance to their cultural identity. Proponents of Greek independence relied on the strong rhetoric that the inhabitants of the land where Greek civilization flourished were the descendants of the ancient Greeks. Their liberation was therefore not an act of radical revolution, but rather the restoration of an ancient status quo, an idea bequeathed to Europe by the Renaissance and the Enlightenment (Tsaoussis, 2001; Beaton, 2021).

One question that took on special significance for the newly founded Greek state was whether Greece was a Western-oriented country that adopted and integrated Western values and principles, or whether it was a special case, a separate space between West and East with its own political, social and cultural characteristics. Although local structures resisted, Greek politicians gave priority to the Westernization of the Greek state. Thus, Greece's state building officially began with the War of Independence of 1821 (also known as the Greek Revolution) and continued along Western lines – with 20th-century French administration, the German legal system and British parliamentary practice (Rozakis, 2001; Koliopoulos and Veremis, 2010).

The widespread introduction of basic cultural elements from Europe into Greek reality since the foundation of the Greek state, which continued in the following centuries, should therefore not surprise us as a phenomenon, since it was initiated from the beginning by the introduction of legal and political institutions based on the European model, but also by the assimilation of European production methods and imported goods. However, the adoption of foreign standards never meant the loss of national identity and Greek autonomy.

While the transition from elite tourism to a more popular form of travelling was taking place in Europe, creating the conditions for the development of the modern tourism industry, Greece faced various obstacles. These were related to: (i) the armed conflicts and the turbulent situation in the Balkans; (ii) the country's geographical location outside the obvious radius of tourist activity in the Central European world; (iii) an inadequate road network; (iv) poor public and private infrastructure; (v) the lack


DOI: 10.1079/9781800626034.0015

of promotional strategies on the part of locals; and (vi) the absence of dynamic state support for tourism. All these factors jeopardized the development of a tourism industry in Greece.

The image of Greece, and in particular the tourist image of Greece, was initially characterized by 'organic' images of tourism, in which the media acted as autonomous mediators in the construction of destinations.

At a time when the means of communication were not as developed as they are today, the image of Greece was mainly characterized by literary books and travelogues, news in the press, works of art and photographs. In any case, postcards and some press adverts for hotels, transport companies and travel agencies also appeared, which can be seen as products of a rudimentary tourism business.

As early as 1791, a book entitled *Modern Geography* was published in Vienna in Greek, in which the ancient name 'Hellas' was used for the southern region of the Balkan Peninsula and the islands of the Aegean Sea; the people living there were called 'modern Hellenes', the descendants of the ancient Greeks (Beaton, 2021).

At this point it should be noted that the name 'Hellenes', which obviously played an important role in the construction of the identity of the Greeks and the image of the newly founded state, was accepted by the various revolutionaries as a national characteristic from the outset and became the property of all the combatants. The adoption of this name was not an expression of a newly acquired national consciousness in the sense that the national consciousness had already existed since the time of Ottoman rule and Byzantium. It can therefore be assumed that a kind of 'cultural memory' or at least a 'sense of the past' already existed among the population (Kakridis, 1963; Tsaoussis, 2001).

In fact, as Assmann (2011) argues, collective identity is based on the sharing of a common knowledge and memory through the use of a common language, or rather, through the use of a common symbol system, and in this case, national identity needed symbols and norms to be strengthened. All this had to be created through selective memory processes that could reduce certain events and concepts to timeless values.

In any case, the Greek language and Christianity were the fundamental components of the Greek community and were passed on by the Orthodox Church, as it was a powerful educational medium at that time.

Be that as it may, the fact is that the newly founded Greek state wanted to ensure the integration and unity of the various ethnic populations into the national body, and the creation of a unified national consciousness was one of its most important goals in this regard. Of particular importance to our study is that many elements of the tourist image of Greece as we know it today emerged in the 19th century with the establishment of the modern Greek state.

Europeans visited Greece mainly because they admired the ancient Greek culture and the Mediterranean landscape; sometimes even many of them became philhellenes and fought alongside the Greek revolutionaries. Many others, on the other hand, visited this place to conduct trade business or diplomatic talks. The general image of Greece at that time was thus the result of its long history, but also of the interaction of modern Greeks as they projected it through their media and the works and publications of those who travelled to the place. An accurate evaluation of the construction of this image is, of course, a very complex task, but we will try to give some simple indications that go in this direction.

## 15.1 Factors That Influenced Development of Greece's Tourism Image

### 15.1.1 Contribution of scholars, writers and poets

First of all, we should refer to some names of important Greek scholars, writers and poets of that time, whose works contributed decisively to the construction of the Greek identity and image. These include Adamantios Korais (1748–1833), Rigas Feraios (1757–1798), Theoklitos Farmakidis (1784–1860), Spyridon Trikoupis (1788–1873), Dionysios Solomos (1798–1857), Panagiotis Soutsos (1806–1868), Alexandros Rangavis (1809–1892) and Achilleus Paraschos (1838–1895). Certainly, there are many others who contributed in one way or another to the idea of the new Greek state, but it would take pages to list them all.

Memoirs and historical texts written by people who fought or lived at the centre of the great historical events of their time were also of particular importance in shaping the image of Greece. Such texts were written (or dictated to others to write) by Theodoros Kolokotronis (1770–1843), Palaion Patron Germanos (1771–1826), Emmanuel Xanthos (1772–1852), Panagis Skouze (1776–1847) and Ioannis Makrygiannis (1797–1864). And of course it would be unthinkable not to mention historians such as Spyridon Zampelios (1813–1881) or Konstantinos Paparrigopoulos (1815–1891), whose works have shaped the perception of the history of modern Greece.

Moreover, as mentioned above, a great wave of modern philhellenism developed during this period, which contributed significantly to the acceptance abroad of the image of an independent Greek state. The term 'philhellenism' refers to the intellectual and political movement that developed, especially in European countries and in America, with the aim of strengthening Hellenism both morally and materially against Ottoman rule during the independence revolution. It is obvious that the aims, roots and motivations of philhellenism were influenced by the past, by the love for the ancient Greek civilization. They were also influenced by the new breath that the principles of humanism and the American and French Revolution brought to Western societies, as well as the cultural, aesthetic and artistic movements of European classicism and romanticism.

The philhellenic movement was very strong in Europe, especially in Germany, but also in the UK, France and Russia, and drew the attention of the whole world to modern Greece, so that it played an important role in building the image of the newly founded Greek state.

Particularly noteworthy is the contribution of some important foreign literary figures who often treated Greek themes in their works. These included, among others, the Italian writer and poet Ugo Foscolo (1778–1827), the British Romantic poet Lord Byron (1788–1824), the Russian poet and novelist Alexander Pushkin (1799–1837) and the French Romantic writer, poet and politician Victor Hugo (1802–1885).

Lord Byron is considered one of the main actors who helped change public opinion in Britain in favour of the rebellious Greeks. In fact, he supported them both financially and through his work. Byron was a passionate traveller, and Samuel Taylor Coleridge (1772–1834), the Romantic poet, once called him a 'picturesque tourist'. Byron's descriptions and poems about Greece were used by many publishers for travel guides for many years after his death. It is worth noting that his friend and London publisher John Murray developed the guidebook industry in Britain and his publishing house was one of the first to publish a systematic guidebook on Greece (Kokkinos, 1978; Dritsa, 2019).

Translations and editions of collections of excerpts from traditional folk poetry by other philhellenes, such as the Frenchmen Claude Charles Fauriel (1772–1844) and Ernest Renan (1823–1892), to name a few, also contributed to the promotion of Greek culture of the time. Even Johann Wolfgang von Goethe (1749–1832), the great German poet, is said to have been inspired to write by Greek folk poetry.

It is also worth mentioning the book Itinerary from *Paris to Jerusalem* (1811) by the eminent French diplomat and writer François-René de Chateaubriand (1768–1848). In this book he reports on his trip to the Middle East in the summer of 1806, during which he greatly appreciated ancient Greece and its monuments, but he sometimes adopted a rather arrogant attitude towards the inhabitants of this region, with the typical feeling of superiority of the West over other peoples. Nevertheless, Chateaubriand is considered one of the main actors of the philhellenic movement in France at the time, who repeatedly expressed his sympathy and admiration for occupied Greece (Kokkinos, 1978).

Occasionally, however, philhellenes living in Greece, embittered by the prevailing situation, sent negative comments to the foreign press. After the establishment of the Greek state, for example, George Finlay (1799–1875), a Scottish philhellene who had problems with his property in Attica, sent critical articles about Greece to *The Times* of London and Edinburgh's *Blackwood's Magazine*. He portrayed Greece as a medieval country infested with outlaws of all kinds and unable to be among the civilized countries of Europe (Koliopoulos and Veremis, 2010).

Philhellenes or not, many people came to Greece during this period, including various diplomats, military officers, scientists, artists and other important personalities who left their writings about this place and its inhabitants. Among many others, these included: (i) French diplomat and historian Louis Félix-Auguste-Beaujour (1765–1836); (ii) British nobleman and diplomat Thomas Bruce 7th Earl of Elgin (1766–1841); (iii) Irish painter and writer Edward Dodwell (1767–1832); (iv) German architect, archaeologist and art historian Carl Haller von Hallerstein (1774–1817); (v) British archaeologist and illustrator Sir William Gell (1777–1836); (vi) British diplomat, antiquarian and writer William Martin Leake (1777–1860); (vii) German educator Friedrich Thiersch (1784–1860); (viii) Danish archaeologist, Peter Oluf Brøndsted (1780–1842); (ix) Baltic archaeologist, writer and painter Otto Magnus Freiherr von Stackelberg (1786–1837); (x) British politician Sir John Hobhouse (1786–1869); (xi) British architect, archaeologist and writer Charles Robert Cockerell (1788–1863); (xii) French diplomat Lodoïs de Martin du Tyrac de Marcellus (1795–1861); (xiii) German educator Karl Otfried Mueller (1797–1840); and (xiv) French diplomat and archaeologist Charles Champoiseau (1830–1909) (Tzermias, 1997; Vingopoulou, 2005).

In many cases, visitors to Greece left behind texts, photographic documents and works of art that not only describe the country's attractiveness, but also point out the destruction caused by the War of Independence, the lack of infrastructure, and the poor housing conditions. Many of these foreign visitors recognized the modern Greek culture and sympathized with the poor and oppressed people who lived there, while others took advantage of the prevailing situation to rob the cultural heritage of the place and impose their interests at the expense of the locals.

Unfortunately, the phenomenon of looting antiquities increased considerably in the 19th century, in parallel with the intensification of excavation activity. Ironically, but very realistically, it can be said that the stolen antiquities exhibited in the museums and palaces of Europe aroused the admiration of Europeans for ancient Greek culture and in a way improved the mood towards Greece. However, it can also be said that this fact gave Europeans the impression that the local authorities and the inhabitants of these areas were either completely ignorant and indifferent to their cultural heritage or opportunistic and corrupt towards money.[1]

### 15.1.2 Artworks

Themes from Greek mythology continued to influence the European art world well into the 19th century, in painting with works such as: (i) *The Ambassadors of Agamemnon in the Tent of Achilles* (1801) and *Apotheosis of Homer* (1827) by Jean-Auguste-Dominique Ingres (1780–1867); (ii) *The Death of Leonidas at Thermopylae* (1814) and *The Wrath of Achilles* (1819) by Jacques-Louis David (1748–1825); and (iii) *Medea about to Kill her Children* (1838) and *Perseus and Andromeda* (1853) by Eugène Delacroix (1798–1863).

At the same time, other artists preferred to paint the Greek landscape. This included Carl Wilhelm Kolbe (*c.*1757–1835) with *Auch ich war in Arkadien* (1801), Hugh William Williams (1773–1829) with *Mountainous Landscape with a Party of Travellers* (1829), Joseph Cartwright (1789–1829) with *Vathi, Ithaca* (1821), and Thomas Eakins (1844–1916) with *Arcadia* (1883), and many others.

Another source of inspiration for painters was the War of Independence of 1821, some of whom followed the trend of romanticism. Among many others this included:

- Eugène Delacroix, who has already been mentioned above, with the works *The Massacre at Chios* (1824) and *Greece on the Ruins of Messolonghi* (1826);
- Lodovico Lipparini (1800–1856) with *Palaion Patron Germanos raises the flag of independence* (1838) and *Death of Markos Botsaris* (1841);
- Louis Dupré (1789–1837) with *Nikolakis Mitropoulos Raises the Flag at Salona* (1827);
- Peter von Hess (1792–1871) with *Alexander Ypsilantis Crosses the Pruth* (nd) and *The Execution of Patriarch Gregory V of Constantinople* (nd);
- Ambroise Louis Garneray (1783–1857) with *The Naval Battle of Navarino* (1827);
- Ary Scheffer (1795–1858) with *The Women of Souli* (1827);

- François-Émile de Lansac (1803–1890) with *The Last Moments of Messolonghi* (1827);
- Vasily Grigorievich Khudyakov (1825–1871) with *Massacre of the Turks in the Streets of Chania* (nd);
- Giuseppe Mazzola (1748–1838) with *The Attack of Ibrahim Pasha on Messolonghi* (nd);
- Martinus Rørbye (1803–1848) with *Conversation on the Parthenon* (1844); and
- Carl Haag (1820–1915) with *Greek Warrior* (1861).

Adam Friedel (*c.*1780–1868) deserves a special mention. He was a Danish military man who travelled to the Ottoman Empire after the Napoleonic Wars, where he served as an officer and personally met many fighters of the Greek Revolution, such as Theodoros Kolokotronis, Andreas Miaoulis and Laskarina Bouboulina. He supported the revolution, but also painted many portraits of its heroes, which were distributed by philhellenic associations in Paris and London between 1825 and 1826, in order to publicize the cause of the War of Independence in Europe.

By putting the themes of the Greek War of Independence on the agenda of the art world, all these works of art undoubtedly contributed to advancing the discussions of Europeans about the current situation in Greece and to developing the idea of establishing a modern Greek state.

Let us take a closer look at one of these paintings to illustrate how a work of art from this period can contribute to the formation of the image of a rebellious people whose goal was the creation of an autonomous state, and which, by disturbing the viewer's emotions, can ultimately support the legitimization of the Greek state in the future. The painting of the then-new French Romantic painter Eugène Delacroix, *The Massacre at Chios* (1824), undoubtedly puts the Greek theme on the agenda of the discussions of art lovers, who at that time came mainly from the elite (Fig. 15.1). The painting, more than 4m high, shows the fear and pain of the civilian population of Chios in the face of the bloodthirsty Ottoman warriors. In the foreground of the painting, we see a half-naked, bloodied wounded man lying handcuffed on the ground, while around him are tragic figures of women, children and other immobilized men, unable to defend themselves and ready to be trampled by a mounted Ottoman soldier. Fear and despair are reflected in the faces of the women. If you look closely, you can see other similar scenes in the background, such as people being rounded up for killing or enslavement, shootings of civilians, and burning villages in the distance. This painting was a metonymy for the massacre of at least 20,000 inhabitants of the island of Chios in the summer of 1822 and the mistreatment, enslavement and expulsion of the remaining inhabitants. It illustrated to Europeans the Ottoman Empire's war practices as acts of supreme barbarism against a Christian population. The liberal romantics of Europe seized upon this, and similar atrocities committed by the Ottomans against the defenceless Greek communities, drew inspiration from them, and created works to show the conservative governments of their day that they could no longer remain indifferent. The painting was exhibited in 1824 at the Salon, the official annual art exhibition in France, and then at the Musée du Luxembourg. In 1874, it was moved to the Musée du Louvre, where it drew the public's attention to the Greek theme and provided an image of Greece at that time.

It is remarkable that today the painting has become one of the iconic signs associated with the word 'slaughter' internationally. Of course, one may wonder what all this has to do with tourism, and yet it does have something to do with it. First of all, this painting, which was created at a time that was still reminiscent of the 'Grand Tour', would certainly have motivated the philhellenes to travel to Greece. Then, in our time, we have to consider that someone who sees the painting in the Louvre or on the Internet will naturally learn about a Greek island called Chios. Therefore, this painting could easily be positioned as one of the 'core images' of the social representation of the destination Chios. Finally, we can assume that alongside the promotion of the traditional 'mastic' of Chios and the tourist posters with beaches, medieval castles and windmills, a painting of the Romantic movement can be an element that composes the contemporary tourist image of the island.

Meanwhile, there were also Greek painters who were inspired by the themes of the Greek Revolution, such as D. Zografos and I. Makrygiannis, who depicted the great conflicts graphically, but also Theodoros Vryzakis (1814–1878) with works such as *The Exodus from*

**Fig. 15.1.** *The Massacre at Chios* (1824) by Eugene Delacroix. Public Domain. The artistic representations of the Greek War of Independence played a role in shaping the image of Greece, along with various other factors. This in turn may have contributed, at least to some extent, to the development of the tourism industry in Greece in the following century.

*Messolonghi* (1853) and *The Battle of Dervenakia* (1860). Others painted subjects related to the history and landscape of the place, as well as to the customs and traditions of the Greeks, in addition to depictions of war. This included: (i) Nikiforos Lytras (1832–1904) with *Execution of Patriarch Gregory V of Constantinople, Carols* (1872); (ii) Nikolaos Gyzis (1842–1901) with *The Orphans* (1871), *The Betrothal of the Children* (1877), *The Secret School* (1886) and *After the Destruction of*

*Psara* (1898); (iii) Konstantinos Volanakis (1837–1907) with *The Inauguration of the Corinth Canal* (1893) and *Women Washing in the River* (nd); (iv) Georgios Jakobides (1853–1932) with *The Little Refugee* (nd); (v) Symeon Savvidis (1859–1927) with *Washerwomen in the Lake* (1888); and (vi) Polychronis Lembesis, (1848–1913) with *The Children Stealing Apples* (1888). These and many others, with their works, directed the imagination of both Europeans and the inhabitants of the newly founded Greek state towards the formation of core images on which Greece was to be built.

Kokkinidis (2001) lists some features of 'Greekness' in art that were more or less adopted by many Greek painters of the time. These were: (i) in terms of colours, strong red, ochre, turmeric, black and white, as well as the greys and blonds of the Greek natural landscape; (ii) in terms of form, the clarity of surfaces and clean contours; (iii) in terms of pictorial space, the counter-perspective arrangement without false sensual tricks of the Western tradition; and (iv) in terms of representational perception, extreme but undistorted representations. All these, however, are not privileged features of Greek art, as we find them in almost all peoples of the Mediterranean.

Various foreign archaeological schools began to take an interest in the region at the southern end of the Balkan Peninsula at this time, and opened branches in the newly founded Greek state. This included the French archaeological school, the German archaeological institute, and various others founded by the Americans, the British and the Italians. They not only carried out systematic archaeological investigations and uncovered numerous ruins, but also recorded the monuments, first with sketches and later with photographs, which gave a tangible picture of Greek antiquities. All this work, of course, helped to increase the number of foreign visitors to Greece, as the finds became very attractive to foreign travellers when news of the discoveries was spread throughout Europe by the foreign press.

### 15.1.3 Contributions of newspaper journalists and the press

As far as the description of the war situation in Greece by the foreign press is concerned, it is obvious that the events that took place in the region during that period were valuable material for the journalists. Certain topics concerning the uprisings against the Ottomans and certain battles, the foreign interventions, the establishment of the Kingdom of Greece, the territorial expansion of the new Greek state and many other political events were frequently reported in foreign newspapers and were of great interest to Europeans. In fact, European newspapers were divided into pro-Hellenic and pro-Ottoman, with both sides attacking and ridiculing each other in their publications. Newspapers that were positive about Greek independence included the Swiss *Gazette de Lausanne*, the British *Courier of London*, the *Morning Chronicle* and the *Morning Post*, among others (Loules, 1983).

In many cases, the articles were not free from an emotional disposition due to the romanticism of the era, which shifted the reference to ancient Greece, the 'cradle of civilization', as well as to Christianity and the oppressive regime of the Ottomans. However, it must be taken into account that all important events that could affect the interests of the ruling class of the European countries were treated by the press according to the goals and ideology of the newspaper's patrons, so that there were always backsliding and contradictions in the general line of the newspaper. Other newspapers expressed mainly negative views about the struggle, such as the Austrian *Oesterreichischer Beobachter*, and the French *Gazette de France* and *Drapeau Blanc* (Dimakis, 1968; Terencio, 1990; Argyrakos, 2020).

In addition, certain events that were a consequence of the war, or of the political-economic conditions of the time described by the foreign press, sometimes contributed negatively to the construction of a tourist image of Greece. For example, the bankruptcy of the Kingdom of Greece in 1893 was an event that left a negative impression through its coverage in the foreign press. Also, the murder of a group of travellers, three upper-class Britons and an Italian diplomat, in the area of Delesi in Boeotia, in 1907 by a gang of robbers was a major defamation of Greece, especially by the English press (Clogg, 2021).

Often well-known personalities of the European continent, who loved and visited Greece, supported the country through positive reports in the press, which contributed to the image of the country. It is worth mentioning

Elizabeth of Austria and Queen of Hungary (1837–1898), often referred to as 'Sisi', who truly loved Greece. She was a prominent figure of the 19th century who had a strong influence on fashion and culture among the European public. Her visits to Greece in the mid-19th century coincided with a period of growing interest in classical antiquity, fuelled by archaeological discoveries and the revival of classical art and architecture. Her travels were widely reported in newspapers of the time, and her fascination with Greek culture, history and mythology contributed to a romantic notion of Greece as a land of beauty and ancient wisdom.

Another element that influenced Europeans' ideas about Greece was political caricatures. As Kokkinos (1978) reports, many of the caricatures drawn by foreign artists aimed to denigrate the official policies of Western countries, which were initially negative towards the Greek Revolution. For example, one cartoon that circulated in the territory of the German Confederation and aroused the ire of the Austrian Minister of Police, symbolically depicted the disaster of the Greeks in Constantinople and the friendly policy of the West towards the Ottomans. The cartoon shows Ottomans with swords slaughtering the Greek population, one of them holding a severed head by the hair in one hand; a little closer is Gregory V, the hanged Ecumenical Patriarch of Constantinople, with 'Agia Sophia', the Christian cathedral of Constantinople from the Byzantine Empire period, converted into a mosque by the Ottomans in 1453, in the background. To the right, the Sultan is depicted, before whom the British ambassador gently curtsies and sticks out his tongue at the slaughtered; while further to the right, Joseph von Pilat, secretary to the foreign minister of the Austrian Empire, Klemens von Metternich (1773–1859), observes the proceedings with a telescope and serious satisfaction (Kokkinos, 1978).

At this time, printed Greek newspapers also began to circulate within the borders of the Greek state, such as the *Salpigx Elliniki* (Kalamata, 1821), published by Theoklitos Farmakidis (1784–1860) and considered the first printed newspaper in Greece. Others included the *Ellinika Chronika* (Mesolonghi, 1824–1826), *Efimeris Athinon* (Athens, 1824–1826), *Filos tou Nomou* (Hydra, 1824–1827) and *Geniki Efimeris tis Ellados* (Nafplio-Aegina-Poros, 1825–1827), which continued the path of the Greek pre-revolutionary press in the now free Greece (Balta, 2005; Koumarianou, 2005).

Be that as it may, Greek newspapers maintained the great importance they attached to politics rather than to the profit motive of their editors, while commercial advertisements to cover production costs began to appear tentatively towards the end of the 1860s.

At this point, it should be noted that the communications landscape in Europe changed after the middle of the century. This was due to a series of factors related to technological development and international affairs, such as the optimization of means of transportation and the invention of the electric telegraph, which enabled communication between Europe and America. Gradually, the great telegraph networks connected the different corners of the Earth and changed the conception and perception of space and time.

In this regard, people, books, newspapers and ideas began to move and spread more easily and quickly than before. The speed of news gathering and dissemination, as well as the creation and gradual expansion of press agencies (Havas, Wollf, Reuter, etc.) globalized the news market to a certain extent. Press culture began to change as information now became associated with speed, and the concept of 'news commodity' began to change the practices, uses and attitudes of journalism. During these years, many other Greek newspapers were founded, such as: (i) *Kairoi* (1872) by Petros Kanellidis (*c.*1848–1911); (ii) *Efimeris* (1873) by Dimitrios Koromilas (1850–1898); (iii) Acropolis (1883) by Vlasis Gavriilidis (1848–1920); (iv) *Skrip* (1895) by Grigorios Efstratiadis (1864–1950) and Evangelos Kousoulakos (1861–1903); and (v) *Empros* (1896) by Dimitrios Kalapothakis (1865–1921) (Balta, 2005; Koumarianou, 2005).

### 15.1.4 Photographs, advertisements and postcards

There were major differences in the design and projection of images by the new established newspapers. Older articles that relied mainly on the development of ideas were replaced by articles that provided information about immediate events and places in the world, which were more descriptive and much more effective for the formation of images. Photographs were added to the newspapers later, some of which showed foreign places. In this way, readers' imaginations were stimulated to form images of various faraway places, and their curiosity and desire to visit these places up close were increasingly aroused. Over time, newspaper advertisements were added that combined description, image, emotion and information to make travel offers attractive. Various studies of the advertising material of newspapers in Athens and other regions of Greece, such as Samos, from this period show that advertisements for hotels and other activities belonging to the tourism industry accounted for only slightly more than 5% of all advertisements (Kariatoglou, 2005; Vasilenas, 2005).

Advertisements at the time took the form of a simple announcement in a square frame. For example, in the 2 November 1883 issue of the *Acropolis* newspaper, an advertisement for tourism at the Grand Hotel Hellas read as follows:

> In Ermou Street, above the confectionery Akrivopoulou, the 'Grand Hotel Hellas' managed by Iliopoulos, has rented the young 'Hotel Europe' in the same street and opposite the 'Hotel Agglia', to satisfy its numerous clientele. The rooms are spacious, the service is good, the prices are very reasonable.
>
> (*Acropolis*, 1883, p. 4)

In another issue of the *Acropolis* of 1 April 1884, an event that takes place every week in Berlin is tantalizingly described by the author 'Mitzos', exciting the tourist imagination of the reader:

> Last week began in Berlin the so-called Corso, an event that takes place in the Tiergarten, in a large square where every Thursday two military bands play loud music. There, horses, pedestrians and private carriages drawn by Arabian horses and driven by gentlemen of the aristocracy and the demi-monde move from one square to another, admired and also envied by those who have no money. This event lasts two hours, then the people disperse into the forest and return to Berlin . . .
>
> (Mitzos, 1884, p. 3)

The first daguerreotypes (early form of photograph) were taken in Greece in the first half of the 19th century by Pierre Gaspard Jolly-de Lotbiniere (1798–1865), a Geneva merchant and photographer who worked for Parisian publisher, optician and daguerreotypist Noël Paymal Lerebours (1807–1873). Lerebours, who viewed photography as a multifaceted imaginary journey with commercial benefits, sent photographers to various parts of the world to compile a monumental travel book titled *Excursions Daguerriennes* (1841–1842).

In the years that followed, more and more foreigners visiting Greece, or travelling through the country to reach the Middle East, brought large-format-view cameras with them. Many visitors, who were either professional photographers or amateurs, roamed the country immortalizing monuments, people and their lives – sometimes natural, sometimes staged. Upon their return, they published special luxury albums or put their photos on the market as postcards.

It was a time of change for the visual perception of travellers. The pictorial stereotypes with which the monuments of ancient Greece had been depicted for so many years by visitors in drawings and paintings remained only as background in the imagination, and now the more realistic photographic representations came to the fore. In this sense, photographs made the tourist destinations and everything that was considered authentic much more real than the paintings of the past. It was the time when the modern 'tourist gaze' was established in Greece.

It is believed that all this photographic record of 19th-century visitors' impressions had a catalytic effect on the way subsequent generations of foreign visitors saw Greece, shaping the expectations of many who chose the same prescribed itineraries and creating timeless stereotypes, many of which are still valid today.

At this point, it should be noted that the first postcards were circulated in Greece around 1893, while their 'golden age', as in other countries, was at the beginning of the 20th century (Xanthakis, 2008; Bonarou, 2012).

### 15.1.5 Contribution of women

Finally, in order to emphasise the role of Greek women and their contribution to such a complex phenomenon as the image of a newborn country in modern times, and bearing in mind the specificity of the inclusion of the female dimension in all the factors we have discussed above (literature, art, press, etc.), let us dwell separately on the contribution of Greek women to the newly founded Greek state. We will not speak too much on the women who participated in the Greek Revolution, mentioning only Laskarina Bouboulina (1771-1825) and Manto Mavrogenous (1796-1840). Anyway right after, apart from Kalliroi Parren (1861–1940), the scholar and journalist who was the editor of the Greek women's periodical: Efimeris ton Kyrion (1887-1917), as well as Aimilia Ktena the editor of the women's periodical Evridiki (1870-1873), many other important Greek women, each with her work, contributed to the construction of the image of the new Greek state in the 19th century and the early beginning of the 20th. Personalities such as the writer Elizabeth Moutzan-Martinegou (1801-1832), the painter Eleni Boukoura-Altamoura (1821-1900), the political and social activist Katerina Botsari (1827-1872), the poetess Aganice Ainianos Mazaraki (1838-1892), the poetess Foteini Oikonomidou (1849-1883), the poetess Agathoniki Antoniadou (1854-1928), the scholar Sevasti Kallisperi (1858-1953), writer Alexandra Papadopoulou, (1867-1906), the music teacher, pianist and suffragist Avra Theodoropoulou (1880-1963), and many other pioneers contributed to the creation of a mental conceptual representation of the region. Although the names of many of these women are no longer well known today, they all supported national independence, female identity, and some of them organised the women's movement in Greece. With their energy showed the whole world at the time that the region of the Southern Balkans, the coasts of Asia Minor, and the islands had a strong female potential that could contribute to the image of the new Greek state from the very beginning (Varika, 1987).

## 15.2 Early Development of Tourism in Greece

It must be emphasized that the devaluation of the word 'tourist' in Europe dates back to this period. In the 19th century the term 'tourist' no longer referred to the participants in the 'Grand Tour' of the previous centuries. Paradoxically, tourists were now those who travelled en masse, guided by tour operators such as Thomas Cook, and took package tours without seeking cultural enrichment or anything other than their personal pleasure and a superficial imitation of intellectuals.

On the other hand, the 'travellers' who travelled alone, on horseback, by ship, etc., were independent and free intellectuals, sensitive at every moment of the journey to everything they saw in their surroundings, seeking the authentic pleasure of travel, motivated by their intention to develop their consciousness. Eventually, the notion crystallized that travellers were cultured and educated people, as opposed to tourists who represented the masses (Buzard, 1993; Mazierska and Walton, 2006; Hionidis, 2014).

Within this logic, as Hionidis (2014) argues, in the case of Greece at that time, one can speak exclusively of travellers, enlightened people who travelled, since there were no railroads in the region, nor did Thomas Cook initially organize trips, and there were no popular 'entertainment' venues or clubs for foreigners in Greece.

In fact, Cook began organizing tour groups to Greece in the late 19th century. In addition to the services of Thomas Cook, and other travel agencies that emerged in this century, the services of 'dragomans', locals who helped visitors, were also used.

The duties of the dragomans included not only helping the foreigners in their daily communication by translating or driving them to the sights, but also: (i) protecting them; (ii) suggesting and guiding them to the sights; (iii) informing them about the history, culture and natural environment of the place; (iv) cooking for them or guiding them to restaurants; (v) organizing accommodation; and (vi) giving them tips on how to interact with the locals. In short, the role of the dragoman was also, in part, to make the trip palatable to the foreign visitors and to protect them from the difficulties and inconveniences of travelling in a foreign country and culture. This suggests that what visitors learned about their visit was mediated not only through the filters of their own observations and perceptions, but also in large part through the explanations and interpretations of their dragoman (Dritsa, 2019).

However, as Dritsa (2019) notes, it is not so certain that the dragomans were aware of the importance of their role in shaping the image of their country, while it is certain that, at least for visitors, they reflected the image of the average modern Greek of the time. Be that as it may, it can be argued that many dragomans had a much better education than the average modern Greek. In order to practice this profession, they had to know at least one foreign language, have historical knowledge, know the geography of the place, and have cooking and organizational skills.

In fact, many of the dragomans had lived or studied abroad, like the famous Greek dragoman Panagiotis Ghiolmas, who had lived in London and even worked in the office of Thomas Cook. Later, Ghiolmas founded the first tourist office owned by a Greek in Syntagma in 1902. He was an excellent organizer of excursions and had a great knowledge of communication, he realized the importance of advertising and did a lot to build a good reputation (Dritsa, 2019).

Overall, the establishment of dragomans can be considered an important step towards the commercialization of tourism and the development of the tourism industry in Greece. In general, it can be said that in those years the dragomans were the main means to shape the image of the country as soon as the tourist entered the country. Already at the beginning of the 20th century, the term 'tourist guide' was used instead of the oriental term 'dragoman'.

Be that as it may, in the early years of the Greek state, one cannot speak of a comprehensive tourism industry that was self-confident enough to strategically promote a specific tourist image of Greece.

### 15.2.1 Guidebooks

Greece was included in Murray's *A Handbook for Travelers* as early as 1840, while Baedeker Publishing's guidebooks in 1880 included places that are now part of Greece (Corfu), and as mentioned earlier, Thomas Cook & Son offered trips to Athens and Constantinople beginning in 1868 (Murray, 1840; Baedeker, 1880; Harlan, 2009).

All of these guidebooks began to market Greece as a tourist product, praising the ancient sites and often the landscape, while often devaluing the traditions, the people of the country, and even the women, who were considered very inferior in terms of beauty and stature (Dritsas, 2006).

Thus, one could say that these first guidebooks reflected a part of the travel literature in which foreign travellers approached Greece with an exaggerated love for the ancient monuments and a distaste for the state of modern Greece (Travlou, 2002).

As we can see, the rich Greek cultural heritage presented through channels such as history, literature and art (i.e. as organic tourist representations) first had to be transformed into tourist products to be understood and recognized as such before it could be consumed. This function was initially performed by the foreign tourist media, while the exploitation of the scarcely existing group travel to Greece was also performed by foreign agencies. Of course, we must not overlook the fact that there were indeed some basic infrastructures for tourism, operated by private or state companies.

The first systematic and comprehensive guide to trade and commerce, which included lists of hotels, restaurants, pastry shops, cafés, etc. in Athens and other major cities, was written and published in 1875 by Miltiadis Boukas so that, as the preface states, 'the foreigner can get a real idea of Greece', although ironically it was written in Greek (Boukas, 1875).

In any case, the concept of tourism as an industry in its own right had not yet been fully understood, even in Europe. Moreover, due to historical circumstances, the newly established Greek state was perhaps more focused on its political construction and development, and therefore unable to implement measures that it considered secondary, such as the promotion of tourism.

As for domestic tourism, towards the end of the century, when the extra-urban space became less dangerous due to the de-escalation of conflicts and robberies, several travelling associations were created, generally supported by the bourgeoisie, which developed a remarkable activity that attracted many people. Regular excursions, visits to archaeological sites, the study of flora, the publication of maps, guidebooks, historical and scientific publications, etc. were among their main activities. Another leisure activity that developed from the end of the 19th century was related to the numerous Greek thermal springs (Kyllini, Hypati, Kaiafas, etc.). It is believed that the reputation of the European elites' preference for such forms of recreation, as well as promotional brochures with medical information, increased the Greeks' attraction to these places (Vlachos, 2016).

### 15.2.2 International exhibitions

Finally, it should be mentioned that the promotion of Greece's image abroad was also sought through the state's participation in numerous global and international exhibitions, beginning in 1851 with the Great Exhibition of Industrial Works of All Nations (Crystal Palace Exhibition) in Hyde Park in London. Among others, the first Olympic Games of the modern era, held in 1896 in the Panathenaic Stadium in Athens, was an event of great interest for the image of Greece and was attended by 70,000 spectators; this fact could be a foreshadowing of what mass tourism can mean in Greece.

## 15.3 Summary

To summarize, we can say that Greece underwent a major cultural transformation in the 19th century, when attempts to create a modern nation state with Western ideals began. However, the image of ancient Greece combined with the state of the country at the time sometimes attracted travellers and sometimes led to negative portrayals in the foreign press. Be that as it may, the philhellenic movements and art presented Greece in a positive way and created a favourable climate for its image in international perception. At the same time, various archaeological excavations and discoveries increased the country's attractiveness and paved the way for modern tourism. The media, including newspapers and photography, certainly played a crucial role and formed the basis for the tourist image of Greece as a cultural destination that began to emerge at this time. Finally, it can also be said that Greece's participation in international events strengthened the country's image and heralded the beginning of the mass tourism that was to come in the second half of the next century.

## Note

[1] In any case, it should be emphasized that many of the above-mentioned people who visited Greece (e.g. Thomas Bruce, the 7th Earl of Elgin, Charles Robert Cockerell, Lodoïs de Martin du Tyrac de Marcellus, Charles Champoiseau, Otto Magnus von Stackelberg, Carl Haller von Hallerstein and many others) either stole the ancient monuments directly or took them abroad with the consent of the Ottoman central authorities, or even with the consent of the poor and illiterate local population lured with money (Simopoulos, 2003; Matthaiou and Hatzidimitriou, 2012).

## References

*Acropolis* (1883) Grand Hotel Hellas advertisement. *Acropolis*, 2 November, p. 4. Digital Library of Newspapers and Periodicals, National Library of Greece. Available at: http://efimeris.nlg.gr/ns/pdfwin.asp?c=126&dc=2&db=11&da=1883 (accessed 7 February 2024).

Argyrakos, G. (2020) Philhellenism as a European cultural phenomenon and the role of newspapers during the Greek Revolution of 1821. Society for Hellinism and Philhellinism. Available at: https://www.eefshp.org/en/philhellenism-as-a-european-cultural-phenomenon-and-the-role-of-newspapers-during-the-greek-revolution-of-1821/ (accessed 12 September 2022).

Assmann, J. (2011) *Cultural Memory and Early Civilization: Writing, Remembrance, and Political Imagination*. Cambridge University Press, Cambridge, UK. DOI: 10.1017/CBO9780511996306.

Baedeker, K. (1880) *Italy: Handbook for Travellers*. K. Baedeker, Leipsic, Germany.

Balta, N. (2005) The evolution of news in Greek newspapers in the 19th century (1830 onwards). In: Droulia, L. (ed.) *The Hellenic Press, 1784 to Today*. Athens Institute for Neohellenic Research/ National Hellenic Research Foundation, Athens, pp. 63–70.

Beaton, R. (2021) *The Greeks: a Global History*. Basic Books, New York.

Bonarou, C. (2012) *Visual Culture and Tourism. Representations of Greece on Tourist Postcards*. Papazisi Publications, Athens. (in Greek).

Boukas, M. (1875) *Commercial, Geographical and Historical Guide to Most of the Great Cities of Greece in 1875*. Elliniki Anexartisia, Athens. Anemi, The Digital Library of Modern Greek Studies. (in Greek). Available at: https://anemi.lib.uoc.gr/metadata/f/4/6/metadata-01-0000271.tkl (accessed 10 September 2023).

Buzard, J. (1993) *The Beaten Track. European Tourism, Literature, and the Ways to Culture, 1800–1918*. Clarendon Press, Oxford, UK. DOI: 10.1093/acprof:oso/9780198122760.001.0001.

Clogg, R. (2021) *A Concise History of Greece*. Cambridge University Press, Cambridge, UK. DOI: 10.1017/9781108953924.

Dimakis, J. (1968) La guerre de l'indépendance Grecque: vue par la presse Française (période de 1821 à 1824). In: *Contribution à l'étude de l'opinion Publique et Du Movement Philhellénique En France*, Vol. 102. Institute for Balkan Studies, Belgrade. (in French).

Dritsa, M. (2019) *Mosaics of Greek Tourism*. Economia Publishing, Athens. (in Greek).

Dritsas, M. (2006) From travellers' accounts to travel books and guide books: the formation of a Greek tourism market in the 19th century. *Tourismos* 1(1), 27–52.

Harlan, D. (2009) Travel, pictures, and a Victorian gentleman in Greece. *Hesperia* 78(3), 421–453. DOI: 10.2972/hesp.78.3.421.

Hionidis, P. (2014) Travelling and the shaping of images: Victorian travellers in nineteenth-century Greece. *International Journal of Cultural and Digital Tourism* 1(2), 30–40.

Kakridis, T. (1963) *Greek Lights. University Lectures*. Estia, Athens. (in Greek).

Kariatoglou, E.I. (2005) Advertisements. In: Droulia, L. (ed.) *The Hellenic Press, 1784 to Today*. Athens Institute for Neohellenic Research/National Hellenic Research Foundation, Athens. (in Greek), pp. 196–200.

Kokkinidis, D. (2001) Conception and expression of Greekness in plastic arts. In: Tsaoussis, D.G. (ed.) *Hellenism and Greekness*. Estia Bookstore, Kolaros and Co., Athens, pp. 211–216.

Kokkinos, A.D. (1978) *History of Modern Greece*. Melissa, Athens. (in Greek).

Koliopoulos, J.S. and Veremis, T. (2010) *Greece – the Modern Sequel. From 1821 to the Present*. Wiley-Blackwell, Oxford, UK.

Koumarianou, A. (2005) Historical path of the Greek press, 1784–1922. In: Droulia, L. (ed.) *The Hellenic Press, 1784 to Today*. Athens Institute for Neohellenic Research/National Hellenic Research Foundation, Athens, pp. 55–61.

Loules, D. (1983) *The Greek Revolution and the British Press*. University of Ioannina Publications, Ioannina, Greece. (in Greek).

Matthaiou, S. and Hatzidimitriou, A. (eds) (2012) *Emigrated Greek Antiquities. Starting Points and Routes*. Institute of Modern Greek Research, Athens. (in Greek).

Mazierska, E. and Walton, J.K. (2006) Tourism and the moving image. *Tourist Studies* 6(1), 5–11. DOI: 10.1177/1468797606070583.

Mitzos (1884) From Berlin, Bismarck-the Corso-the heir. *Acropolis*, 1 April, p. 3. Digital Library of Newspapers and Periodicals, National Library of Greece. Available at: http://efimeris.nlg.gr/ns/pdfwin.asp?c=126&dc=1&db=4&da=1884 (accessed 7 February 2024).

Murray, J. (1840) *A Handbook for Travelers in the Ionian Islands, Greece, Turkey, Asia Minor, and Constantinople; Being A Guide to the Principal Routes in those Countries, Including A Description of Malta; with Maxims and Hints for Travelers in the East*. John Murray, London.

Rozakis, C. (2001) Greece in the international arena. In: Tsaoussis, D.G. (ed.) *Hellenism and Greekness*. Estia Bookstore, Kolaros and Co., Athens. (in Greek), pp. 91–120.

Simopoulos, K. (2003) *Foreign Travelers in Greece: Public and Private Life, Popular Culture, Church and Economic LIfe, from Travel Chronicles*, Vol. 3. Stachy, Athens. (in Greek).

Terencio, Y. (1990) The struggle of the Greeks and European press in 1821. (in Greek). *Istoria* 261, 36–47.

Travlou, P. (2002) Go Athens: a journey to the centre of the city. In: Coleman, S. and Crang, M. (eds) *Tourism: Between Place and Performance*. Berghahn Books, Oxford, UK, pp. 108–127. DOI: 10.1515/9780857457134.

Tsaoussis, D.G. (2001) Hellenism and Greekness, the problem of the new Greek identity. In: Tsaoussis, D.G. (ed.) *Hellenism and Greekness*. Estia Bookstore, Kolaros and Co., Athens, pp. 15–26.
Tzermias, P.N. (1997) *The Image of Greece in the Foreign World*. Sideris Publications, Athens. (in Greek).
Varika, E. (1987) *The Ladies' Revolt: The Genesis of a Feminist Consciousness in Greece 1833-1907*. Research and Education Foundation of the Commercial Bank of Greece, Athens. (In Greek).
Vasilenas, S. (2005) Advertisements: the diary of a place at the turn of two centuries (Samos 1895–1914). In: Droulia, L. (ed.) *The Hellenic Press, 1784 to Today*. Athens Institute for Neohellenic Research/ National Hellenic Research Foundation, Athens. (in Greek), pp. 201–214.
Vingopoulou, I. (2005) *The Greek World Through the Eyes of Travelers (15th–20th Century): an Anthology from the Dimitrios Kontominas Collection: Exhibition Catalogue, Benaki Museum, February 8, 2005–March 6, 2005*. Benaki Museum Exhibition, Kotinos, Athens. Available at: https://hdl.handle.net/10442/14283 (accessed 12 August 2022).
Vlachos, A.F. (2016) *Tourism and Public Policies in Modern Greece 1914–1950*. Corfu-Economia Publishing, Athens. (in Greek).
Xanthakis, A. (2008) *History of Greek Photography 1839–1970* Papyros, Athens. (in Greek).

# 16 Tourism Enters the Greek World as an Economic Activity

**Abstract**

The formation of a tourist image of Greece at the beginning of the 20th century took place with elements that emerged organically from the worldwide collective memory of ancient Greek civilization, the media of the time, literature, painting, the press, cinema, etc., but also from the efforts of a rudimentary tourism industry. It can be said that in the first half of the 20th century, tourism in Greece was only very slowly taken seriously as an activity that could generate foreign exchange for the state and thus be part of the country's economy. The Greek National Tourism Organisation and other state institutions that were established as an alternative played an important role in this direction.

In the first decades of the 20th century, the emergence of tourism in Greece was often referred to as the 'industry of foreigners'. Despite this phraseology, the development of the tourism sector in Greece lagged behind that of other Western nations. Both the private sector and the Greek state did not evolve significantly in their perception of tourism during this period, and a coherent and consistent tourism policy was lacking. In the initial phase, the Greek state formulated a rudimentary tourism policy, developed it further, abandoned it again and reorganized it. This process was repeated in response to changing circumstances and the ebb and flow of economic and political developments.

However, the responsibility for this situation cannot be attributed only to the Greek state, but various inhibiting factors have contributed to it, such as the Balkan wars, the First World War, the Asia Minor catastrophe, the Second World War, the Greek Civil War (1944–1949), political instability, the disagreement of Greek politicians on tourism strategies, bureaucracy and the poor living conditions. The separation of the region from the other Central European countries due to its geographical periphery and difficult access to the mountain border by land, as well as the underdeveloped transport infrastructure, constituted another fundamental obstacle. Moreover, since the 19th century, the prevailing image among Western Europeans was that Greece was associated with remarkable archaeological monuments but not with a developed tourist infrastructure with modern facilities. Therefore, various aspects and situations prevented the development of Greek tourism with a concrete long-term state economic policy, as was the case in many other European countries.

## 16.1 A Rudimentary Tourism Industry

It cannot be said that in those early years there were no professionals in the field of tourism, no tourism organizations or tourism associations, or no interest in tourism at all on the part of the Greek state. This is because as early as 1914, several tourism organizations were gradually established, for example the Office for Foreigners and Exhibitions, which published promotional material such as an official guide in French entitled *Guide officiel des Communications de Grece*. Furthermore, the Hellenic Association of Tourist and Travel Agencies (HATTA) was founded in 1927 and the Greek National Tourism Organisation (GNTO) began its work



 *Communication and Tourism: Reflecting on the construction of the tourist image of Greece* (Michael Tsangaris)
DOI: 10.1079/9781800626034.0016

in 1929 under the supervision of the then Greek Ministry of Economy. In 1931 the State School of Tourist Guides was established, and in 1935 the Greek Tourist Police as well as the Panhellenic Federation of Tourist Hotels Organisation (which took the name Hellenic Chamber of Hotels in 1946). The School of Tourist Professions was officially founded at the end of the 1930s. Furthermore, in the 1920s, the Hiking Association and the Hellenic Touring and Automobile Club (ELPA) were founded, and as early as 1935, there were specialist magazines about the tourism industry in Greece, such as *Touristiki Hellas* (Moussa, 2012; Katsigiannis, 2017; Dritsa, 2019).

For many years, foreigners visited Greece mainly as cultural enthusiasts interested in the archaeological sites, rather than as beach vacationers and mass tourists, as they did in the second half of the 20th century. That is, they visited the country not as masses flooding entire beach areas in the summer, as in the 1980s, but as groups of observers whose goal was primarily to visit archaeological sites in the Peloponnese, Delphi, Attica, etc., to hike in the Greek countryside, and to study Greek nature. Occasionally, however, they also visited the thermal springs (Kyllini, Hypati, Kaiafas, Edipsos, Loutraki, etc.) or spent their summer vacations by the sea (Corfu, Rhodes, Mykonos, etc.). It should be pointed out that in the first four decades of the 20th-century, tourism in Greek thermal springs assumed a larger scale than in the previous century, but was never as successful as in other countries (Vlachos, 2016).

In the first years of this period, despite the initiatives of some individuals and organizations, such as the hiking and nature-lovers associations and the thermal spas, there were few plans for domestic tourism in Greece: it can be said that its development was rudimentary. However, it cannot be said that there were no Greeks who travelled, visited places and enjoyed the amenities there. It should also not be forgotten the considerable number of lower-class people who left their homes during the summer months and settled in improvised shelters (huts, sheds, etc.) or in tents on the coast. However, this type of issue was only marginally studied, no statistical data was provided, and it was not included in the context of tourism because it was not considered part of the tourist development of the tourism industry in terms of the prevailing European capitalist model.

So, we can say that at that time there were signs of the phenomenon of tourism and that the Greeks travelled, but not in the same way and to the same extent as the Western Europeans. We must also keep in mind that there were quite a number of Greeks who lived abroad and visited Greece. Nevertheless, perhaps out of a desire to match the Westerners, or perhaps out of an inner need, the number of Greeks who visited the thermal baths or, more generally, travelled according to the Western European model, slowly increased.

The fact that foreign travel guides were translated into Greek, printed and distributed, such as the one published in 1901 by 'D.G. Efstratiou', also testifies to this. It was a Baedeker travel guide that included a variety of excursion routes throughout Greece, as well as timetables for the Athens Tramway, the Attica Railway, the Athens–Piraeus Railway, the Piraeus–Athens–Peloponnese Railway, the Northwest Greece Railway, and the Thessaly Railway, with their respective fares.

Another publishing house that issued tourist publications in Greek very early on was Eleftheroudakis, which published *Attic Walks* in 1922. It was written by the Greek storyteller, poet and folklorist Costas Pasayiannis (1872–1933), one of the pioneers of modern Greek travel literature.

A year later, Eleftheroudakis published a *Map of Attica* in Greek. However, the first comprehensive travel guide to Greece in Greek was published by the same publisher in 1926 in three volumes with maps, colour and black-and-white plans of cities and archaeological sites. In 1930 an expanded travel guide appeared in eight volumes, divided into Greek regions as follows: (i) Athens and Attica; (ii) North-west Greece; (iii) Thessaly; (iv) Crete; (v) Macedonia; (vi) Islands; (vii) Peloponnese; and (viii) Main Greece (Vlachos, 2016; Katsigiannis, 2017; Dritsa, 2019).

## 16.2 Emergence of the Tourist Image of Greece

It was during this period that Greece began to link tourism more closely with the economy and, albeit unconsciously, with the country's tourist image building. In an era marked by significant advances in mass communication technologies and practices, such as the cinema, the telegraph, the telephone, the press and the radio, which had a profound impact on the way information and culture were disseminated and consumed by the masses, the image of Greece gradually began to be shaped by the combination of more organic and less induced destination images.

It should also be added here that, according to Greek tourism scholars, the first commissioning of a project to promote tourism by the Greek government was in 1911 by Greek Prime Minister Eleftherios Venizelos (1864–1936) himself. After meeting with Swiss philhellene photographer Frédéric Boissonnas (1858–1946), who had already photographed Greece, the prime minister commissioned him to visit the Cyclades and Crete with his friend Daniel Baud-Bovy (1870–1958), director of the Geneva School of Art, to photograph landscapes and snapshots of everyday life to be used for tourism promotion (Katsigiannis, 2017).

Boissonnas, both on his own initiative and on behalf of the Greek state, continued to photographically reproduce the Greek image of the time, not only to promote tourism, but also as a means of political propaganda. With the series of photo albums *L'image de la Grèce*, he not only reproduced a simple photographic impression of the areas that belonged to the geographical or administrative boundaries of the state, but he tried to contribute to its territorial claims by including those areas whose inhabitants, as he said, felt themselves to be 'Greeks at heart'. The series of these photo albums united memories of antiquity, the unknown treasures of the museums, but above all it showed the recently liberated provinces and those that were Greek but 'subject to the unjust decisions of diplomacy'. His sons Edmond and Henri, who continued his work, contributed to the publication of other albums, but also to the organization of exhibitions on Greece, reflecting Greekness, the ancient monuments, but also the modern face of the country, the Asia Minor campaign, etc. (Girardin, 1998; Boudouri, 2003).

From all this, it is clear that there was some kind of tourism promotion in Greece since the beginning of the century through organizations, programmes and structures, but the situation was volatile and their plans changed from one day to the next. There was no stable organization and financial base; they did not follow the rules of a rational economic policy and were not coordinated in a concerted economic sector as required by the modernized capitalist model of the West. Perhaps it would be better to say that in those years, due to Greece's geographically isolated location and geopolitical problems rooted in the past, as well as the instability of Greek governments, the foundations of the modern Greek tourism industry were laid in a hesitant and poorly organized manner compared with the Western models.

On the other hand, core images of Greek antiquity and civilization, as well as the military successes of the Greeks (first in the expansion of the Greek state and later against the Italians and the Germans during the Second World War, which aroused admiration and great positive publicity abroad) were important aspects for a positive predisposition in favour of the country. All this was reported in foreign newspapers, newsreels and radio broadcasts, creating a complex and sometimes contradictory image of Greece abroad, which nevertheless attracted tourism.

### 16.2.1 Newspapers inspired travel

Mayer (1959) refers to the first decade of the 20th century as the golden age of Greek newspapers, since the rapid development of technical means and the progress of communication radically affected their number, content and appearance. The newspapers of this period improved greatly from the point of view of information, as the number of editors increased and the means of communication with home and abroad were developed. The numerous correspondents, the telegraph, telephone, telex, telephotography and the cooperation with worldwide news agencies contributed to inform the Greek people in a relatively short time about everything that happened in all parts of the world. Its spread increased with the growth of the Greek population due to the refugees arriving from Asia Minor, and the decline of illiteracy. Many of the newspapers founded in the previous century continued their production, but some new ones were also founded. These included: (i) *Astrapi* and *Neo Asty* both in 1901; (ii) *Esperini* and *Chronos* in 1903; (iii) *Patris* in 1905; (iv) *Makedonia* in 1911; (v) *Nea Hellas*, *Dimokratia* and *Ethnos* in 1913; (vi) *Rizospastis* in 1917; (vii) *Efimeris tis Elladas* and *Kathimerini* in 1919; (viii) *Eleftheros Typos* and *Eleftheron Vima* in 1922; (ix) *Eleftheros Logos*, *Vradyni* and *Apogevmatini* in 1923; (x) *Athinaika Nea* in 1931; and (xi) *Eleftheria* in 1944 (Mayer, 1959).

As the pages grew, new concepts were introduced, including events in faraway places, travel by celebrities, and general affairs in distant lands that caught the public's attention. Photo reports were often included that piqued readers' curiosity and created a strong desire or impulse to travel, explore new destinations, and learn about other cultures. Tourism advertisements still account for no more than 5% of all adverts in newspapers and mainly concern transportation companies, hotels and travel agencies (Kariatoglou, 2005).

In an article on the front page of the newspaper *Empros* of 5 October 1910, the editor tells of his visit to a grand old French country hotel near Paris, where one could see how much the world had changed. In the hotel's magnificent and ancient courtyard, where the bulky 'diligences', the great four-wheeled public stagecoaches, once pulled in, now parked the machine of the century: the automobile. He continued to wonder where the charioteers from the novels were: where were the mysterious men on horseback, wrapped in great cloaks, drinking wine, sleeping and setting out on a crime or duel? The car had erased all that. Now the occupants of the cars rush into the old hotel, the men in flat caps, big glasses and boots, the women just getting in, looking in the mirror and frantically fixing their hair. When they get back into the cars to drive away, the women stand up for a while and put on tank tops with aesthetic gestures . . .

In descriptions like these in newspaper articles, it is clear how images were organically constructed for tourists. By describing a new reality and modern trends in 1910s' France as he himself experienced them on his trip, the narrator creates a compelling and immersive travel experience in the minds of potential tourists. Indeed, effective narratives such as this can evoke emotion, arouse curiosity and create a desire to visit a particular destination. Be that as it may, at that time only the elite of Greek society could afford to travel abroad.

### 16.2.2 Writers promoted modern Greece

During these years several Greek writers were inspired to work with the aim of aestheticizing the politics of national culture and promoting the autochthonous aspects of Greek culture. At the same time they were trying to bring modern Greece closer to the cultural scene of Western Europe. This included personalities such as Konstantinos Kavafis (1863–1933), Nikos Kazantzakis (1883–1957), Angelos Sikelianos (1884–1951), Stratis Myrivilis (1890–1969), Giorgos Seferis (1900–1971), Andreas Embirikos (1901–1975), Elias Venezis (1904–1973), Angelos Terzakis (1907–1979), M. Karagatsis (1908–1960), Dido Sotiriou (1909–2004), Yiannis Ritsos (1909–1990) and Odysseas Elytis (1911–1996). They have contributed decisively to the formation of a modern Greek cultural identity with their work and became famous during this period, but their fame grew greatly in the second half of the century. Of course, there are many differences between them, but what interests us here are their common meanings, which are an associative link of a common tendency that contributes to the formation of the image of Greece.

Many of them are usually associated with the advent of modernism in Greece, which has close ties with Europe. Notwithstanding the various differences between its Greek and European representatives, Greek modernism can be considered European in spirit and letter. Many of its adherents studied in Europe and were deeply involved in European literature and aesthetics (Mackridge, 2004).

But, in general, for all the authors mentioned above, the issue of tradition and cultural roots, the relationship between the nation state, national culture and cosmopolitanism, modernization, language and education were some of the objects of their interest in their efforts to identify, conceptualize and concretize the national identity and the Greek specificity (Leontaritis, 2001).

Most of them, influenced by similar tendencies of the time in the West, seek and define the elements of 'Greekness'. However, what interests us is that they are another means to establish this place as a symbol of cultural continuity and highlight the natural beauty of the land, organically building a tourist image. Moreover, they motivated many groups of neo-Greeks to take long walks in nature, as another component to gild the national identity. In this context, the creation of hiking clubs and nature-related associations experienced an unprecedented boom, a development that was also observed in the rest of Europe during the interwar period.

It should also be mentioned that many of the works of these intellectuals were translated and published in other languages. The famous works of most of these literati and poets, which were published in large numbers and read by thousands of people all over the world, also contributed to shaping the image of Greece abroad.

In addition, there were also some well-known foreign writers of the 20th century who creatively participated in this process of constructing the tourist image of Greece with their works. For example, Henry Miller (1891–1980) with *The Colossus of Maroussi* (Miller, 1941) captures the spirit and charm of the country during his visits in the 1930s, and Robert Byron with his book *The Road to Oxiana* (1937), which is mainly about his travels in Persia, but also contains sections about Greece.

Let us now consider the complicated relationship between travel literature and travel journalism in Greece during this period. As mentioned in Chapter 9, this volume, an important difference between travel literature and travel journalism is in approach. While journalism focuses on the practical and informative side of travel, literature is about emotional, subjective and even imaginative accounts. It is therefore rather difficult to distinguish between Greek travel literature and Greek travel journalism in the early period, as many Greek literary figures were active as journalists and foreign correspondents, such as P. Nirvanas (1866–1937), Z. Papantoniou (1877–1940), S. Prokopiou (1883–1957), N. Kazatzakis (1883–1957) and S.E. Chiliadakis (1890–1955). They expressed their personal feelings in their works and helped the Greek public to form a picture of other countries. It should be noted that many other authors also described Greek destinations in their journalistic works and shaped the image of Greek places in their own way. It can be assumed that all these articles by the above-mentioned authors belong to travel literature rather than travel journalism.

Finally, we should not forget to mention some names that contributed to the historiography of Greece, such as Spyridon Vellinis (1818–1887), Yanis Kordatos (1891–1961), Ioannis Kakridis (1901–1992), Konstantinos Dimaras (1904–1992), Apostolos Vakalopoulos (1909–2000), Milton Anastos (1909–1997), Michael Sakellariou (1912–2014) and Nikos Svoronos (1911–1989). Greek historians played an important role in shaping the tourist image of Greece by highlighting its history, culture, natural beauty and cultural heritage. Their writings created a sense of fascination and admiration for Greece and attracted travellers from all over the world to explore this ancient and culturally rich country.

### 16.2.3 Contribution of modern art

The various currents of modern Greek art since the beginning of the 20th century have also helped to establish the notion and concept of 'Greekness', associated with aspects of what is considered folklore and a genuine form of Greek national identity. In this context, the art of Theophilos Chatzimichail (1870–1934) should be highlighted, which is very characteristic with works such as *Athena and Artemis* (1927), *Constantine Palaiologos* (1932) and *Alexander the Great*.

Theophilos completed an apprenticeship as a plasterer, but worked mainly as a day labourer, decorating the walls of cafés, taverns, small stores and houses with his painting. For most of his life, his art remained unrecognized, but a few years before his death, the Parisian art critic Tériade (Stratis Eleftheriadis 1897–1983) discovered him, promoted him and made him known to a wide audience (Clogg, 2021). The paintings of Theophilos were mainly inspired by life and nature, Greek landscape, Greek mythology, history and religion. In this sense, the Swiss-French architect and painter Le Corbusier, in an exhibition of Theophilos' paintings after his death in Paris in 1936, remarked, among other things, that in his work one sees the landscape and the people of Greece in the characteristic colours, the pines and olive groves, the sea and the mountains of the gods in a dangerous calm that lends itself to the prickly skirmishes of the soul (Makris, 1975).

Of course, at this point we should also mention some other painters and their works from the many that contributed to the concept of 'Greekness' and the formation of the image of Greece. Among many others, these included:

- Eleni Prosalenti (1870–1911) with works like *Sunrise*;
- Olga Prosalentis-Papadimaki (1870–1930) with *Riga's School in Zagora* (1925);
- Thalia Flora-Karavia (1871–1960) with *Landscape with Poppies* (nd) and *Bizani* (nd);
- Konstantinos Parthenis (1878–1967) with *Kaisariani* (1909) and *Around the Acropolis* (1909);
- Konstantinos Maleas (1879–1928) with *Monemvasia* (1923) and *Santorini* (1925);
- Stavros Kantzikis (1885–1958) with *On the Beach* (nd) and *King Constantine in the Balkan Wars* (1915);
- Georgios Roilos (1867–1928) with *The Poets of the Generation of 1880* (nd) and *Greek Sailors Collect the Body of Patriarch Gregory V in the Bosphorus* (nd);
- Giorgos Gounaropoulos (1890–1977) with *Fishing Boat* (nd) and *Prometheus* (1952);
- Spyros Papaloukas (1892–1957) with *Arsanas on Mount Athos* (1924) and *Pantokratoros* (1924); and
- Photis Kontoglou (1895–1965) with *Taygetus* (1927) and *Meteora* (1927).

Some European painters still painted the Greek cityscape and landscape, as well as subjects from recent Greek history, such as the Russian painter Alexander Barkoff (1870–1942) with *Thessaloniki* (1930) and *Athens Street View* (1936), and the French painter and illustrator Georges Bertin Scott (1873–1943) with *King Constantine I of Greece during the Balkan Wars* (nd) and *The Battle of Kresna* (nd).

It is also very important to mention here that in art circles it is discussed that famous foreign artists of the 20th century, who completely revolutionized the art world of their time, were ultimately inspired by Greek and Roman art. Such artists included Auguste Rodin (1840–1917), Henri Matisse (1869–1954), Amedeo Modigliani (1884–1920), Pablo Picasso (1881–1973), Giorgio de Chirico (1888–1978) and Alberto Giacometti (1901–1966). In this sense, then, we can always claim that they passed on something of the Greek spirit with their works (Prettejohn, 2012).

As for the photographic representations of the period, the Greek landscape is increasingly shown alongside the representations of the ancient monuments. In fact, in the illustrated magazines widely distributed at the time, many articles on tourism and travel are accompanied by photographs of landscapes and excerpts from the works of authors such as Karagatsis, Venezis, Kazantzakis and so on (Bonarou, 2012).

### 16.2.4 Music and the cultural image of Greece

During this period in Greece, in addition to traditional Greek music (dimotiko) and some composers of 'serious music' who followed the tradition of Western classical music, various popular music genres developed, such as rebetiko, eptanesian, Athenian and minore (Karagiorgis, 2012).

Accordingly, we can mention some personalities who have brought honour to the Greek classical music scene, such as the composers Manolis Kalomiris (1883–1962), Dimitris Mitropoulos (1896–1960), Giannis Constantinidis (1903–1984) and Nikos Skalkottas (1904–1949). Their work showed that in Greece, despite the limited economic and technical development, the creative expression of classical music fell on fertile ground and benefitted the cultural image of the country. Although these few composers may not have had the same direct impact on tourism promotion as other figures in Greek history and culture, their contributions to the world of classical music and their glorification of Greek identity and heritage have certainly helped to shape the country's cultural image and increase its attractiveness as a destination for art and culture lovers.

Among the popular musical genres that developed in Greece at this time, worthy of mention for our study is the modern bouzouki, whose origins can be traced back to the past (i.e. to similar stringed instruments of the ancient Greeks and Byzantines). However, the great spread of the bouzouki in its present form began in modern times with the rebetiko. The rebetiko was a kind of folk song of an urban subculture, in which musical instruments such as the bouzouki and the baglamas played an important role. It emerged in the second half of the 19th century and developed into its familiar form in the first three decades of the 20th century. This music was played mainly in the wider working-class environment and spread after 1922 with the arrival of refugees from the Asia Minor catastrophe by the Turks, and with the professionalization of musicians in the 1930s and the subsequent gradual commercialization of the genre. Originally, these songs were performed in the closed environment of marginalized groups that retained some common social characteristics of deviance associated with their subculture (Kotaridis, 1999; Tsertos, 2020).

Musicians such as Markos Vamvakaris (1905–1972), Michalis Yennítsaris (1917–2005), Giannis Papaioannou (1913–1972) and Dimitris Gogos (1903–1985) were instrumental in preserving and promoting this type of Greek music. Their work helped to publicize Greece's rich cultural heritage, which in turn attracted tourists who wanted to experience authentic Greek culture. However, the bouzouki enjoyed great popularity and became a symbolic element of Greek tourism a few years later, in the 1960s.

### 16.2.5 Role of cinema, films and theatre

As we have shown in Chapter 10, this volume, cinema, together with music, is a fundamental means of forming tourist images and mobilizing tourist demand. Although early cinema is one of the least explored areas of encounter between cinema and the Greco-Roman era, there are a large number of films related to antiquity that date from the beginning of the 20th century. A few examples are: (i) *Jupiter's Thunderballs* (1903 Georges Méliès); (ii) *The Return of Ulysses* (1909, André Calmettes and Charles Le Bargy); (iii) *The Fall of Troy* (1911, Luigi Romano Borgnetto and Giovanni Pastrone); and (iv) *Homer's Odyssey* (1911, Francesco Bertolini, Giuseppe de Liguoro and Adolfo Padovan). Thus, it can be said that all these works show that cinema's contribution to the formation of representations about Greece and Italy was already great in its early years. They laid the foundation for Western moviegoers to become future generations of travellers who wanted to personally explore these historical sites as destinations (Michelakis, 2017).

The first foreign film shot in Greece is said to have been an Austrian production entitled *Der Kosak und die Nachtigall* (1935, Phil Jutzi), a romantic thriller starring Jarmila Novotna, Iván Petrovich and Gerda Maurus. It was followed sporadically by other foreign productions, mainly documentaries such as *La Grèce, problème Mondial* (1948, François Villiers), *Autels des Dieux* (1948, François Villiers), but also feature films such as *Unter Heissem Himmel* (1936, Gustav Ucicky) and *Fronttheater* (1942, Arthur Maria Rabenalt), which definitely contributed to the foreign film audience forming a certain image of Greece (Katsigiannis, 2017).

After the Second World War and the Greek Civil War (1946–1949), which levelled the place to the ground, Dr Herbert Arnold Wilkinson, head of the Travel Division of the US Office of International Trade based in Washington, made a proposal to Greek politicians as part of the Marshall Plan and the strengthening of tourism in Greece. He proposed a collaboration with Spyros P. Skouras (1893–1971), then president of 20th Century Fox (1942–1962), who was interested in producing films with vivid depictions of everyday Greek life. However, nothing significant came of it (Athanasiou and Alifragis, 2012).

On the basis of this policy, it is clear that the idea of film tourism had already been invented in those years in the USA, while the Greek state practically ignored this promotional strategy at first and paid little attention to the importance of Greek cinema in general. It appears that proponents of Greece's economic development and many government officials were rather sceptical of cinema as a means of promoting Greek tourism, although the major foreign film production companies wanted to use themes about Greece and Greek mythology because they were attractive at the time and sold well at the box office. Anyway, the use of film tourism did not occur until many years later.

Despite this, surprisingly, some factors of the Greek state sometimes saw the intention to cooperate with foreign film productions to promote Greece, as indicated by Alifragis (2012) whose study of the subject revealed that at least 16 foreign productions and co-productions were filmed wholly or partially in Greece in the 1950s.

To continue our discussion in the field of performing arts, we should first mention that there was a revival of ancient Greek drama in Europe at the beginning of the 20th century (Puchner, 1993). In the spirit of this rebirth, ancient theatre became an important tourist attraction and has remained so to this day. It is worth mentioning that in 1927 and 1930 the Delphic Festival was held in the ancient theatre of Delphi, which was a unique cultural event for Greece at that time and attracted the interest of a large number of foreigners and Greeks visiting Delphi.

The transfer of the ancient drama to the cinema inspired numerous Greek and foreign producers and directors already in the interwar period. Since many of these productions participated in international film festivals and were shown worldwide, they were also groundbreaking for the tourist representations of the image of Greece in the world.

*Prometheus Bound* by Dimos Vratsanos is considered the first silent film made in Greece to treat the ancient drama. The film is based on a screenplay by Greek poet and playwright Angelos Sikelianos, inspired by the tragedy of the same name by ancient Greek tragedian Aeschylus. Filming began in 1928 and included numerous scenes of the Delphic Festival and ended in 1929. It can be argued that cultural productions such as *Prometheus Bound* have contributed to Greece's cultural image, which could have a long-term impact on tourism; in fact, ancient Greek theatre and mythology continue to be important attractions for tourists visiting Greece today (Soldatos, 1995).

### 16.2.6 Other forms of promotion of the tourist image

Although there were efforts to establish some radio stations in Greece since the early 1920s, a national radio station was established and operated in Athens in 1938, while the National Broadcasting Foundation (EIR) was established in 1945. During the dictatorship of Ioannis Metaxas (1936–1941), the organization of lectures and radio broadcasts to promote tourism was among the official tasks of the administration of the Deputy Ministry of Press and Tourism (Vlachos, 2016).

It should be emphasized that promotional activities were particularly important in those years. Such activities included the distribution of advertising brochures, film screenings, lectures, translated thematic tributes to ancient Greek theatre and popular architecture, and the presentation of tourist material on overseas ships. Tourist stands at exhibitions abroad were also an important contribution to a coordinated orientation of the 'national image'.

The invocation of the glorious past to spread and establish a good reputation for the present was common in many Western countries at this time, including the modern Greek state, as we have seen above. The same practice, however, was often insidiously employed by totalitarian regimes that looked deep into history for legitimacy for their abuses. Nazi Germany developed a myth of origin from the ancient Greeks by associating them with the Germans and attributing the same origin to both peoples. This was probably one of the ways the Nazis tried to overcome the humiliation of the Germans at Versailles in 1918 by pretending to have a much more glorious past. Thus, the Nordic origin of the ancient Greeks was confirmed by Nazi historians and ethnographers and published in various journals, not always in a scientific manner.

Several Nazi theorists such as Hans Friedrich Karl Günther (1891–1968) and Alfred Rosenberg (1893–1946) were the architects of this myth.[1] This narration was not only taught in schools in Germany, but also appeared in public ceremonies, architecture, sculptures and movies, to ensure that Nazi ideology was rooted in the past. However, as far as modern Greeks were concerned, the purity of their race was questioned by the Nazis (Tzermias, 1997; Chapoutot, 2016).

Be that as it may, it could be argued that the symbolic association of Italian fascism and German Nazism with ancient Roman and Greek civilizations through architecture, sports, and other cultural practices and products can be seen as a negative point in terms of international collective memory and the image of ancient Greece.

In addition to the typical association of Greek tourism with monuments, the then rudimentary Greek tourism industry later began to develop another feature of the image of Greece. Through the contribution and collaboration of various artists, an operation began that could be called the 'construction of Aegeanicity' (i.e. the characteristic metonymic representation of all Greek tourism with an image that presents the white houses of the Cyclades, the sea, the sand and the sun). These two metonymic representations of Greek tourism, the monuments of antiquity and the landscape of the Aegean, formed the strongest myth of Greek tourism, which is still valid today (Vlachos, 2016).

## 16.3 Promotion of Tourism by the GNTO

As far as the organized promotion of tourism and the shaping of Greece's tourist image by the tourism industry is concerned, the following discussion will focus only on the work of the GNTO, which made a decisive contribution in this direction, without, of course, claiming that the promotional activities of the private sector were irrelevant.[2] Undoubtedly, there have been numerous creative promotional activities for Greek tourism, both by Greek and by foreign tourism companies, aimed at attracting tourists to Greece by highlighting the country's natural beauty, historical sites and cultural attractions.

Anyway, since its foundation in 1929, the GNTO has made great efforts to promote tourism, with great artistic sensitivity, since one of its main objectives was to observe and influence the image of the country abroad. In the same year, it printed the first advertising poster highlighting the Propylaea and the Parthenon, an outstanding work by the internationally renowned Greek photographer Elli Sougioultzoglou-Seraidari (Fig. 16.1).[3]

The artist continued her collaboration with the GNTO, and a few decades later, in 1939, the Deputy Ministry of Press and Tourism commissioned her to decorate the Greek pavilion at the New York World's Fair with her photographs, one of the first significant efforts to promote Greek tourism in the USA (Katsigiannis, 2017).

One of the main objectives of GNTO was the systematic allocation of works to recognized artists of the time, with the aim of assembling a large collection of various artistic representations (photographs, paintings, films, etc.) for the promotion of Greece through the Greek embassies, foreign travel agencies, large foreign shipping companies, chambers of commerce, newspapers and magazines.

In addition, very early on, the organization created an archive of films that it acquired from various production companies, but also produced with its own film crew after acquiring the appropriate equipment.

In those early years, GNTO assisted film crews with their shoots on various occasions and participated in international film events and exhibitions. The organization's film archive gradually acquired small promotional footage of the Delphi Festival, the Balkan Games, and many other events and characteristic landscapes of the country.

Moreover, in 1932, GNTO contributed decisively to the realization of a large cinematographic sound film with footage from all the typical tourist areas of the country, focusing on folkloric elements, monuments and natural attractions. The following year, the American production company Fox Movietone News released the film in 600 theatres in the USA (Vlachos, 2016).

The GNTO posters initially depicted mainly picturesque Greek landscapes, antiquities, figures from Greek mythology, folk themes and historical monuments. They were created by eminent Greek painters such as Celest Polychroniadis (1904–1985), Mimis Vitsoris (1902–1945) and Michael Papageorgiou (1896–1987), and impressively decorated tourist offices abroad. In general, it can be said that the GNTO used the tourist poster intensively over a long period of time, which made it one of the most important tools for shaping Greece's tourist image.

It can be argued that the Greek tourist poster, despite the dimensions of its lifespan, was able to evoke a strong emotional involvement in foreign viewers due to its great artistic quality. Because art products can convey broader meaning in a symbolic way, form tourist myths, and lend great prestige to a destination, the

**Fig. 16.1.** Greek National Tourism Organisation (GNTO) poster, *Athens-Parthenon* (1929) photograph by Nelly's. The Parthenon has always been a key symbol of Greek tourism, featuring prominently in promotional materials and cultural representations. Its iconic status emphasizes Greece's rich history, attracts tourists and shapes the country's cultural image. Image used with permission from the GNTO.

relatively easy-to-produce GNTO posters were an ideal tool for building Greece's image abroad. In this sense, the GNTO posters became one of the most important artistic means to establish the tourist myth of Greece as a country of summer vacation and cultural tourism, and some of them are considered as works of art by many people today.

Very soon, the organization also published information brochures in four foreign languages, English, French, Spanish and German, which contained general information about travelling in Greece, hotels, spas, roads, etc.

## 16.4 Summary

To summarize, the emergence of a rudimentary tourism industry in early 20th-century Greece, initially referred to as the 'industry of foreigners', had to contend with geopolitical hurdles and a fragmented political landscape. Despite concerted efforts, Greece struggled with a lagging tourism infrastructure. The media, including newspapers and travel literature, had a major influence in shaping Greece's image, luring tourists with its cultural and historical opulence. The seminal works of famous modern Greek artists epitomize 'Greekness' through folkloric elements and promote a sense of national identity. In addition, media platforms such as photography and cinema reinforce Greece's appeal and increase its attractiveness to potential tourists. At the same time, the GNTO, together with other state agencies, played a crucial role in shaping Greece's profile as a desirable tourist destination in the first decades of the 20th century.

It could therefore be argued that the GNTO, in conjunction with the international collective memory of ancient Greek civilization and the strong image that the media organically built up in the first half of the 20th century, played a crucial role in establishing Greece as a popular and enduring tourist destination.

## Notes

[1] Hitler wrote in his book *Mein Kampf* that there was a racial unity between Greeks, Romans and Germans and that these three peoples were together in the fight against the same millennial war (Chapoutot, 2016).
[2] In 1914, the Office for Foreigners and Exhibitions was first established in the Ministry of National Economy, and in 1919 it was transformed into an independent Service for Foreigners and Exhibitions. In 1929, the Greek National Tourism Organisation (GNTO or EOT; the latter being the Greek abbreviation for Ellinkos Organismos Tourismou) was established under the supervision of the Ministry of National Economy. In 1936–1937, the GNTO was abolished, and the Ministry of Press and Tourism was established. In 1941, the Deputy Ministry was dissolved, and the Department of Spas and Tourism was established in the Ministry of National Economy. In 1945, the General Secretariat of Tourism was established. In 1950–1951, the GNTO was finally re-established. Today in Greece, the Ministry of Tourism is responsible for state tourism planning and tourism policy formulation, as well as tourism development. The GNTO is a legal entity under public law that reports to the Ministry of Tourism (Katsigiannis, 2017).
[3] Elli Sougioultzoglou-Seraidari (1899–1998), known by her nickname Nelly's, was a Greek photographer of international fame. She carried out one of the most discussed photo shoots of the Acropolis in the history of Greece, in which Mona Paeva, the dancer of the Comédie-Française, danced half-naked on the Parthenon. This was published in the French weekly of Paris *L'Illustration*.

## References

Alifragis, S. (2012) From the bright light to the hot sun II: the existing light, 1941–1969. Tourist and travel documentaries, EOT's tourist films. In: Athanasiou, A., Alifragis, S., Moussa, M. and Fokaidis, P. (eds) *Postwar Modernism, Architecture, Politics and Tourism in Greece 1950–1965*. PEVE, Athens. (in Greek), pp. 365–399.

Athanasiou, A. and Alifragis, S. (2012) Americanism and Greek tourism I: 1948–1955. The Marshall Plan and the pre-education of the Greeks in modernism. In: Athanasiou, A., Alifragis, S., Moussa, M. and Fokaidis, P. (eds) *Postwar Modernism, Architecture, Politics and Tourism in Greece 1950–1965*. PEVE, Athens. (in Greek), pp. 19–53.

Bonarou, C. (2012) *Visual Culture and Tourism: Representations of Greece on Tourist Postcards*. Papazisi Publications, Athens. (in Greek).

Boudouri, I. (2003) Photography and foreign policy (1905–1922): the contribution of the Boissonnas family. In: Stathatos, J. (ed.) *1st Conference on the History of Greek Photography*. Museum of Photography/Kythera Photographic Encounters, Thessaloniki, Greece. (in Greek).

Byron, R. (1937) *The Road to Oxiana*. MacMillan and Co., New York.

Chapoutot, J. (2016) *Greeks, Romans, Germans: How the Nazis Usurped Europe's Classical Past*. University of California Press, Berkeley, California. DOI: 10.1525/california/9780520275720.001.0001.

Clogg, R. (2021) *A Concise History of Greece*. Cambridge University Press, Cambridge, UK. DOI: 10.1017/9781108953924.

Dritsa, M. (2019) *Mosaics of Greek Tourism*. Economia Publishing, Athens. (in Greek).

Girardin, D. (1998) Frédéric Boissonnas: between tradition and modernity. *History of Photography* 22(3), 281–283. DOI: 10.1080/03087298.1998.10443891.

Karagiorgis, E. (2012) *The Greek Song of the 20th Century*. Ministry of Education and Culture, Nicosia.

Kariatoglou, E.I. (2005) Advertisements. In: Droulia, L. (ed.) *The Hellenic Press, 1784 to Today*. Athens Institute for Neohellenic Research/National Hellenic Research Foundation, Athens (in Greek), pp. 201–214.

Katsigiannis, K.X. (2017) *Hellenic Tourism Organization: Travel Through Time*. Private Edition, Athens. (in Greek).

Kotaridis, N. (1999) *Rebetes and Rebetiko Song*. Plethron, Athens (in Greek).

Leontaritis, G.V. (2001) Nationalism and internationalism: politics and ideology. In: Tsaoussis, D.G. (ed.) *Hellenism and Greekness*. Estia Bookstore, Kolaros and Co., Athens. (in Greek), pp. 27–36.

Mackridge, P. (2004) *Contemporary Greek Fiction in a United Europe: From Local History to the Global Individual*. Legenda, Oxford, UK.

Makris, K. (1975) Theofilos. In: Rayas, G. (ed.) *The Greek Painters*. Melissa Publications, Athens (in Greek), pp. 450–487.

Mayer, K. (1959) *History of the Greek Press II*. G. Serbinis-I. Makrides, Athens. (in Greek).

Michelakis, P. (2017) Greece and Rome on screen: on the possibilities and promises of a new medium. In: Pomeroy, A.J. (ed.) *A Companion to Ancient Greece and Rome on Screen*. Wiley, New York, pp. 15–35.

Miller, H. (1941) *The Colossus of Maroussi*. Colt Press, San Francisco, California.

Moussa, M. (2012) Policies for tourism: the realization of the vision for tourism through the GNTO building project. In: Athanasiou, A., Alifragis, S., Moussa, M. and Fokaidis, P. (eds) *Postwar Modernism, Architecture, Politics and Tourism in Greece 1950–1965*. PEVE, Athens. (in Greek), pp. 151–221.

Prettejohn, E. (2012) *The Modernity of Ancient Sculpture: Greek Sculpture and Modern Art from Winckelmann to Picasso*. I.B. Tauris, London.

Puchner, W. (1993) The impact of ancient Greek drama on Europe from the Renaissance to the early 20th century. In: *Greek Classical Theatre: Its Influence in Europe*. Municipality of Athens Cultural Centre, Athens. (in Greek), pp. 53–66.

Soldatos, G. (1995) *Greek Cinema: a Brief History*. Capricorn, Athens. (in Greek).

Tsertos, G.B. (2020) *Discography and live performance: the options of a professional status for bouzouki players in the decade of 1930*. Master's degree thesis. Panteion University of Social and Political Sciences, Athens, Greece. (in Greek). Available at: http://pandemos.panteion.gr/index.php?op=record&pid=iid:20394&lang=en (accessed 3 May 2023).

Tzermias, P.N. (1997) *The Image of Greece in the Foreign World*. Sideris Publications, Athens. (in Greek).

Vlachos, A.F. (2016) *Tourism and Public Policies in Modern Greece 1914–1950*. Corfu-Economia Publishing, Athens. (in Greek).

# 17 The Phenomenon of Mass Tourism in Greece

**Abstract**

After the mid-20th century, when the number of travellers to Greece was still relatively low, several state and private actors began to show some interest in developing tourism as an industry that could support the weak Greek economy. In this atmosphere, organic tourist images from literature, painting, the press, cinema and television, as well as the advertising of the developing tourism industry and the Greek National Tourism Organisation, helped to build Greece's tourist image. By the end of this period, however, mass tourism had become established in the southern part of the Balkan Peninsula and on the islands, and a relatively strong tourist image for Greece had finally emerged that satisfied the demands of tourists.

## 17.1 Transformation of Greek Tourism

To illustrate the enormous transformation that Greek tourism underwent during this period, it is worth noting that on the tourist map of a Thomas Cook advertising brochure from 1951, which listed all the European countries to which the travel agency systematically offered organized tours, Greece was not even mentioned. However, towards the end of the 20th century it was difficult to find a European travel agency that did not advertise Greece as a desirable destination for a summer holiday (Nikolakakis, 2013).

Therefore, we can clearly see that at the beginning of this period Greece was not yet one of the main destinations of the European tourism industry, if we can put it that way.

The development of Greek tourism showed a slow upward trend after 1950, when the infrastructure of communication and transport networks was relatively improved, the organized tourism business began to grow indecisively, destinations multiplied and expanded, state investment in the sector continued, and the Greek National Tourism Organisation (GNTO) expanded its work.

Due to these facts and some important socio-economic changes in post-war Europe in general, certain elements such as the sun, the sea and the pleasure of the beach were given special emphasis in the international tourism media and had a decisive influence on the image of Greece. Examples of these socio-economic changes include: (i) the emergence of an enlarged middle class accompanied by rising incomes and more leisure time; (ii) the spread of consumer culture; (iii) the continuous development of the international tourism market; (iv) the emergence of the summer holiday; and (v) the increase in narcissistic-exhibitionist behaviour on the beach.

Gradually, tourism areas expanded to other coastal regions and islands, and eventually the summer holiday model based on 'sun, sea and sand' prevailed to the detriment of cultural tourism (Tsartas, 2010).

Thus, the form of seasonal mass tourism and organized travel through travel agencies gradually took hold. At the same time, however, domestic tourism also increased and numerous private holiday homes were built for the local population, mostly near the sea.

In addition to hotel establishments, organized campsites and other accommodation increasingly sprang up. However, many room rentals were operated illegally, which was detrimental to the general image of the country's tourism services (Tsartas, 2010).


 *Communication and Tourism: Reflecting on the construction of the tourist image of Greece* (Michael Tsangaris)
DOI: 10.1079/9781800626034.0017

## 17.2 Discourse Projected by the Greek Mass Media

Already in the middle of this period, we can speak of a normative discourse projected by the Greek mass media (newspapers, books, cinema) beyond advertising campaigns, which shaped the new patterns of consumption and especially the practices of bathing in the sea, summer holidays, nightlife entertainment and so on. All of this, of course, reflected the values of Greek society at the time, including the rules of masculinity and femininity based on traditional stereotypes. However, once this process was completed by the Greek media, there was no lack of criticism of the 'subordinate' culture of mass entertainment as well as the ecological destruction caused by mass tourism (Nikolakakis, 2013).

### 17.2.1 Newspapers' influence on tourism imagery

The circulation of Greek newspapers increased steadily, reaching its peak in the mid-1980s and 1990s. Newspapers now had many more pages than before, and more topics were covered, first only in columns, then in full pages, then in sections, then in supplements and finally in embedded magazines. The supplements and magazines covered topics such as health, lifestyle, television, history, comics, real estate, books, professions, studies and of course holidays and tourism.

Thus, not only did the press become a platform for organic forms of tourist imagery with articles and photos of various faraway places, but tourism advertising itself developed in its pages to a much higher level than before. As early as the 1970s, tourism advertising even appeared on the front page of serious newspapers. An example is the 10 May 1978 issue of *Kathimerini* with the announcement of 'package tours' inviting Greeks to get to know 4000-year-old China, and with the indication that this fascinating historical country, which had only recently welcomed travellers, would offer them countless impressions.

### 17.2.2 Literature and the appeal of Greece

Over time, as intellectual contexts evolved differently and the conditions for the production and reception of literary works also changed, new generations of writers appeared in Greece. They wrote in various forms and modes of the Greek language about the dynamics of life and the power of existence, sometimes with a strong political character, sometimes poetically, describing near and far places and interesting journeys. As mentioned in the previous chapter, there were many Greek writers who became famous with their works and managed to be translated into different languages and read all over the world. Some of them were honoured with the Nobel Prize for Literature during this period, while many others were considered contenders for this prize for a while, and although they did not receive it, they were recognized in one way or another with other awards as bearers and shapers of Greek culture.

Nevertheless, it is important to emphasize the exceptional Greek writers and poets whose fame at this time transcended national borders. Giorgos Seferis (1900–1971) received the Nobel Prize for Literature in 1963 in recognition of his exceptional lyrical achievements, while Odysseas Elytis (1911–1996) was also honoured with this prestigious award in 1979. Nikos Kazantzakis (1883–1957), on the other hand, did not receive this honour during his lifetime, although he was nominated for the Nobel Prize for Literature several times.

Seferis is a very famous Greek poet with a relatively rich literary presence, translated several times into many languages; he was a Nobel laureate and had a varied activity. One of the main themes of his poetic work is the Greek historical tradition. The recourse to local traditions and the historical past with strong influences of modernity and the connection of Greek mythology with contemporary Hellenism are the reference points of his poetry. Seferis may not have played a direct role in promoting Greek tourism in the traditional sense, but his poetic works have contributed to the cultural and literary appeal of the country. His writings continue to attract people interested in Greek culture, history and literature to explore the beauty and importance of Greece as a tourist destination.

But Elytis, another in the long list of Greek national poets whose entire oeuvre reflects the image of the modern Greek, also deserves special mention. In the 1970s, Elytis became known outside Greece when translations of his poems appeared in many languages. This culminated in the award of the Nobel Prize for Literature in 1979, and his creative genius became for Greeks a symbol of their cultural identity. Elytis' poems are dominated by the sun, the sea, ancient monuments and the landscape of the Aegean, which reinforce the image of the region where the new Greek state was founded. For Elytis, however, Greece is, among other things, a lively and dynamic feeling, a way of conceiving and expressing life that is crucial for the modern individual (Carson, 2004; Linardaki and Aslanides, 2020).

Kazantzakis is indeed one of the most internationally recognized Greek writers. He was a tireless explorer of time and space, but also a passionate explorer of the mind and soul. Before his main career as a literary author, he began writing travel novels. In his travel writings, he describes places, people, events, stories and situations, focusing on ideological, cultural, historical, anthropological and social themes of each country to illustrate the complex interactions between place, time and the individual. As he used to say, every new country we enter can and should be an opportunity for our soul to expand (Mathioudakis, 2013).

Kazatzakis played a crucial role in shaping the image of modern Greek culture in Western countries and internationally. His novel *The Life and Times of Alexis Zorba* (Kazatzakis, 2017), to take just one example, is one of the most famous Greek books. It won the prize for the best foreign novel in France in 1954 and was on the bestseller list in many countries.

Zorba, following the urge to exist, is the type of true man who ecstatically lives the meaning of life, full of curiosity, always on the move because he is constantly taking a new look at himself and the whole world. According to the French critic of *Le Monde* newspaper, Marcel Brion, Zorba is a child of nature, overflowing with dance and life, embodying the wanderer without being a social rebel (Agathos, 2017).

The charm of the Mediterranean landscape with its white houses, enchanting sea and the tantalizing scents of the region, coupled with the rich history of Crete, its vibrant culture and diverse people, who are sometimes friendly, sometimes talkative, sometimes reserved, sometimes angry, leads the reader of *The Life and Times of Alexis Zorba* to a nuanced understanding that the portrayal of Greece is complicated and full of contradictions. In this literary work, every facet, from the content of the text to the settings mentioned to the characters and cultural nuances of the inhabitants, contributes significantly to painting the multifaceted picture of Greece at the time.

But if we think about it, the personality of Zorba himself, based on the insatiable desire for life and movement, on the dominance of the senses and on curiosity, which in a certain sense are directly related to spontaneous wandering and travelling for pleasure, is also as such part of the general spirit of tourism with the meaning we give to this word in this book.

Of course, there are many other Greek writers and poets whose works were read abroad during this period, such as: (i) Stratis Myrivilis (1890–1969); (ii) Melpo Axioti (1905–1973); (iii) Yiannis Ritsos (1909–1990); (iv) Stratis Tsirkas (1911–1980); (v) Miltos Sachtouris (1919–2005); (vi) Eleni Vakalo (1921–2001); (vii) Tasos Leivaditis (1922–1988); (viii) Alki Zei (1925–2020); (ix) Manolis Anagnostakis (1925–2005); (x) Costas Taktsis (1927–1988); (xi) Dinos Christianopoulos (1931–2020); (xii) Kiki Dimoula (1931–2020); (xiii) Nikos Dimou (1935–); (xiv) Giorgos Ioannou (1927–); and (xv) Petros Markaris (1937–). All of these contributed in their own way to shaping the image of Greece.

In the last decades of the 20th century, there was a tendency among some writers, such as Dimitris Nollas, Neni Efthymiadi and Alexis Stamatis, to develop special relationships between Greeks and foreigners in some stories that go beyond the usual geographical boundaries of Greece. As part of this tendency, some books were written about journeys to the Balkans, to Europe, but also to the rest of the world. However, these journeys did not take place within the framework of a sterile cosmopolitanism aimed at impressing Greek readers with foreign places, because the characters' experiences abroad usually seem to be the result of a deep contact with the 'other' and not just a simple tourist contact. Europe, in particular, was not treated in most of these narratives as a foreign destination, but as a larger homeland where the Greek characters felt physically, culturally and ideologically at home. In these narratives, the Greek protagonists feel that they belong to Europe, know its culture well and, as descendants of the ancient Greeks, feel connected to the people who all contributed to European civilization (Mackridge, 2004).

At this point we should also mention some foreign literary figures who characterized Greece in their works and contributed to shaping the international image of the country at that time.

Quite well known are the travel books of the British writer Lawrence Durrell (1912–1990), who, among other places of special beauty, vividly described the Greek islands. In his books *Prospero's Cell* (1945), *Reflections on a Marine Venus* (1953), *Bitter Lemons* (1957) and *The Greek Islands* (1978), Durrell portrays Corfu and Rhodes, where he lived for years with his wife, as well as the other Greek islands. His descriptions of the islands contributed significantly to the formation of the foundations of travel stereotypes about Greece and enriched the early mental models that later led to the elementary tourist representations of this part of the Mediterranean (Butler, 1990).

Durrell was not the only one to speak of the beauty of the Greek region at this time. Rather, there were a number of well-known writers of the 20th century who creatively participated in the construction of the tourist image of Greece with their works. This includes: (i) Jacques Lacarrière (1925–2005) with *L'été Grec: une Grèce Quotidienne de 4000 Ans* (Lacarrière, 1975); (ii) Charmian Clift (1923–1969) and George Johnston (1912–1970) with *The Sponge Divers* (Clift and Johnston, 1992); (iii) Leigh Fermor (1915–2011) with *Mani: Travels in the Southern Peloponnesus* (Fermor, 2004); and (iv) John Fowles (1926–2005) with *The Magus* (Fowles, 1965). Examples of more recent works are Richard Stoneman's *A Literary Companion to Travel in Greece* (1994), Christa Wolf's *Cassandra: a Novel and Four Essays* (1984) and Storace (1996) *Dinner with Persephone: Travels in Greece*, some of which attempt to demystify the legacy of the Greek 'ideal' as the origin of the West and rewrite it under other aspects such as gender, orientalism and environment (Kolocotroni and Mitsi, 2008).

### 17.2.3 Cultural and natural beauty perceived through art and music

Greek painters continued to experiment with themes related to Greek nature, the city, Greek mythology, Greek history, Greek identity and modernization. Artists, such as those listed here, contributed from their perspective to the composition of the multifaceted image of Greece:

- Nikos Egonopoulos (1907–1985) with *Au rendez-vous Allemand* (1963) and *Olympia* (1970);
- Nikos Hadjikyriakos-Ghikas (1906–1994) with *Party by the Sea II* (1970–1979) and *Propylaea* (nd);
- Yannis Tsarouchis (1910–1989) with *Small Coffee House in Athens* (1955) and *Landscape with a Neoclassical House and a Sailor* (1965);
- Yiannis Spyropoulos (1912–1990) with *Mykonos* (1955);
- Yiannis Moralis (1916–2009) with *Ancient Athens* (1966) and *Aegean Landscape* (1976);
- Dimitris Mytaras (1934–2017) with *View of Greece* (1970) and *Composition in Ionic Style Composition* (1972); and
- Alekos Fassianos (1935–2022) with *Athenian Panorama* (1989).

Even if these artists did not primarily want to promote tourism, their artistic contributions have played an important role in the perception of Greece and its cultural and natural beauty.

As for photographic representations, the focus shifted to the sea and in particular to the Aegean, and 'Aegeanicity' was now decisively introduced into international tourist thinking.[1]

At the same time, the photographic representation of Greece as the cradle of Western civilization continued, with images of ancient monuments such as the Acropolis, Olympia, the Temple of Poseidon at Sounio and Delphi superimposed on the screen. In the late 20th century, however, photographers began to reject stereotypical representations of picturesque Greece, and showed a technical and aesthetic sophistication and experimentation in depicting the Aegean landscape in terms of architectural details, geometric shapes, abstract frames and intense colour contrasts.

In addition to the various types of Greek music already mentioned in Chapter 16, this volume (dimotiko, rebetiko, minore, etc.), there were major musical developments during this period with genres such as laika, entechno, neo kyma and Greek rock.

The bouzouki in particular enjoyed great popularity among Greeks from the 1960s onwards and later also became a symbolic element of Greek tourism, appreciated and sought after by tourists and an essential part of Greece's tourist image. Nowadays, the bouzouki plays an important role in promoting Greek tourism by serving as a cultural symbol, providing entertainment and offering unforgettable experiences to tourists. Its connection with Greek identity and traditions makes it an integral part of the cultural landscape, attracting visitors from all over the world.

Of course, there are many high-quality songs that have become tourist hits, such as 'Never on Sunday' (Hatzidakis 1960) and 'Zorba's Dance' (Theodorakis 1964). These are from the films *Never on Sunday* (1960) and *Zorba the Greek* (1964), respectively (discussed below in Section 17.2.4 'Greek films increase tourist interest'). Also many lighter songs, such as 'Disco Bouzouki', written by Daniel Vangarde and Jean Kluger, have had great success and are now strong tourist indicators for Greece.

Be that as it may, we should mention some of the personalities in the various music scenes in Greece, as they have in their own way made an important contribution to Greek music and shaped the image of Greek tourism. Among many others, these are: (i) Iannis Xenakis (1922–2001); (ii) Maria Callas (1923–1977); (iii) Vassilis Tsitsanis (1915–1984); (iv) Mimis Plessas (1924–); (v) Manos Hatzidakis (1925–1994); (vi) Mikis Theodorakis (1925–2021); (vii) Nana Mouskouri (1934–); (viii) Giannis Spanos (1934–2019); (ix) Stavros Xarchakos (1939–); (x) Vangelis Papathanassiou (1943–2022); (xi) Dionysis Savvopoulos (1944–); (x) Demis Roussos (1946–2015); and (xi) Thanos Mikroutsikos (1947–2019). Also rock bands have had an influence, such as Aphrodite's Child (1967–1972), Pavlos Sidiropoulos with Spyridoula (1976–1979), Sharp Ties (1980–1992), The Last Drive (1983–1992, 2007–) and Trypes (1983–2001). In all these cases, the artists have left their mark on the musical and cultural landscape of Greece, helping to create a tourist image that showcases the country's rich history, traditions and contemporary creativity through music and art. Their influence has been instrumental in attracting tourists who want to experience Greek culture first hand.

### 17.2.4 Greek films increase tourist interest

The period from 1955 to 1973 is often referred to as the 'golden age' of Greek cinema, as not only did film production boom during this period, but also many of the most important Greek films were produced.

Greek cinema, which was aimed at broader sections of society, not only provided material for dreaming, but also reflects and reveals today the behaviours and attitudes that functioned in the context of Greek society at that time (Rouvas and Stathakopoulos, 2005).

Soldatos (1995), a scholar in the field of Greek cinema, emphasizes the important role played by the Greek film industry in the promotion of Greek tourism worldwide. Despite the remarkable contributions of the film industry, Soldatos notes that there has been a lack of proactive efforts by the state to strengthen it through essential measures and sufficient subsidies.

Be that as it may, there were certainly some notable actions that helped to attract international attention, such as the Greek Film Festival of Thessaloniki, founded in 1960, and many others that followed, such as the International Festival of Short Fiction Film (1978) and the Mediterranean Festival of Young Filmmakers of Larissa (1992). These organizations were undoubtedly important in attracting tourists from Greece and abroad and contributed to the international promotion of the country.

Summarizing the history of Greek cinema, one can highlight some of the Greek productions that have enriched the image of Greece through their participation in international film festivals or their nomination for the Academy Award for Best Foreign Language Film. Characteristic examples are the films:

- *The Counterfeit Coin* (1955, Giorgos Tzavellas);
- *Stella* (1955, Michael Cacoyannis);

- *A Girl in Black* (1956, Michael Cacoyannis);
- *The Lake of Thinking* (1958, Giorgos Zervos);
- *Girls in the Sun* (1968, Vassilis Georgiadis);
- *The Travelling Players* (1975, Theodoros Angelopoulos);
- *Iphigenia* (1977, Michael Cacoyannis);
- *The Man with the Carnation* (1980, Nikos Tzimas);
- *From the Edge of the City* (1998, Constantine Giannaris); and
- *Peppermint* (1999, Kostas Kapakas).

Nevertheless, it could be argued that some other Greek films seem more 'touristy' because they were shot in Greek hotels and on sandy beaches, such as: (i) *Marina* (1947, Alekos Sakellarios), shot in Hydra; (ii) *Vacations in Aegina* (1958, Andreas Lambrinos), shot in Aegina; (iii) *Rendezvous in Corfu* (1960, Dimis Dadiras), shot in Corfu; (iv) *Some Like It Cool* (Giannis Dalianidis, 1963), shot in Athens; (v) *Kiss the Girls* (1965, Giannis Dalianidis), shot in Athens, Rhodes, Hydra and New York; (vi) *Mermaids and Tramps* (1968, Giannis Dalianidis), shot in Hydra; (vii) *Apollo Goes on Holiday* (1968, Giorgos Scalenakis), shot in Athens, Epidaurus, Nafplion, Corfu and Andros; and (viii) *The Countess of Corfu* (1972, Alekos Sakellarios), shot in Corfu.

As Papadimitriou (2000) notes, the images that characterize the many popular Greek films of the 1950s and 1960s are those of blue skies and clear Greek seas, sailboats, taverns, sirtaki dancing men, bikini-wearing women, white-washed villages, narrow streets, sunsets and nightclubs (Fig. 17.1).

However, the film *Never on Sunday* (1960), a romantic comedy starring Melina Mercouri (1920–1994) and directed by Jules Dassin (1911–2008), became a legend that fired the international imagination of modern Greece. The film won the award for best female performance at the 1960 Cannes Film Festival and received the Oscar for best song for its musical theme, namely the song 'The Children of Piraeus' composed by Manos Hadjidakis. This song, 'The Children of Piraeus', became something like the national melody of Greece, a poetic composition with music that evoked the image of Mercouri and consequently the image of Greece in the minds of foreign listeners. All these awards, as well as the international success of the film, contributed to its international recognition as a symbol of post-war Greece.

Another film production that made an important contribution to the international perception of Greece was the film *Zorba the Greek* (1964). The film was a Greek-British production, inspired by Kazantzakis' book *The Life and Times of Alexis Zorba* (2017; first published in 1946) and directed by Michael Cacoyannis. Several famous actors took part, but the main roles were played by Anthony Quinn, Alan Bates, Lila Kedrova, Irini Pappas, Sotiris Moustakas and George Foundas. The film was a huge success worldwide, winning three Oscars and is considered the film that brought Crete and Greece in general to the forefront of international tourism interest. The film music composed by Mikis Theodorakis, to which Giorgos Provias choreographed the 'sirtaki', is now considered one of the most characteristic dances of modern Greece and one of its international tourist landmarks.

According to Eva Stefani (2001), the two films mentioned above, *Never on Sunday* (1960) and *Zorba the Greek* (1964), not only created a stereotypical image of Greece, but also helped to demarcate the country as a periphery against an imaginary but all-powerful Western centre. Stefani (2001) explores this consideration through three discourses expressed in the films' texts: (i) Greece as a place of pleasure; (ii) Greeks as barbarians; and (iii) Greeks as children. In both films, the premise is similar: a Westerner visits Greece and encounters a culture that is different from his own. This encounter causes the stranger to form a picture of contemporary Greek culture and confirms the diversity of his own national identity, which belongs to the powerful Western centre.

At this point, it should be noted that despite the big words of politicians and the relevant references in the laws supporting foreign film producers shooting in Greece, there was still a certain state disinterest in the country's film industry and even more so in the concept of 'film tourism'. It was only in the 1980s, with Law 1597/1986, which refers to the protection and development of cinema and the strengthening of Greek cinematography, that the Greek state began to show a somewhat more serious interest in the film industry. Before that, since 1961, there was a corresponding law *On Measures for the Development of Cinematography in Greece*

**Fig. 17.1.** Greek National Tourism Organisation (GNTO) poster, *Euboea* (1976). The images commonly associated with popular Greek films from the 1950s and 1960s, with blue skies, clear seas, sailboats, tavernas, sirtaki dancing, women in bikinis, whitewashed villages, narrow streets, sunsets and nightclubs, blend seamlessly with the advertising images used by the tourism industry to create the enduring tourist myth of Greece. These common images not only reflect the cinematic portrayal of Greek culture and lifestyle during this period, but also illustrate the influential role that film and tourism imagery has played in shaping and perpetuating the idealized perception of Greece as a picturesque and welcoming destination. Image used with permission from the GNTO.

(Law 4208), which theoretically allowed foreign productions to shoot in Greece, as it was assumed that they would make the country known abroad, but this did not work as it should (Sotiropoulou, 1989).

It was only much later, with Laws 4487/2017 and 4582/2018, that the relationship between cinema and tourism was effectively supported by strengthening the production of audio-visual works in the country, creating investment incentives to attract international producers, and codifying the producers of cinema and television tourism.

Spectacular films anchored in ancient Greece such as *Ulysses* (1954, Mario Camerini), *Alexander the Great* (1956, Robert Rossen), *Helen of Troy* (1956, Robert Wise) and *Le fatiche di Ercole* (1958, Pietro Francisci) made up a large part of Hollywood and international film production in the 1950s, regardless of how accurately they portrayed the ancient world according to academic standards. Although they were not filmed in Greece, they all show a strong sense of how the Greek image was constructed (Nikoloutsos, 2017).

After the films mentioned above, a plethora of foreign productions emerged, which gradually shaped the image of Greece. These include films such as:

- *Boy on a Dolphin* (1957, Jean Negulesco) and *Celui qui doit mourir* (1957, Jules Dassin);
- *The Guns of Navarone* (1961, J. Lee Thompson), *Atlantis, the Lost Continent* (1961, George Pal) and *Colosso di Rodi* (1961, Sergio Leone);
- *The 300 Spartans* (1962, Rudolph Maté) and *It Happened in Athens* (1962, Andrew Marton);
- *In the Cool of the Day* (1963, Robert Stevens);
- *The Moon-Spinners* (1964, James Neilson) and *Topkapi* (1964, Jules Dassin);
- *Vaghe Stelle dell' Orsa* (1965, Luchino Visconti);
- *The Road to Corinth* (1967, Claude Chabrol) and *Edipo Re* (1967, Pier Paolo Pasolini);
- *Lebenszeichen* (1968, Werner Herzog), *Oedipus the King* (1968, Philip Saville) and *The Magus* (1968, Guy Green);
- *The Greek Tycoon* (1978, J. Lee Thompson), *Fedora* (1978, Billy Wilder) and Escape to *Athena* (1979, George Cosmatos);
- *For Your Eyes Only* (1981, John Glen), *Tempest* (1982, Paul Mazursky), *Summer Lovers* (1982, Randal Kleiser), *The Big Blue* (1988, Luc Besson), *Pascali's Island* (1988 James Dearden) and *Shirley Valentine* (1989, Lewis Gilbert); and
- *Mediterraneo* (1991, Gabriele Salvatore) and *Tsatsiki, morsan och polisen* (1999, Ella Lemhagen).

The cinematographic representation of authentic Greek sites or other sites modelled on the originals in all these films inscribed them in the consciousness of international audiences and formed a network of sites of ancient and contemporary Greece as visitor attractions. This network included Athens, Sparta, Marathon, Hydra, Thermopylae, Olympia, Rhodes and Crete, sites that were at the same time systematically promoted by the Greek state.

The film *Boy on a Dolphin* (1957, Jean Negulesco), starring Sophia Loren, contributed greatly to the spread of the Greek image abroad with its shots of Athens, Delphi, Corinth, Meteora, Hydra, Poros, Rhodes, Delos, Mykonos and Santorini. It became a pioneer and legend of cinema tourism for Greece. Needless to say, a statue erected in Hydra by sculptor Giorgos Xenoulis depicting a boy with a dolphin in reference to the film could also be one of Greece's first steps towards cinema tourism (Alifragis, 2012).

### 17.2.5 Television inspires travellers

The first official television broadcast in Greece took place in 1960 on the occasion of the International Exhibition in Thessaloniki, but the two state broadcasters, the National Radio Foundation (EIR, later EIRT and ERT) and the Armed Forces Television (TED, later YENED), started systematic television broadcasts between 1968 and 1969 (Hairetakis, 1998; Valoukos, 2008).

In addition to the various documentaries that presented both Greek and foreign destinations, one of the first pure travel programmes in Greece was *Tourist Greece*, which was broadcast by YENED every Sunday from 1974. This programme was intended to inform viewers about the different regions of Greece and contribute to their entertainment in an educational way.

Later other programmes in a similar style were the music programme *Musical Journey with Domna Samiou*, which was broadcast by ERT and the programme *One City – One Story*, which was broadcast by YENED from 1977. Subsequently, other programmes with similar content were broadcast on state television, such as *Another Greece* (1982), *Unknown Greece* (1983) and *Mediterraneo* (1995). *Mediterraneo* was an international co-production broadcast by European public broadcasters, including RAI (Italy), France 3 (France), TVE (Spain) and ERT (Greece), that presented themes from different countries in the Mediterranean region to show the intercultural influences resulting from the coexistence of peoples in this particular region, the cultural and social mixing and the way of life of the people there. Characteristic of this period of tourist television programmes is that they were preferably filmed with outdoor scenes (Pliakas *et al.*, 2018).

The liberalization of broadcasting in Greece after 1989 brought significant changes to the landscape of mass communication with the introduction of numerous private television channels such as Mega, Ant1, New Chanel and SKAI. It seems that the programme *Akounda Mantata* (Mega) in 1998 was the first tourism programme on private television in the strict sense (i.e. excluding documentaries about monuments, festivals and places produced by the private sector). This programme moved away from the more serious style of state broadcasts towards infotainment (Pliakas *et al.*, 2018).

Since 1999, with programmes such as *Travelling with Maya Tsoli* (ERT), a new style has developed that can be described as travel journalism, albeit in a hybrid form that can always be described as infotainment.

In the area of television advertising for tourism, it should be emphasized that in the initial phase, commercials showing the Greek landscape and culture were broadcast sporadically in various countries. These commercials, which were mainly from travel agencies or transport companies, played a crucial role in promoting Greek tourism and shaping the country's tourist image. However, it is important to mention that these efforts lacked a consistent and organized approach as they were carried out sporadically instead of forming a continuous and systematic promotional strategy.

A prime example of the ingenuity of these television adverts is an Olympic Airways commercial that was broadcast in various countries in 1967. This captivating advert, which was repeated in various versions over the following decade, depicts a fascinating scenario. A charming stewardess serves authentic Greek cuisine, wine and fruit to a distinguished gentleman on board, creating an atmosphere of genuine pleasure. As the gentleman becomes increasingly animated and breaks out into a spontaneous sirtaki dance, lively Greek music plays in the background. The narrator's voice accompanies the scene and says:

> Every night, Olympic Airways' super-fan jets fly to Europe and the Middle East. On board there's plenty to do, we planned it that way . . . after all, Olympic is the airline of Greece: there's 21 club meals, one of the largest wine cellars in the air, lively passengers, first-class movies and seven stereo channels, including one with vibrant Greek music. If you are flying our way, call your travel agent or Olympic Airways, it's a party all the way to Paris, Rome, Athens, Cairo or Tel Aviv, but just one thing we ask you: please no dancing in the aisles.
>
> (Olympic Airways, 1967)

Greek television also regularly broadcast adverts inviting viewers to 'discover Greece' or take part in various events, including the wine festivals in Dafni, Attica or Alexandroupoli in Thrace and the festival of Epidaurus, to name but a few. Although it is difficult to quantify the exact impact of these audio-visual messages on the public, it is undeniable that, in conjunction with cinematic films showing people enjoying their holidays in glamorous or picturesque settings, they had a significant impact on the development of society's perception of certain types of holidays. These portrayals in the media played a crucial role in the formation of enduring social and cultural stereotypes that have remained virtually unchanged to this day.

## 17.3 Advertising by the Greek Tourism Industry

As far as the overall advertising of the Greek tourism industry is concerned, as early as 1978 the category 'Tourist Travel' was among the ten product/service categories with the highest advertising expenditure, while TV accounted for 10% of advertising expenditure in the tourism sector (Hairetakis, 1997, p. 437, p. 441).

In the late 1990s, tourism-related businesses such as hotels, travel agencies and tour operators began to present themselves on the Internet through websites and simple online advertising. These early websites often contained basic information about accommodation, destinations and contact information. However, online advertising, social media marketing and other digital marketing techniques as we know them today were not yet mature.

As far as government tourism advertising efforts are concerned, the first properly organized advertising campaign for Greek tourism abroad was unveiled in 1951. This is when the GNTO, at the suggestion of the Economic Cooperation Administration's Special Mission to Greece under the Marshall Plan, hired the American advertising agency Cone and Belding Co. to place advertisements in magazines and the daily press in Europe and Egypt. In the same year, the GNTO's first paid advertising article appeared in the *New York Times*, followed by several more in various European newspapers and magazines in the following years (Athanasiou and Alifragis, 2012; Katsigiannis, 2017).

The tourist image of Greece that the emerging tourism industry tried to convey in the post-war years can be very well explained by the words of the head of the GNTO's Advertising Department (1955), Costas Dimitriadis. He prescribed that Greece's tourist advertising publications must, above all, have something of the warmth of the Mediterranean sun, something of its deep blue sky, something of the aroma of thyme and of the ancient Greek spirit, so that they immediately attract the special interest of the foreigners who leaf through them (Vlachos, 2016).

The basic idea was that state advertising should target the category of wealthy visitors from Europe and the USA. Visiting the country was to flatter those who wanted to be seen as rich and civilized, and the advertising was to encourage those who wanted to be seen as such. In other words: in the absence of a mass tourism infrastructure, the first advertisements were aimed at a few but wealthy visitors; they thus used the logic of conspicuous consumption and reflected the social and psychological dynamics of the growing consumer society, in which individuals demonstrate wealth and superiority within a social hierarchy through their purchasing decisions. Nevertheless, this practice was not successful in the case of GNTO, as Greece is now internationally classified mainly as a middle- and low-income country, mainly due to the summer holidays (Tsartas, 2010; Katsigiannis, 2017).

During the same period, the production of GNTO's advertising posters reached remarkable heights thanks to the contributions of renowned artists and visual designers. These included Georgios Vakirtzis (1923–1988), Yiannis Moralis (1916–2009), Louisa Montesantou (1917–2007), Michalis Katzourakis (1933–), Agni Katzourakis (1936–), Freddy Karabott (1924–2011), Dimitris Mytaras (1934–2017), Panagiotis Tetsis (1925–2016), Alekos Fassianos (1935–2022) and Theodoros Pantaleon (1945–). Along with numerous others, they built on the themes of the artists mentioned in Chapter 16, this volume, but also created something new. It is noteworthy that the foreign-language posters of this era first appeared in English and French and later in German. Many of these posters introduced progressive graphic concepts and embraced the aesthetic principles of modernism and abstraction. They featured modern, abstract and flat compositions that cleverly blended ancient, Byzantine and folk traditions with minimalism and occasionally included detailed colour photography.

In addition to all the GNTO's posters, the organization has also issued colour maps of Greece for tourists, calendars with beautiful photos in line with its tourism policy, promotional stamps, records with typical Greek music for tourists and much more (Katsigiannis, 2017).

As early as the 1960s, the GNTO published the album *Tourism in Greece* every year, which, in addition to artistic photographs of archaeological sites, typical landscapes and everyday life in the country, also contained an appendix with useful information (e.g. road maps, information about museums and archaeological sites, useful telephone numbers and artistic events). This annual edition was discontinued in the early 1990s. However, the organization continued to publish various thematic albums (Katsigiannis, 2017).

As far as film productions are concerned, it can be said that the GNTO generally either bought ready-made films to promote Greece from foreign or Greek producers and directors, partially financed some productions, or eventually commissioned someone to produce them and fully supported them.

By the mid-1960s, the GNTO had already begun to work systematically in the field of television and film promotion. Their short films testify to the organization's creativity in promoting tourism. Among others, these short films include:

- *Skiathos* (1963, Yiannis Panagiotopoulos);
- *Argolis* (1964, Roussos Koundouros);
- *Crete* (1964, Roussos Koundouros);
- *The Girl Game* (1967, M. Stavrou and V. Drimaropoulos);
- *Memories from Greece* (1967, Francis Carabott);
- *The Fairies of Greece* (1968, Francis Carabott);
- *Festivals in Greece* (1969, Aristides Karydis-Fucks);
- *Thunder of the Gods* (1969, Ann Todd-Pox Simpson);
- *Thassos* (1970, Takis Kanellopoulos);
- *Athens, the City of Eternal April* (1971, Angelos Stefanou);
- *Greece, the Affordable European Vacation* (1972, William Wurtzel); and
- *Spleil d' Hiver* (1974, Giuliano Tomei).

It is also worth mentioning the organization's support for the production of two films by Jacques Cousteau (1910–1997), *Calypso's Quest for Atlantis* Parts 1 and 2 (1977), as part of his work *The Cousteau Odyssey*, the projection of which was very important in building the image of the country as surrounded by the sea.

Most of these films show the sun, the sea and the natural beauty of the Greek landscape, the monuments of antiquity, the picturesque life of the people in the countryside, but also the attractiveness of modern Athens (Alifrangis, 2012).[2]

In the last decade of the century, the promotion of Greek tourism gradually became better organized and more coordinated advertising campaigns emerged, such as 'Greece: Chosen by the Gods' (1991–1993) or 'Greece: the Authentic Choice' (1997). These were presented in a uniform way in terms of slogan, theme or colours in press releases, posters, spots on TV, radio spots and, a little later, on their websites.

## 17.4 Summary

To summarize, art and literature have played a crucial role in shaping the image of Greece as a tourist destination, an image that has lasted well into the 20th century. Writers such as Seferis, Elytis and Kazantzakis, as well as painters and photographers, have expressed the appeal of Greece through cultural narratives and visual representations. The ancient cultural richness of Greece, the Mediterranean landscape, combined with the evolving literary and artistic expressions, increased the global appeal of Greece and crystallized the typical tourist stereotypes of the country. The influence of the media on Greece's tourist image was also profound, particularly through music, cinema and television. The bouzouki, which became popular in Greek music from the 1960s onwards, became a symbol of Greek tourism. Films such as *Never on Sunday* (1960) and *Zorba the Greek* (1964) played an important role and created a lasting perception of Greece abroad. Television programmes, documentaries and advertising campaigns continued to shape the image of the country by showcasing its natural beauty, excellent weather, cultural heritage and modern attractions. These media portrayals have been instrumental in attracting tourists and creating lasting stereotypes about holidays in Greece.

## Notes

[1] In general, a change took place in the representation of the Greek tourist space in the promotional publications following the 'construction of Aegeanicity', which gradually foregrounded the sun, the sea, the Greek beaches and the white island houses, pushing the historical monuments into the background (Nikolakakis, 2013).

[2] However, the limited number of these films cannot be compared with the large number of non-archival productions about Greece at the time that were made unofficially by other institutions or amateurs; or even the audio-visual material produced by Greek and foreign cultural organizations to document the historical monuments of the country. In particular, there were numerous foreign film productions on the subject of Greek culture. These included: (i) educational and scientific documentaries by British Pathé; (ii) productions by the United Nations Relief and Rehabilitation Administration; (iii) educational films by the American Secret Service; (iv) propaganda films for the Marshall Project (European Film Unit); and (v) other productions by foreign archaeological schools operating in Greece. Much of this material was shot by internationally renowned directors such as John Fernhout, Carl Dudley, Basil Wright and others (Alifragis, 2012).

## References


Agathos, T. (2017) The timeless charm of Kazantzakis' Zorba. In: Kazatzakis, N. (ed.) *The Life and Times of Alexis Zorba*. Ekdoseis Kazatzaki, Athens, Athens (in Greek), p. XIII–XLVII.

Alifragis, S. (2012) From the bright light to the hot sun II: the existing light, 1941–1969. Tourist and travel documentaries, EOT's tourist films. In: Athanasiou, A., Alifragis, S., Moussa, M. and Fokaidis, P. (eds) *Postwar Modernism, Architecture, Politics and Tourism in Greece 1950–1965*. PEVE, Athens. (in Greek), pp. 365–399.

Athanasiou, A. and Alifragis, S. (2012) Americanism and Greek tourism III: 1947–1964. Road Tourism and Modern Architecture. In: Athanasiou, A., Alifragis, S., Moussa, M. and Fokaidis, P. (eds) *Postwar Modernism, Architecture, Politics and Tourism in Greece 1950–1965*. PEVE, Athens. (in Greek), pp. 107–149.

Butler, R.W. (1990) The influence of the media in shaping international tourist patterns. *Tourism Recreation Research* 15(2), 46–53. DOI: 10.1080/02508281.1990.11014584.

Carson, J. (2004) Introduction. In: Elytis, O. (ed.) *The Collected Poems of Odysseus Elytis*. Johns Hopkins University Press, Baltimore, Maryland, pp. xv–xliv.

Clift, C. and Johnston, G. (1992) *The Sponge Divers*. Angus and Robertson, Sydney, Australia.

Durrell, L. (1945) *Prospero's Cell*. Faber, London.

Durrell, L. (1953) *Reflections on a Marine Venus*. Faber and Faber, London.

Durrell, L. (1957) *Bitter Lemons*. Faber, London.

Durrell, L. (1978) *The Greek Islands*. Viking, New York.

Fermor, P.L. (2004) *Mani: Travels in the Southern Peloponnese*. John Murray, London.

Fowles, J. (1965) *The Magus*. Jonathan Cape, London.

Hairetakis, M. (1997) *Television and Advertising: the Greek Case*. Sakkoulas, Athens. (in Greek).

Hairetakis, M. (1998) The penetration of TV and its audience in Greece. In: Navridis, K., Dimitrakopoulos, G. and Paschalidis, G. (eds) *Television and Communication*. Paratiritis, Thessaloniki, Greece. (in Greek), pp. 197–208.

Katsigiannis, K.X. (2017) *Hellenic Tourism Organization: Travel Through Time*. Private Edition, Athens. (in Greek).

Kazatzakis, N. (2017) *The Life and Times of Alexis Zorba*. Ekdoseis Kazatzaki, Athens. (in Greek).

Kolocotroni, V. and Mitsi, E. (eds) (2008) *Women Writing Greece: Essays on Hellenism, Orientalism and Travel*. Rodopi, Amsterdam. DOI: 10.1163/9789401206440.

Lacarrière, J. (1975) *L'été Grec: une Grèce Quotidienne de 4000 Ans*. Plon, Paris. (in French).

Linardaki, C. and Aslanides, A. (2020) From poem and song to cultural diplomacy: challenges and opportunities for place branding and tourism promotion. *Place Branding and Public Diplomacy* 16(4), 304–315. DOI: 10.1057/s41254-020-00176-1.

Mackridge, P. (2004) *Contemporary Greek Fiction in a United Europe: From Local History to the Global Individual*. Legenda, Oxford, UK.

Mathioudakis, N. (2013) A literary journey to England – introduction. In: Kazatzakis, N. (ed.) *Travelling – England*. Ethnos Publications, Athens. (in Greek), pp. 5–25.

Nikolakakis, M. (2013) Tourism and Greek society in the period 1945–1974, Doctoral dissertation, University of Crete, Rethymno, Greece. (in Greek).

Nikoloutsos, K.P. (2017) The resurgence of epics in the 1950s: classical antiquity in post-war Hollywood. In: Pomeroy, A.J. (ed.) *A Companion to Ancient Greece and Rome on Screen*. Wiley, Hoboken, New Jersey, pp. 91–117. DOI: 10.1002/9781118741382.
Olympic Airways (1967) Olympic Airways, Nightly to Europe and the Middle East. Television advertisement. Available at: https://www.youtube.com/watch?v=BPkk3gjBVjQ (accessed 21 June 2023).
Papadimitriou, L. (2000) Traveling on screen: tourism and the Greek film musical. *Journal of Modern Greek Studies* 18(1), 95–104. DOI: 10.1353/mgs.2000.0015.
Pliakas, T., Triantafyllaki, V., Galatsopoulou, F. and Kenderelidou, K. (2018) Fifty years of television with travel impressions and experiences. In: Vamvakas, V. and Paschalidis, G. (eds) *Fifty Years of Greek Television*. Epikentro, Thessaloniki, Greece (in Greek), pp. 407–422.
Rouvas, A. and Stathakopoulos, X. (2005) *Greek Cinema Volume B*. Greek Letters, Athens. (in Greek).
Soldatos, G. (1995) *Greek Cinema: a Brief History*. Capricorn, Athens. (in Greek).
Sotiropoulou, C. (1989) *Greek Cinema 1965–1975: Institutional Context – Economic Condition*. Themelio, Athens. (in Greek).
Stefani, E. (2001) Cinema's ideological construction of the periphery. In: Moutsopoulos, T. (ed.) *The Hybrids of Globalization, City and Mass Culture in the Periphery*. Futura/a-station, Athens. (in Greek), pp. 217–230.
Stoneman, R. (ed.) (1994) *A Literary Companion to Travel in Greece*. Getty Publications, Los Angeles, California.
Storace, P. (1996) *Dinner with Persephone: Travels in Greece*. Pantheon Books, New York.
Tsartas, P. (2010) *Greek Tourism Development*. Kritiki Publications, Athens. (in Greek).
Valoukos, S. (2008) *History of Greek Television*. Aigokeros, Athens. (in Greek).
Vlachos, A.F. (2016) *Tourism and Public Policies in Modern Greece 1914–1950*. Corfu-Economia Publishing, Athens. (in Greek).
Wolf, C. (1984) *Cassandra: a Novel and Four Essays*. Farrar, Straus and Giroux, New York.

# 18 The New Face of Tourism in Greece

**Abstract**
At the beginning of the second millennium, there were many changes in the media landscape and in international tourism. In Greece in particular, events such as the Olympic Games, the victory of the Greek football team in the 2004 European Football Championship, the opening of the new Acropolis Museum, the economic crisis and the negative portrayal of the country by some European media, as well as the advance of new communication technologies and social media, have had a major impact on Greece's image. Moreover, some media that traditionally provided 'organic tourist images' seem to be cooperating with the marketing of the tourism industry to promote the new tourist image of Greece. In this context, the Greek National Tourism Organisation has taken a dynamic and admirable course for the development of the image of Greek tourism through the use of new media technologies.

## 18.1 Changes in Media and International Tourism

After the beginning of the new millennium, the changes in the communication landscape and the use of new media applications by the tourism industry and tourists had a direct impact on shaping the image of Greek tourism and on the functioning of the Greek tourism industry in general.

The new conditions had a decisive impact on who is involved in the construction of tourist images and in what way, but also on the idea of mass tourism and led to the emergence of new types of tourism forms and businesses in the industry.

The development of information and communication technologies and the decentralization of tourism programmes from tourism agencies to the tourists themselves, combined with other factors (e.g. the changing preferences of travellers, the environmental and ethical considerations of society, and the desire for more diversity and meaningful travel experiences) have led many mass tourists to become 'alternative tourists'. As a result, alternative forms of tourism have been introduced and promoted in Greece, albeit with inadequate infrastructure and relatively unqualified staff.

Although the state has made many efforts to promote alternative tourism, this has been difficult as there are constant problems concerning the quality of the tourist product in Greece due to the seasonality, the uneven geographical distribution of tourist activities, the lack of infrastructure, and the relatively low level of training in this field. This fact could, of course, lead to a general deterioration of Greek tourism in an environment of international competition, especially as other tourist destination countries appeared much more prudent and adaptable to the changing tourism market.

Despite the above shortcomings, communication technology was a valuable tool for Greece to promote and support alternative tourism by raising awareness, facilitating bookings, improving sustainability practices and connecting travellers with unique and culturally enriching experiences.

While technology has enhanced the travel experience and provided travellers with a wealth of information, it has not eliminated all sources of uncertainty in international tourism, which can still be affected by a variety of factors such as economic crises, health and safety concerns, natural disasters, geopolitical tensions, issues related to immigration and refugee movements.

It is fair to say that tourists' preferences and perceptions of Greece are now very diverse, and the country appeals to a wide range of travellers, making it a popular and welcome destination for many.

Be that as it may, although upscale accommodation and amenities for higher-income

 *Communication and Tourism: Reflecting on the construction of the tourist image of Greece* (Michael Tsangaris)
DOI: 10.1079/9781800626034.0018

tourists have emerged in some regions and on some islands, and opportunities for alternative tourism have been organized, Greece remains an affordable destination for middle- and low-income summer holidaymakers, especially considering the variety of options available.

## 18.2 Greece's Tourism Image After the Turn of the Millennium

The construction of Greece's tourism image after the turn of the millennium was influenced by various events and promotional activities, but also by other facts that do not belong to the tourism industry, although in this period we see a kind of contact between the sources that generate organic and induced tourism images.

Some important factors that contributed to Greece's tourism image are: (i) the 2004 Olympic Games; (ii) the construction of the new Acropolis Museum; (iii) the trend of promoting cultural heritage and sustainable tourism, film and media; (iv) the emergence of social media; and (v) digital marketing by state tourism agencies.

Greece's entry into the Eurozone in 2001 harmonized the country's image in the European landscape more than ever, while the hosting of the 2004 Olympic Games had a positive impact on the country's international image.

Most people believed that the euro would bring benefits such as price stability, cheaper and safer transactions for tourism businesses and tourists and, above all, convenience, as tourists could move around a very large area without having to face the complications and uncertainties of changing currencies, which was a positive aspect for Greek tourism.

### 18.2.1 Hosting the 2004 Olympic Games and foundation of the new Acropolis Museum

In connection with the hosting of the Olympic Games, the organizing committee commissioned research on the 'image of Greece' from various companies (e.g. TNS ICAP in 2003 and the cooperation of MRB, VPRC and Research International in 2004). In this context, it is clear that the issues related to reputation, international public opinion and the 'image of Greece' were taken very seriously by the state after all.

Against this background, the country's tourism officials finally began to closely examine concepts related to image communication with scientific rigour and methodological precision (Fola, 2010).

The praise for the organization of the Olympic events by the international media, which had previously taken a very critical stance during the preparations, played an important role in creating a positive international climate with regard to Greece's image for some time at the beginning of the period we are dealing with.

In addition, as part of the 2004 Olympic Games, a programme was implemented to promote artistic and cultural events of all kinds beyond the period of the global sports festival, with the aim of combining sport and culture with the support of the mass media. The 'Cultural Olympiad' events were very diverse and lasted an unusually long time – before the Games, during the Games and after the end of the Games. Several of them took place outside Greece, from Europe and America to Japan and Australia.

In Athens, new facilities and stadiums were built and famous architects such as the Spaniard Santiago Calatrava, who has created works of high aesthetics, were invited to design the infrastructures for the events. While progressive ideas for communication were introduced and dynamic initiatives were taken for the organization of the Olympic Games, the Greek cultural heritage was undoubtedly the background that promoted the image of Greece (Papanikolaou, 2015).

Incidentally, in the summer of 2004, the Greek national football team won the European Championship and thus caused one of the biggest surprises in European football. This fact brought Greece even more fame and certainly played an important role in creating a positive perception and promoting the country as a tourist destination.

A few years later, in 2009, the new Acropolis Museum, a reconstructed archaeological museum in state-of-the-art architecture focusing on the finds on the Acropolis of Athens, was officially inaugurated and opened to the public. It was promoted by the media as an international event for human culture and was another great contribution to Greece's tourist image.

### 18.2.2 Greek contemporary art and tourism

As for the influence of Greek painting on tourism during this period, it is worth noting that the famous Greek artists continued to contribute voluntarily to the formation of the image of Greece with their works. For example Alekos Fassianos greeted the Olympic Games with his work *The Spirit of the Olympic Champion* (2001), and many other active artists of his generation made similar actions. At the same time a large number of other visual artists emerged who also contributed to the formation of the image of Greece, be it with their works, their words or their origins. Such artists include Venia Bechrakis, Tasos Chonias, Anna Fafaliou, Nikos Floros, Aggelika Korovessi, Maria Maragkoudaki, Maria X, Yiannis Melanitis, Yiannis Mouravas, Anastasios Nyfadopoulos, Athina Rachel Tsangari, Mina Papatheodorou-Valiraki, Erietta Vordoni, Sofia Petropoulou, Danae Stratou and Lydia Venieri.

In the context of artistic representation, it is interesting to note that a few years before the 2004 Olympic Games in Athens, there was a massive graffiti clean-up operation in the urban environment because of the upcoming international event that would attract thousands of tourists. The state was determined to project a safe and organized image of the city so that the thousands of visitors who would flock to Athens for the Olympic Games could feel safe. Paradoxically, during the Olympics, the 'Festival of Graffiti' took place in Chromopolis, organized by the 'Cultural Olympiad' in collaboration with the Crape Dien group, which was one of the starting points in Greece for a relative acceptance of these alternative communication practices. The organizers of the event officially invited 16 well-known international artists to Greece to work on public walls.

In this respect, we can clearly see that the Greek state used co-optation strategies to transform a truly radical practice of rebellious subcultural aesthetic expression into a promotional activity for the mainstream celebration of a cultural-commercial event such as the Olympic Games.

Thus, the Athenians' opinion of graffiti and street art gradually changed, and many Greek graffiti writers and street artists gained international fame with their works created in Greece but visible worldwide via social media. Their art attracted tourists to the respective areas of Greece where they worked. In this respect various local travel agencies promoted a new form of tourism, street art tourism, that was already developing in major metropolises around the world such as New York, Berlin and Melbourne. To name just a few of the artists who have worked in Greece: (i) Billy Gee; (ii) Dizi; (iii) Fikos; (iv) iNO; (v) N_Grams; (vi) Sonke; (vii) Stamatis Laskos; and (viii) Tonia Ainot.

Accordingly, many tour operators have sooner or later developed programmes to showcase the city's most famous graffiti, stencils, stickers, street art and traditional murals. For a fee, tourists on foot or by bicycle can view and photograph these works of art, learn more about their significance and how each one relates to Athens' current social and economic situation.

While tourists now take selfies with stencils of political activists and graffiti in the background, it seems that these products of an alternative communication practice that was used by many social groups in the subcultural sphere to express their protest have finally become a tourist attraction and have perhaps lost their power to simultaneously arouse astonishment and vigilance.

The rise of digital photography and social media has not only made it easier to capture and communicate travel moments, graffiti and other attractions. It has also significantly changed the way people engage with, share, enjoy and understand travel images. Before the instant sharing of images on social media platforms, it was photography in print, on postcards, in travel albums and in many other ways, such as art photography, that played the most important role in shaping people's perceptions of the world and inspiring their wanderlust and imagination.

Be that as it may, even though the Internet has reduced the sending of postcards by traditional mail, tourists often choose to buy postcards only as souvenirs or gifts from their travels. Thus, Greek postcards continued to be produced during this time. Typically they featured images that evoke a sense of nostalgia, such as landmarks (e.g. the Parthenon), mesmerizing landscapes, sunny beaches, the beauty of Santorini, the famous 'shipwreck of Zakynthos', local culinary delights, snapshots of daily village life with its elderly inhabitants and charming details of traditional island architecture – sometimes even accompanied by a cute cat. Postcards can thus still somehow capture the imagination of tourists, in a kind of formal way carried out by the tourism industry (Bonarou, 2012).

### 18.2.3 Contemporary literature and literary tourism

As far as the influence of contemporary Greek literature on the image of Greece is concerned, the low presence of new Greek authors on the international book market during this period is a central issue that has led to many important discussions in this field.

It can be said that Greece remains one of the few European countries that does not propose a specific national book policy, as other countries have been doing for some years, in order to achieve systematic distribution of its books on the international market (Stamou, 2015).

Be that as it may, while the established poet of the old generations, Elias Simopoulos, was nominated for the Nobel Prize in 2011, many poets, writers and literary personalities have emerged during this period. In their own way they have contributed to shaping the image of Greece, such as Christos Ikonomou, Amanda Michalopoulou, Kallia Papadaki, Sofia Nikolaidou, Yannis Palavos, Maria Mitsora, Kostas Akrivos, Thanassis Valtinos, Zyranna Zateli and Nikos Dimou.

At this point, mention should also be made of some books by foreign authors whose works contribute in one way or another to painting a more up-to-date picture of everyday life in Greece. This includes books such as:

- Tom Stone's *The Summer of My Greek Taverna: a Memoir* (2002);
- John Mole's *It's All Greek To Me: a Tale of a Mad Dog and Englishman, Ruins, Retsina and Real Greeks* (2004);
- Sofka Zinovieff's *Euridice Street: a Place in Athens* (2005);
- Victoria Hislop's *The Island* (2005), which also became a television series in Greece, and *One August Night* ( 2020);
- Aurelia Smeltz's *Labyrinthine Ways* (2012);
- Rebecca Hall's *Girl Gone Greek* (2015);
- Marissa Tejada's *Chasing Athens* (2015);
- Jennifer Barclay's *An Octopus in my Ouzo* (2016);
- Sharon Blomfield's *Sifnos Chronicles: Tales From a Greek Isle* (2016);
- Leah Fleming's *A Wedding in the Olive Garden* (2020);
- Marjory McGinn's *How Greek Is Your Love?* (2020);
- Suzi Stembridge's *No Ordinary Greek Odyssey* (2020); and
- Jeffrey Siger's *A Deadly Twist* (2021).

This is only a small sample of the large number of authors who have written about Greece from a kind of visitor's perspective, for most of them have travelled Greece. They share their experiences and describe various aspects of current Greek culture and everyday life, whether in the form of detective stories, love stories, novels, memoirs or personal adventures, enriched with historical truths and myths.

We should also mention contemporary Greek literary tourism, which motivates people to travel to Greece in connection with literary festivals, competitions, conferences, exhibitions, places described in books, the residences of classical authors, and so on.

In Greece, activities related to literary themes have always been at the centre of cultural life, but the connection of these events with a broader international environment has developed in recent decades due to the globalized world.

In addition to the tourist attraction of ancient Greek writers and their connection to various events and tourist products in general, events related to new literature are taking place throughout the country. Examples of this movement are some international literature and poetry festivals: (i) the Athens World Poetry Festival, which has been held in Athens every year since 2013; (ii) the annual International Book Fair in Thessaloniki, one of the few state-sponsored international events related to books;

(iii) the International Literature Festival, which has been held on the island of Tinos since 2010; and (iv) many other international literature events in various cities such as Patras and Chania (Linardaki and Aslanides, 2020).

### 18.2.4 Importance of films promoting Greek tourism

Throughout this period, cinema remained a powerful tool for promoting Greek tourism, as it skilfully portrayed the country's natural splendour, cultural richness and historical importance on the world stage.

Many notable Greek film productions and Greek co-productions in collaboration with foreign companies continue to contribute to the enrichment of Greece's image. This includes films such as: (i) *A Day in August* (2001, Constantine Giannaris); (ii) *A Touch of Spice* (2003, Tassos Boulmetis); (iii) *Brides* (2004, Pantelis Voulgaris); (iv) *Uranya* (2006, Costas Kapakas); (v) *El Greco* (2007, Yannis Smaragdis); (vi) *Dogtooth* (2009, Yorgos Lanthimos); and (vii) *Attenberg* (2010, Athina Rachel Tsangari).

Likewise, foreign film productions in Greece have helped to improve the country's image and promote it as a tourist destination, to name a few according to their year of production:

- *Captain Corelli's Mandolin* (2001, John Madden);
- *Boat Trip* (2002, Mort Nathan) and *The Bourne Identity* (2002, Doug Liman);
- *Lara Croft Tomb Raider: the Cradle of Life* (2003, Jan de Bont);
- *The Sisterhood of the Traveling Pants* (2005, Ken Kwapis);
- *Mamma Mia* (2008, Phyllida Lloyd);
- *My Life in Ruins* (2009, Donald Petrie) and *Eden à l'Ouest* (2009, Costa-Gavras);
- *Before Midnight* (2013, Richard Linklater);
- *The Two Faces of January* (2014, Hossein Amini);
- *The Lost Daughter* (2021, Maggie Gyllenhaal); and
- *Glass Onion: a Knives Out Mystery* (2022, Rian Johnson), *Triangle of Sadness* (2022, Ruben Östlund), *Crimes of the Future* (2022, David Cronenberg) and *The Enforcer* (2022, Richard Hughes).[1]

A 2014 report by the Foundation for Economic and Industrial Research on international productions in Greece showed that some foreign films had a positive impact on tourist flows. In particular, a significant increase in tourist arrivals at hotels on Kefalonia, Santorini and Skopelos was observed following the international screening of the films *Captain Corelli's Mandolin* (2001, John Madden), *Lara Croft Tomb Raider: the Cradle of Life* (2003, Jan de Bont) and *Mamma Mia* (2008, Phyllida Lloyd), which were filmed on these islands, respectively (Athanasiadis *et al.*, 2014).

However, there was general disappointment, both among local tourism operators in the areas where films such as *Captain Corelli's Mandolin* (2001, John Madden) were shot, and among researchers of Greek tourism, that no appropriate measures were taken by the state to implement a film tourism programme. Instead the sets were completely destroyed after filming and there was nothing left for visitors to see.This fact is another indication of how long it took for the Greek state to recognize and use film tourism in the most conventional way (Hudson and Ritchie, 2006; Assadourian, 2011).

At this point, however, it is interesting to note that after the success of the film *My Big Fat Greek Wedding* (2002, Joel Zwick), which was not filmed in Greece but obviously has a connection to the country, the Greek National Tourism Organisation (GNTO) planned a unique promotional programme in the USA for tourism. This was to take advantage of the positive impact of the film's success by placing advertisements in *Modern Bride* magazine and distribute free DVDs of the film with the magazine. At the same time, a contest was held for married couples to win a prize: a trip to Greece and 'A Big Fat Greek Wedding', which greatly enhanced Greece as a tourist destination (Assadourian, 2011).

In this atmosphere, the National Centre of Audiovisual Media and Communication and the Greek Film Centre with its Hellenic Film Commission began to intensify their actions to promote foreign productions or Greek co-productions with other countries and, in general, to make every effort to ensure the attractiveness of Greece for foreign film producers. These institutions could easily recognize the great importance of cinema tourism for the country's tourism development due to the nature of their activities and their knowledge of what is being done abroad in this field.[2]

In addition, the National Centre of Audiovisual Media and Communication has taken on the task of establishing and operating the 'National Network of Audiovisual Production Facilitation Offices'. In this context, film offices have been established in 13 regions of Greece to promote their areas as desirable filming locations and to help foreign filmmakers to shoot there (National Centre of Audiovisual Media and Communication, 2018; Komi *et al.*, 2021). We can therefore assume that the serious and organized actions to promote a central programme for film tourism in Greece began during this period.

### 18.2.5 Contribution of television

Together with film production, it can be said that television production also contributed sufficiently to the formation of tourist images by exploring the theme this time with the perception of the Greek audience in mind. Apart from the imported tourist TV programmes and documentaries, Greek productions of this kind experienced a boom during this period. In fact, there was a proliferation of travel programmes in most Western countries in the 2000s.

Different television channels, sometimes in combination with advertising on the Internet, produced a wide range of travel-related content. A number of new travel journalism and infotainment programmes were broadcast on Greek television, such as *Beautiful World* (Mega) and *Images* (Alpha). Over time, as new technologies conquered media culture and changed the style and way mass communication was perceived, most travel programmes on Greek television introduced elements of audience interaction, initially through e-mail and text messaging. Many programmes even began to broadcast live directly from the regions or studios, such as *We stay in Greece* (ERT1).

At the same time, some tourist programmes continued to show the daily life and culture of the people in the different countries and the particular sociocultural elements of each region in a documentary style, such as the programmes *Balkan Express* (ERT3), *The Tales of Water* (ERT3) and *One Way Ticket* (ERT2).

In general, however, there was an increase in programmes that make dynamic use of social media, either to interact with the audience or to promote the programme, such as *Girismata* (Skai TV), *Travelling with Terence Quick* (Star), *I love Gr* (Skai TV), *World Party* (ERT3, Alpha TV), *60s Greece* (Alpha TV) and *Happy Traveller* (Skai TV).

Many of these new programmes also dealt with other aspects of tourism, such as gastronomy, environment, sports or cars. This mixing was sometimes very pronounced, as with the tourist sports and gastronomy programmes, *Live Greece* (ERT3) and *A Thousand Miles of Taste* (Sky). To be more precise, a new type of gastronomic programmes has developed, for example in connection with travel.

Gastronomy is a typical theme that can be linked to tourist narratives, as certain dishes can be associated with the cultural identity of certain places. Thus, programmes such as *Mom's Kitchen* (ERT3, Alpha), *Sunday in the Village* (ERT3), *Food and the City* (Mega), *Boukia kai Sichorio* (Mega), *A Journey to the Flavours* (Epsilon TV) and *No Recipe* (Mega) maintain the general feature of gastronomic tourism. Other programmes which have more to do with tourism promotion, or with cooking recipes, have gradually become a fundamental element of Greek television programming (Pliakas *et al.*, 2018).

Characteristic of the television production aesthetic of this period is the use of various new media tools such as drones, mobile phones, tablets, action cameras and selfie sticks. These tools bring television programmes closer to the idea of social media and citizen journalism (Pliakas *et al.*, 2018).

It could be argued that filming with mobile selfie-stick cameras, as is often done in some Greek travel shows, is a deceptive motivational technique used in travel-show production because most young people nowadays use this method when posting their travel experiences on social media and it is easy to identify with the presenter. In short, when the presenter follows a set script to promote destinations and praise the places, the young viewers identify with him/her and are inclined to judge them positively. They get into a vicious circle of actions and judgements that the presenter leads them into because they can put themselves in his/her shoes, being used to taking selfies all the time.[3]

Other recent tourism programmes combine the style of feature films and documentaries with advertising techniques and present appealing results with cheerfulness and good humour, such as *We have arrived* (ERT 1) and *Van Life* (ERT 2).

## 18.3 Negative Trends for Tourism

We must now point out some negative trends of this period in relation to Greece's tourist image, which we consider important, but which are also interesting in the context of media impact theories.

After the international financial crisis of 2007 and the Greek debt crisis, which among other things led to the country's inability to repay its national debt, Greece's image suffered a severe setback.

The mass media of other countries, such as the German magazine *Focus* and the German newspapers *Bild, die Welt* and *Süddeutsche Zeitung*, published a series of vituperative articles against Greece. These articles surprisingly not only referred to the topic of the crisis (i.e. the concrete economic situation, the economic policy of the Greek government or, in the worst case, political personalities) but they were completely directed against contemporary Greek culture and the Greek people, portraying them as corrupt and taking up pro-racist theories reminiscent of the erroneous theory of Jakob Philipp Fallmerayer.[4] This situation naturally triggered a negative spectrum that affected the international image of Greece in general (Melaniti, 2017; Rigopoulou, 2017).

Some studies on the cultural representation of the Greek crisis in political cartoons, and on the front pages of newspapers and magazines of the foreign press, have shown that the power of the sketches, combined with the ironic, cynical and irreverent humour, made the contemporary mass media privileged instruments of political criticism and public opinion-forming by disparaging Greece's position in the crisis at the time (Chalániová, 2013; Talalay, 2013; Papacharalambous, 2017).

However, the country has made efforts to counter the negative portrayals in the media by highlighting its positive attributes and achievements to overcome these challenges and has succeeded in maintaining a positive tourism image.

At this point, we can refer to the theories on the great power of the media and argue that it is quite difficult for the media to convince the public of certain positions when they are false and malicious, while at the same time there is a strong global social memory on the subject. As in the case of the defamation of Greece by the German media, the image of Greece was not distorted in the end, either because this did not last long or because it was a false slander, and the collective memory worldwide had a different opinion.

According to the United Nations World Tourism Organization (UNWTO) (2020), Greece was the 13th most visited tourist destination in the world in 2019, before the coronavirus disease 2019 (COVID-19) pandemic. However, it is worth noting that even during the pandemic and the restrictive measures imposed by the Greek government from 2020 onwards, the 'Greece from Home' platform was set up to support the protection of public health worldwide and encourage people to stay at home, but also to strengthen the country's image as a tourist destination and bring Greek culture and sights to the world through virtual travel.

## 18.4 GNTO Develop the Image of Greek Tourism

In addition to all these developments in the field of communication in Greece in relation to tourism, what is most important, as already mentioned, is that during this period public institutions gradually began to take seriously the important political issue of the 'image of the country' in general and, of course, in 'economic terms', the tourist image of Greece in particular.

In this context, the GNTO has been very active with a series of campaigns, such as:

- 'Greece: That's life!' (1999–2000);
- 'Greece: Beyond Words' (2002–2003);
- 'Greece: Your Best Time' (2004);
- 'Live Your Myth in Greece' (2005–2006);
- 'Greece: Explore Your Senses' (2007);

- 'Greece: the Real Experience' (2008);
- '5000-Year-old Greece: a Masterpiece You Can Afford' (2009);
- 'Greece: Kalimera' (2010);
- 'You in Greece' (2010–2011);
- 'Greece: a 365 Day Destination' (2017);
- 'Visit Greece Now!' (2018);
- 'Greece is Always in Season' (2020);
- 'Greece from Home' (2020); and
- 'All you want is Greece' (2021).

The truth is that through all these campaigns, the myth of Greece as the ideal holiday destination, based on the promotion of the 'sun-sea-sand' triptych, has generally been recycled further. To some extent, however, some of the later campaigns were enriched with other themes to promote various alternative forms of tourism and began to promote year-round tourism.

It is worth noting that many of GNTO's audio-visual commercials have a special artistic value and some of them have been internationally recognized, such as 'Greece: a 365 Day Destination', which was awarded in 2018 by the International Committee of Tourist Film Festivals (CIFFT), among others.

As far as the organization's use of new media is concerned, the first GNTO website (gnto.gr) was launched at the beginning of the new millennium, which can generally be said to have been initially a formal presence on the Internet rather than a dynamic use of a communication tool to build an image and promote Greek tourism. In 2004–2005, however, a more serious attempt was made to build a promotional website with the greektourism.gr website.

From 2010 onwards, the greatest importance was attached to the GNTO's communication via the Internet. Therefore, the organization developed a new promotion strategy by shifting the focus of its advertising from television, print media, radio and press entries to the Internet and social media (GNTO, 2012). Since then, the new website visitgreece.gr has been the main tool for promoting Greek tourism, while social media (Facebook, YouTube, Twitter, Instagram, Google+, etc.) also play an important role in promotion.

Moreover, in a 2020 study, the Hellenic Tourist Board (SETE) set as its main goal for the future, the wider adoption of new technologies (e.g. artificial intelligence, augmented reality and virtual reality) to automate and enhance the travel experience and provide security through health checks (Ikkos *et al.*, 2021).

Indeed, virtual tours and online applications have become increasingly important tools to promote Greek tourism. These technologies allow travellers to explore destinations and attractions in Greece from the comfort of their own homes, helping to maintain interest in Greek tourism and prepare for future trips.

Of great interest are the various online virtual tours set up during this period by public and private organizations, some of which are free, such as 'Athens, the city is the museum' or the 'Street View Virtual Tour of the Acropolis Museum'. These were created in collaboration with Google and use maps or image menus that organize virtual walks. These virtual tours suggest stopping at particularly important points, antiquities or works of art, all from the comfort of your armchair in your living room. However, there are many other offers like these that are self-guided virtual tours that the user pays for, proving once again that myths, history, art and civilization can be commercialized and turned into superficial, post-modern packages just to make money.

In any case, by offering these diverse and immersive experiences, online apps have helped to present Greece as a multifaceted destination that appeals to a wide range of interests, from history and culture to nature and adventure, thus increasing the country's appeal to potential tourists.

In general, it is now considered that the promotion of Greek tourism through the new digital media and network technologies has reached an excellent level.

In this context, it is worth noting that at the end of the first decade of the new millennium, Greece had achieved a very good position in the global tourism industry, according to the Travel and Tourism Competitiveness Index (TTCI) of the Information Society Observatory. Be that as it may, the promotion of Greek tourism businesses and Greek tourism in general via the Internet, whether by official government bodies, tourism companies or major international agencies, is considered satisfactory (Katsoni, 2017).

Finally, we can mention that there are some alternative surveys and measurement frameworks, such as: (i) the Anholt-Ipsos Nation

Brands Index, which measures a country's 'brand image' through a composite analysis of global public opinion in the areas of exports, governance, culture, heritage, people, tourism, investment and immigration; and (ii) Brand Finance's Nation Brand Annual Report, which provides additional insights into the perception of nations. According to these surveys and measurements, Greece is not in the best position in many respects, but it has a fairly positive image in terms of tourism (Brand Finance, 2021; Ipsos, 2021).

## 18.5 Summary

To summarize, recent literature, art, mass media and the emergence of new applications of digital media have profoundly influenced the perception of the image of Greece and its tourist landscape since the beginning of the new millennium. Events such as the 2004 Olympic Games and various cultural endeavours such as the establishment of the new Acropolis Museum have significantly enhanced Greece's international reputation. Technological developments promoted alternative methods of tourism promotion and marketing that began to overcome disadvantages and successfully overcame the problems caused by the crises (financial crisis, COVID-19) that have hit the country. In addition, cinema, literature and television have further enriched the country's tourist image. Consequently, the evolving communication environment has redefined the paradigms of tourism by utilizing virtual platforms to create enticing images and facilitate tourist transactions. This digital transformation has improved accessibility and made Greece's rich cultural and natural heritage easily accessible on a global scale. It can be said that Greece is embracing these technological changes and is now positioning itself as a modern destination with a dynamic tourism story that emphasizes its timeless attractions in the digital age.

## Notes

[1] In 2009, the GNTO published a booklet on foreign film productions in Greece, containing summaries of 118 films from the period 1935–2009 (Katsigiannis, 2017).

[2] The Hellenic Film Commission (HFC) is a directorate of the Greek Film Centre, under the Ministry of Culture and Sports and subsidized by the state. The main objective of the HFC is to promote and develop Greece as a location for international film and TV productions (Hellenic Film Commission, 2022).

[3] A similar effect has been observed with films shot with earlier, commercially used recording techniques such as the video camera. For example, the film *The Blair Witch Project* (1999, Daniel Myrick and Eduardo Sánchez) was presented as 'found footage' shot with a Hi8 camcorder. Since camcorders were a common consumer product at the time, it was easy for viewers who had experience with video recordings to identify with the person doing the filming, to follow the logic of the film, to be psychologically influenced by the film and to be gripped by great fear.

[4] Jakob Philipp Fallmerayer (1790–1861) was a 19th-century Austrian journalist, politician and historian who believed that the race of ancient Greeks was eradicated from Europe during the Migration Period and replaced by Arvanite, Aromanian, Slavic and Turkic peoples. He claimed that the physical beauty, intellectual brilliance, innate harmony, simplicity, artistic spirit and other qualities of the ancient Greeks, or even their name, had disappeared from the surface of the present Greek continent. He used to say that not the smallest drop of unadulterated Hellenic blood flows in the veins of the Christian population of modern Greece.

## References

Assadourian, R. (2011) Greece: lights, camera ... travel! analysis on the process of attracting foreign filmmakers and the way their films can contribute to strengthen a destination brand: the case of Greece. *Journal of Tourism Research* 2, 19–52.

Athanasiadis, T., Papadakis, E., Paratsiokas, N., Danchev, S. and Tsakanikas, A. (2014) *Motion Picture Production in Greece: Effects on the Economy*. Foundation for Economic and Industrial Research, Athens. (in Greek).

Barclay, J. (2016) *An Octopus in my Ouzo*. Octopus Publishing Group, London.

Blomfield, S. (2016) *Sifnos Chronicles: Tales From a Greek Isle*. The Parea Group, Waterloo, Ontario, Canada.

Bonarou, C. (2012) *Visual Culture and Tourism: Representations of Greece on Tourist Postcards*. Papazisi Publications, Athens. (in Greek).

Brand Finance (2021) Nation Brands 2021. Available at: https://brandirectory.com/rankings/nation-brands/ (accessed 2 September 2022).

Chalániová, D. (2013) Turn the other Greek. How the Eurozone crisis changes the image of Greeks and what visual representations of Greeks tell us about European identity. *Perspectives: Review of International Affairs* 21(1), 5–41.

Fleming, L. (2020) *A Wedding in the Olive Garden*. Head of Zeus, London.

Fola, M. (2010) *Communication strategies for building the image of a country as a brand: the case of the Athens Olympic Games*. Doctoral dissertation, University of Crete, Rethymno, Greece. (in Greek).

Greek National Tourism Organisation (GNTO) (2012) GNTO promotion and promotion strategy on the internet. Available at: https://gnto.gov.gr/wp-content/uploads/2022/04/online_stratigiki_EOT.pdf (accessed 2 May 2021).

Hall, R. (2015) *Girl Gone Greek*. Createspace, Scotts Valley, California.

Hellenic Film Commission (2022). Available at: https://filmcommission.gr/ (accessed 4 May 2023).

Hislop, V. (2005) *The Island*. Headline Review, London.

Hislop, V. (2020) *One August Night*. Headline Review, London.

Hudson, S. and Ritchie, J.B. (2006) Film tourism and destination marketing: the case of *Captain Corelli's Mandolin*. *Journal of Vacation Marketing* 12(3), 256–268. DOI: 10.1177/1356766706064619.

Ikkos, A., Rasouli, G. and Karagounis, A. (eds) (2021) *Tourism in 2040: the Main Trends That Will Shape the Travel Path in Tourism*. Association of Greek Tourist Companies, Athens. (in Greek).

Ipsos (2021) 2021 Anholt-Ipsos Nation Brands Index. Available at: https://www.ipsos.com/en/nation-brands-index-2021 (accessed 2 September 2022).

Katsigiannis, K.X. (2017) *Hellenic Tourism Organization: Travel Through Time*. Private Edition, Athens. (in Greek).

Katsoni, B. (2017) *e-Tourism: Online Transactions in Tourism*. Greenbooks, Athens. (in Greek).

Komi, K., Mini, P. and Tsagarakis, N. (2021) Tangible benefits and people's bonding in international film production: the case of films shot in Crete. *International Journal of Euro-Mediterranean Studies* 14(2), 105–133.

Linardaki, C. and Aslanides, A. (2020) From poem and song to cultural diplomacy: challenges and opportunities for place branding and tourism promotion. *Place Branding and Public Diplomacy* 16(4), 304–315. DOI: 10.1057/s41254-020-00176-1.

McGinn, M. (2020) *How Greek Is Your Love?* Pelagos Press, London.

Melaniti, M. (2017) Greece in the German press: reproduction of stereotypical images in *Bild, die Welt, Süddeutsche Zeitung* and *Focus* magazine. In: Rigopoulou, P. (ed.) *The Image of Greece in the Media, Cultural Pride and Prejudice*. I. Sideris Publications, Athens. (in Greek), pp. 159–190.

Mole, J. (2004) *It's All Greek to Me: a Tale of a Mad Dog and Englishman, Ruins, Retsina and Real Greeks*. Nicholas Brealey Publishing, London.

National Centre of Audiovisual Media and Communication (2018) Film Offices: All across Greece. Available at: https://www.ekome.media/audiovisual-production-invest/film-offices/ (accessed 3 April 2021).

Papacharalambous, D. (2017) Cultural representations of the Greek crisis in political cartoons. In: Rigopoulou, P. (ed.) *The Image of Greece in the Media, Cultural Pride and Prejudice*. I. Sideris Publications, Athens. (in Greeks), pp. 83–158.

Papanikolaou, P. (2015) Cultural institutions, museums, memory and heritage in Greece. Athens 2004 and the institution of the Cultural Olympiad. Doctoral thesis, University of Ioannina, Ioannina, Greece. (in Greek).

Pliakas, T., Triantafyllaki, V., Galatsopoulou, F. and Kenderelidou, K. (2018) 50 years of television with travel impressions and experiences. In: Vamvakas, V. and Paschalidis, G. (eds) *50 Years of Greek Television*. Epikentro, Thessaloniki, Greece. (in Greek), pp. 407–422.

Rigopoulou, P. (ed.) (2017) *The Image of Greece in the Media, Cultural Pride and Prejudice*. I. Sideris Publications, Athens. (in Greek).

Siger, J. (2021) *A Deadly Twist*. Poisoned Pen Press, Scottsdale, Arizona.

Smeltz, A. (2012) *Labyrinthine Ways*. Cosmos Publishing, River Vale, New Jersey.

Stamou, E. (2015) Discussion: On modern Greek literature abroad. *The Reader*. Available at: https://www.oanagnostis.gr/syzitisi-gia-ti-sygchroni-elliniki-logotechnia-sto-exoteriko-tis-eyas-stamoy/ (accessed 23 September 2020).

Stembridge, S. (2020) *No Ordinary Greek Odyssey*. Komboloi Press, Chalandri, Greece.

Stone, T. (2002) *The Summer of My Greek Taverna: a Memoir*. Simon and Schuster, New York.

Talalay, L.E. (2013) Drawing conclusions: Greek antiquity, the economic crisis, and political cartoons. *Journal of Modern Greek Studies* 31(2), 249–276. DOI: 10.1353/mgs.2013.0023.

Tejada, M. (2015) *Chasing Athens*. Harmony Books, New York.

United Nations World Tourism Organization (UNWTO) (2020) *World Tourism Barometer,* Volume 18, Issue 2. UNWTO, Madrid.

Zinovieff, S. (2005) *Euridice Street: a Place in Athens*. Granta Books, London.

# Conclusions

## Means of Communication that Shape our Ideas About Travel

Since time immemorial, non-verbal cues, oral traditions, rituals and symbols have proved to be indispensable means of communication for travelling safely and meaningfully in foreign countries.

Nowadays, the information we receive daily via the media undeniably shapes our ideas about travelling. It influences the images and opinions we form about places near and far, and thus affects our travelling decisions.

Painting and the visual arts have encouraged people's desire to travel through visual representations of distant places, cultures and historical events. Indeed, artworks can evoke emotion, curiosity and a sense of wonder, encouraging individuals to explore the world and experience the places of aesthetic representation in real life.

Like an enchanting key, art opens the door to the imagination, liberates us from the constraints of geography and ignites an unquenchable thirst for freedom and exploration. It is a universal language that transcends borders. It spins stories about distant lands and invites us to embark on a search for the secrets of the world and our inner selves.

Literature also has the power to inspire the imagination, broaden horizons and awaken the reader's desire to travel. It has encouraged people to explore the world, connect with other cultures and embark on journeys to discover themselves. Whether in the form of fiction, non-fiction or poetry, literature has been instrumental in fostering people's love of travelling and exploring the richness of our planet.

In addition, by providing valuable information about destinations, including travel tips, accommodation, restaurants and local attractions, the press and travel journalism have encouraged people to travel and provided them with inspiration and cultural insights. The press not only informs and educates, but also appeals to readers' emotions and makes travelling a more accessible and enriching experience for a global audience.

The invention of photography has played an important role in the conception of the tourist gaze and in the emergence and growth of the tourism industry. It has shaped the perception of tourism by enabling people to explore the world through images and inspiring them to embark on adventures themselves. In fact, photography in brochures and leaflets, travel books, newspapers, magazines, posters or postcards has been one of the most effective tools to promote tourism for many decades.

The dynamics of moving images and radio broadcasts revolutionized the audience's emotional attachment to narratives and encouraged sensual, emotional and intellectually stimulating media consumption.

Indeed, the advent of cinema increased people's thirst for discovery. With its dynamic, moving images, cinema created visual and emotional connections to different places, cultures and experiences. It awakened the urge for mobility, sparked enthusiasm for travelling, inspired concrete plans and encouraged the spirit of discovery – both on screen and in the real world.

Although radio is not as visually powerful as cinema, not as detailed as travel journalism and not as emotional as literature, it conveys a sense of immediacy and has also played a role in promoting tourism through various programmes about destinations and travel talk shows, and more generally through any kind of coverage of a geographical area. In fact, radio is still a valuable tool today for conveying images of places that reach and inspire potential travellers



DOI: 10.1079/9781800626034.CONCL

and provide them with important information for their trips.

Television has played an influential role in promoting tourism through the presentation of destinations, travel experiences, lifestyle features, culinary delights, live events, festivals and cinematic narratives. Through its dynamic messages, seamless continuity and real-time immediacy, television not only informs and entertains, but also serves as a powerful source of inspiration that often sparks an increased interest in travelling. The central role of global television networks extends to shaping the perceptions of international travellers in a unified way and tangibly influencing their decisions as they explore the world.

However, it is clear that the constant advances in communication technology have not only made travelling one of the first topics on the agenda of public discussions, but also more convenient and accessible to a wider audience (i.e. more universal than ever before).

The new digital media that has made communication in tourism easier than ever, with online booking platforms, travel planning resources, traveller reviews and ratings, social media influencers and online travel communities, as well as travel apps, virtual tours and live webcams streaming from destinations, has significantly advanced tourism.

There is no doubt that the advent of new digital media has transformed the travel industry by increasing convenience, personalization, accessibility and participation for all in the creation of tourism images. They have also 'democratized' the creation and distribution of travel content, providing travellers with a wealth of information, inspiration and resources to plan and enjoy their trips more efficiently and meaningfully.

As a result, tourists now have access to a variety of sources to learn about the world and express themselves, as well as tools and resources to enhance their overall travel experience and increase their safety and comfort while travelling.

It is therefore perfectly understandable that the image we form of the world around us and the destinations that attract us is largely shaped by the stories we hear from various sources around us and the images we see in works of art and in the mass media in general, and all this, of course, when we have no personal experience of a previous visit. Undoubtedly, people are not born with innate ideas or knowledge about places in our world, but they acquire their knowledge through sensory experiences and interactions with the outside world and mass media.

## Media Influence on the Tourist Image of a Destination

The interplay of beliefs, ideals and perceptions that individuals or a collective have of a place at a given time forms the 'image of the destination', a decisive factor in the field of tourism, which is significantly influenced by the media. This image, which is an integral part of the tourism phenomenon, can be seen as a social representation – a framework that contains basic elements shared by several social groups and which can also serve as a common basis for cooperation between these groups.

But is the 'image of a destination' the basis for tourism and to what extent can it be a controlled construction?

We can argue that the 'image of a destination' is indeed crucial for the development of tourism, but it is not a necessary condition for the existence of the human need to travel and discover.

But even if we accept that the 'image of a destination' is not the only factor in mobilizing tourists, we should consider another doubt. While governments, tourist boards and the tourism industry can influence and shape this image to a certain extent through various measures and with the support of the mass media, it also depends on external factors and certainly on public perception.

The image of a destination is a complex concept that is constantly evolving. If it is positive, it is an asset that can attract tourists and boost the tourism industry, but it requires constant effort and adaptability to external influences.

However, our question about the relationship between media and tourism becomes even more complicated when we ask ourselves whether it is true that the more we know about a place (or the easier it is to get information and buy tourist services for it), the more we want to visit it.

Although media portrayal can undoubtedly influence the desire to travel, the idea that exposure alone drives wanderlust is a gross oversimplification. Even though the mass media has a significant influence, we must approach the theory of the media's overwhelming influence with caution and nuance.

People's motivations and desires for travelling are very different, and several factors play a role when it comes to visiting a particular place. It is a widely held view that tourists' behaviour is influenced by a variety of factors besides media effects (e.g. their cultural capital, the influence of friends, perception, personality, economic situation, global crises, lifestyle, the allure of the unknown and adventure).

Within this framework, researchers have closely examined the multiple factors that influence destination choice, including age, gender, personality, ethnicity, education, life cycle, income, cost, travel distance and media exposure (Uysal and Hagan, 1993; Crompton and McKay, 1997; Dellaert *et al.*, 1998).

Some researchers are of the opinion that travellers are driven by a dual motivation, namely that of discovery and that of escapism. Their intrinsic desire for new experiences, knowledge and challenges drives them to immerse themselves in different cultures and landscapes in order to grow personally. At the same time, they long for a temporary escape from the routine of daily life and seek relaxation and recreation. This dynamic interplay allows them to quench their thirst for adventure while finding solace in a tranquil environment. The journey becomes a symbolic representation of this dual pursuit – leaving the familiar for the unknown and harmonizing the quest for knowledge with the need for relaxation. This harmony between exploration and escapism makes travelling a profoundly enriching and transformative experience (Iso-Ahola, 1983, 1990).

Our desire to travel to undiscovered places, to hidden gems that are not in the media spotlight and less advertised as tourist hotspots, is very much in keeping with the adventurous spirit of the great explorers of the Age of Discovery. Just like the pioneers who unlocked the secrets of unexplored territories, our desire to visit lesser-known places reflects the curiosity and courage that fuelled their discoveries.

Be that as it may, if we look at things philosophically and from a much more general perspective after this discussion, we can assume that before any media influence, the primal 'instinct of seeking' exists. The main purpose of this is to help the individual satisfy their basic needs, expand their awareness and adapt to their environment, and which therefore also includes the concept of 'escaping' from undesirable situations.

We assume that the 'instinct of seeking' inherent in human beings is a fundamental drive that propels them towards tourism, with communication playing a central role. If we define the human need for mobility and communication as the core premise for modern tourism, we can gain concrete insights into the interplay between tourism and communication and the construction of tourism images under the influence of the media.

Be that as it may, this relationship between communication and tourism, and the way in which the image of a place is created, can be better illustrated if we examine specific places and their historical development in relation to this theme.

## How the Image of a Place is Created: the Case of Greece

The case of Greece, which is analysed in this book, is interesting from several points of view. It is a region within the geographical boundaries of the Western world in which a particular civilization developed early on. Its inhabitants not only possessed an insatiable 'instinct of seeking', but also showed exceptional skills in preserving and spreading their culture by skilfully using the means of communication of their time. The unique geography, characterized by a network of mountains, islands and an extensive coastline with many bays, capes and inlets, left the inhabitants little choice and forced them to sail the seas. As a result, travelling became an integral part of their lives, woven into every medium they used to portray their rich culture.

Although the country had limited natural resources, which led its inhabitants to travel, emigrate and establish colonies far from their homeland, this particular geographical location, together with the sun, the sea and the intellectuality of the dwellings, facilitated trade and contact with neighbouring civilizations. This included Egypt, Persia and the ancient Near East, and produced a remarkable cultural and political landscape encompassing philosophy, science, literature, drama, art and architecture that was admired by the whole world and attracted people to visit this land.

From this analysis, it is clear that the image of a country is a long-term process, made up of the representations in the media of each era and the way the world perceives these representations. And be that as it may, the sum of the beliefs, ideals and impressions that people have of a particular place at a particular time is, as we have often emphasized in this book, very important for tourism.

Stories, performances and representations from ancient Greece have always been fertile ground for popular entertainment and art, so much so that it is almost impossible to find a time in history when the Greek paradigm did not play a role in the popular culture of the Western world. All these works from antiquity have always appealed to many countries and different audiences, from the elite to members of the lowest socio-economic strata.

In this sense, the image of Greece was constructed and developed over centuries through a complex interplay of media representations of historical events, literature, art, philosophy and archaeology, which had various cultural influences in the Western world.

However, it must be emphasized that the image of Greece was shaped by both internal and external factors, by Greek and foreign media, that it varied according to the perception of the respective public and that it went through different phases and interpretations over time.

In antiquity, early literary works, including the writings of ancient Greek poets, philosophers and historians, such as Homer, Plato, Aristotle and Herodotus, emphasized the importance of Greece's contributions to philosophy, art and science. The artistic and architectural achievements of the Classical and Hellenistic periods, including the Parthenon and other symbolic buildings, statues, ceramics, coins, museums, libraries, etc., decisively shaped the image of Greece as the centre of art and culture of the time.

The Romans, who admired Greek culture, contributed significantly to the preservation and reconstruction of the image of Greece with their own resources, thus contributing to the continuity of Greece's historical importance. The influence of the Byzantine Empire on writing, art and religion enriched Greece's cultural heritage.

While in Europe, with the Renaissance and the Enlightenment, ancient Greek art and philosophy were highly valued after their adaptation to the times, the image of Greece under the rule of the Ottoman Empire was also influenced by the struggle of the subjugated Christian peoples and by its important position at the crossroads of Eastern and Western civilizations.

In the 18th and 19th centuries, Western travel literature and exploration reports popularized the landscapes, ruins and culture of Greece for a wider European audience, awakening romantic notions of the 'Grand Tour' and the fascination of the ancient Greek heritage.

One could argue that the perception of Greece in the 19th century was shaped not only by travel literature, but also by various media of the time, such as paintings and newspapers. The depiction of Greece's quest for independence from the Ottomans played a crucial role in shaping the image of Greece as a symbol of national resilience and the struggle for freedom.

After the emergence of modern tourism and the transition to the 20th century, the Greek state slowly began to recognize the hypothetical potential of tourism as an activity that could significantly contribute to the economy of the country. As a result some state agencies for the promotion of tourism were gradually created, which contributed to some extent to the formation of Greece's image abroad.

In the second half of the 20th century, when tourism was recognized as a viable economic enterprise in Greece, the state, local authorities and the private sector turned to investment in tourism infrastructure and services. With the aim of improving the country's image in the world, Greece's historical and archaeological treasures, such as the ancient temples and theatres, were highlighted as major attractions, as were the Mediterranean landscape and climate. Intensive efforts were made to excavate and restore these sites to make them more accessible to tourists. Greece actively participated in international events and used platforms such as world fairs and exhibitions to showcase and promote its growing tourism offer.

News and entertainment reports in the Western press, on radio and television, as well as images from the cinema, but also the advertising campaigns of the Greek National Tourism Organisation (GNTO), in which antiquity, the 'Mediterranean landscape', 'sun, sea, sand' were shown together with excerpts from modern Greek culture, were the most important factors in the formation of the modern image of Greece.

In recent decades, Greece has faced economic difficulties, particularly the financial crisis, which have tarnished the country's image of stability and financial responsibility. However, in the face of these challenges, both the Greek state and private tourism companies have proactively invested in digital marketing measures. They have created engaging tourism platforms and social media channels to spotlight Greece's many archaeological wonders, breathtaking landscapes, delicious cuisine and diverse cultural experiences to inspire tourists.

The shaping and reshaping of Greece's image is therefore an active endeavour in the Greek and international media. Although the reconstruction process involves a variety of undertakings, the core strategies revolve around the appropriation of themes from classical antiquity and ancient Greek culture. At the same time, the iconic 'sun-sea-sand' triptych, a timeless cornerstone in the narrative of Greece's appeal, is still tirelessly promoted (Fig. C.1).

Fig. C.1. Greek National Tourism Organisation (GNTO) poster, *Balcony with sun view* (1959). The 'sun-sea-sand' triptych forms the basis of Greece's appeal to tourists and was used by the GNTO very early on in tourism advertising. This iconic blend embodies the country's enduring splendour and is an essential part of the tourism myth and an eternal magnet for travellers in search of a Mediterranean utopia. Image used with permission from the GNTO.

To summarize, the enduring link between tourism and communication from antiquity to the present day underlines the central role of the media in shaping the perception of travel and tourist images. New media technologies such as the Internet and smartphones have revolutionized travel by democratizing information and enabling personalized experiences. The evolving tourist image of Greece, influenced by ancient and neoclassical elements, literature, art and alternative forms of tourism, shows the dynamic interplay between media and tourism. Despite all the challenges, the GNTO has played a central role in shaping and promoting the image of tourism by combining ancient culture, the Mediterranean sun, the myth of 'Aegeanity', alternative travel narratives and modern technologies. Historical eras and current events promoted by the media, such as the Olympic Games and economic changes, continue to shape Greece's identity as a destination.

## How Tourism Could Have Evolved Differently

Let us now break out of the ordinary and immerse ourselves in a realm of imagination to stimulate creativity and broaden our horizons. Instead of staying in the present, let us dive into the realm of 'what if' and imagine how tourism could have evolved in unconventional ways, but still intertwined with the fascinating world of communication. Referring to these alternative realities can give us new perspectives on the dynamics of travelling, exploration, pleasure and tourism.

The exploration of possibilities and contingencies serves as an intellectual playground, a speculative enquiry that unfolds as a liberating game. In this imaginative space, we encounter a world in which tourism takes unexpected paths, characterized by alternative historical turns, various technological advances and shifting social values.

However, for those who see the current historical paradigm as eternal wisdom and are also unfamiliar with the dimension of contingency that every minute in this world can have, this may seem pointless, exaggerated and paradoxical.

The facets that make up today's definition of tourism, as well as the wide range of services offered by the ever-evolving tourism industry, have their roots in intrinsic human tendencies that have persisted since the earliest epochs of our civilization.

From the basic elements of communication and social interaction to the more nuanced experiences of hospitality – forging new connections, revelling in the wonders of nature, seeking solace and tranquillity in far-flung places, encountering diverse wildlife, savouring culinary delights and thirsting for adventure, embarking on a journey to understand the world, immersing oneself in different cultural landscapes, making religious and spiritual pilgrimages, admiring art and impressive sites, participating in vibrant festivals and events or simply having a good time – these activities have seamlessly blended into the fabric of modern tourism. However, it is important to recognize that they have always been woven into the fabric of human existence and have not only a contemporary, but also a deep historical and cultural significance.

So, we can easily see that modern tourism has integrated many activities that have been part of human culture for centuries. Even though the way these activities are experienced and organized may have changed over time, their roots can be traced back to ancient human traditions.

In any case, we should ask ourselves how we would perceive the concept of tourism if things could have been different throughout history. When we say 'different', we mean that 'travelling for pleasure' could have taken other paths over time, if some important historical turning points had not taken place or had taken place differently. These turning points are the discovery of agriculture, the development of a sedentary culture, the huge technological advances in transport and communication, the emergence of nations, the Industrial Revolution, but above all naming of this phenomenon 'tourism' after the advent of mass capitalist society and, finally, its complete commercialization and formation of an economic sector defined as the tourism industry.

So, the questions are:

- If a different path into modernity had been followed, could the term tourism have a

different meaning today? Or at least would the media have constructed the concept of tourism differently?
- Would tourism have been able to escape the commercialized mass model that it has internalized for so many years, even if it now adorns itself with the possibilities of alternative tourism?
- So, is there any chance at all of expressing a non-commercial element in the concept of tourism today?

In this sense, having developed the theme of tourism, communication and culture, taking into account the specificities of Greece, we can make some conjectures about what could have happened, or what did not happen, in the development of tourism. This would help us understand what actually happened, or better approach this particular reality in which we live and the historical involvement of the media in this construction.

Let us conclude by formulating a brief idea for a different development and explore alternative scenarios or outcomes that deviate from the actual historical timeline in a 'counterfactual historical perspective' (Bunzl, 2004).

Among the thousands of alternatives, there could also be a version of the modern world in which people have no fixed abode and are constantly on the move; in which nomadism is still a staple, and in which technology is at the service of humanity and not dominated by its own logic. This would be a world that is not so commercialized and where the economy is not primarily focused on capital accumulation and the pursuit of profit, but simply on making the best possible use of the Earth's limited resources to meet people's basic needs, all with great respect for the environment.

In other words, we could imagine that we have evolved into a non-sedentary society where the value of individuals is not determined by the things they own and the land they possess. A society in which we only own and consume what we need to fulfil our basic needs, a society in which human creativity is not used to produce useless goods and services that provide us with a false comfort and convenience but ultimately do not contribute to our basic happiness and 'eudaemonia'. A society based on a biocentric culture that considers it painful to take more from nature than we need to live, where we respect life in all its forms, where we positively recognize all differences between humans and respect the rights of other animals. A society based on a rhizomatic form of political organization and thinking.

A world in which tourism is seen as a human condition, just as the mobility of peoples has been for more than three million years before the discovery of agriculture.

So perhaps we can say that the above questions, which were part of our discussion, have already been answered constructively and positively, as has our argument about the relationship between communication and tourism, which has preoccupied us in this book.

Tourism is in itself a form of communication that reflects the innate 'instinct of seeking', and as a 'primitive tribal lore' that links modern societies with prehistoric ones, it may indeed have a different meaning from the one that identifies it today with the tourism industry, which developed strongly with the onset of the Industrial Revolution and the rise of capitalism.

Be that as it may, a world such as we have described above, and which we believe might have been one of the ideal developments, may not be attractive to many people in the world today, just as it is possible that if we lived in such a society, we would dream of something other than this ideal.

## References

Bunzl, M. (2004) Counterfactual history: a user's guide. *The American Historical Review* 109(3), 845–858.

Crompton, J.L. and McKay, S.L. (1997) Motives of visitors attending festival events. *Annals of Tourism Research* 24(2), 425–439.

Dellaert, B.G., Ettema, D.F. and Lindh, C. (1998) Multi-faceted tourist travel decisions: a constraint-based conceptual framework to describe tourists' sequential choices of travel components. *Tourism Management* 19(4), 313–320.

Iso-Ahola, S.E. (1983) Towards a social psychology of recreational travel. *Leisure Studies* 2, 45–56.

Iso-Ahola, S.E. (1990) Motivation for leisure. In: Jackson, E.L. and Burton, T.L (eds) *Understanding Leisure and Recreation: Mapping the Past, Charting the Future*. Venture Publishing, State College, Pennsylvania, pp. 247–279.

Uysal, M. and Hagan, L.A.R. (1993) Motivation of pleasure travel and tourism. In: Khan, M., Olsen, M. and Var, T. (eds) *Encyclopedia of Hospitality and Tourism*. Wiley, Hoboken, New Jersey, pp. 798–810.

# Index

This book is published by **CABI**, an international not-for-profit organisation that improves people's lives worldwide by providing information and applying scientific expertise to solve problems in agriculture and the environment.

CABI is also a global publisher producing key scientific publications, including world renowned databases, as well as compendia, books, ebooks and full text electronic resources. We publish content in a wide range of subject areas including: agriculture and crop science / animal and veterinary sciences / ecology and conservation / environmental science / horticulture and plant sciences / human health, food science and nutrition / international development / leisure and tourism.

The profits from CABI's publishing activities enable us to work with farming communities around the world, supporting them as they battle with poor soil, invasive species and pests and diseases, to improve their livelihoods and help provide food for an ever growing population.

CABI is an international intergovernmental organisation, and we gratefully acknowledge the core financial support from our member countries (and lead agencies) including:

UKaid from the British people

Ministry of Agriculture People's Republic of China

Australian Government Australian Centre for International Agricultural Research

Agriculture and Agri-Food Canada

Ministry of Foreign Affairs of the Netherlands

Schweizerische Eidgenossenschaft Confédération suisse Confederazione Svizzera Confederaziun svizra

Swiss Agency for Development and Cooperation SDC

## Discover more

To read more about CABI's work, please visit: **www.cabi.org**

Browse our books or explore our online products at:
**https://www.cabidigitallibrary.org**

Interested in writing for CABI? Find our author guidelines here:
**https://www.cabidigitallibrary.org/books/authors**

www.ingramcontent.com/pod-product-compliance
Lightning Source LLC
LaVergne TN
LVHW070829220926
848486LV00029B/163